Three-Dimensional Attached Viscous Flow

Ernst Heinrich Hirschel · Jean Cousteix
Wilhelm Kordulla

Three-Dimensional Attached Viscous Flow

Basic Principles and Theoretical Foundations

Springer

Ernst Heinrich Hirschel
Herzog-Heinrich-Weg 6
85604 Zorneding
Germany
e.h.hirschel@t-online.de

Wilhelm Kordulla
Wilhelm-Evers-Str. 2
37120 Bovenden
Germany
wkordulla@gmail.com

Jean Cousteix
158, rue de Pont Vieux
31810 Le Vernet
France
jeancousteix@orange.fr

ISBN 978-3-642-41377-3 ISBN 978-3-642-41378-0 (eBook)
DOI 10.1007/978-3-642-41378-0
Springer Heidelberg New York Dordrecht London

Library of Congress Control Number: 2013950407

Printed on acid-free paper

Springer is part of Springer Science+Business Media (www.springer.com)

Preface

Fluid mechanics is generally considered more or less a two-dimensional affair. This holds for teaching and also for the bulk of text books. Viscous flow usually is treated in the frame of boundary-layer theory and as two-dimensional flow. The popular books on boundary layers give for three-dimensional boundary layers at most the describing equations and, with very few exceptions, some solutions for special cases.

The economical and ecological pressures on all transportation means, in particular on airplanes, are progressively increasing. The demands are large to improve the efficiency of flight vehicles in all speed regimes. Goals for the worldwide civil air transport request for the next decades sizeable reductions of fuel consumption, pollutants, and noise emanation. This requires that also the aerodynamic design must be refined, and that advanced drag reduction measures—for instance by means of laminar flow control and by means of turbulent flow management—must be incorporated into the vehicle design.

All this calls for an increasingly better handling of viscous effects during the flight vehicle definition and development phases. The viscous effects mostly are related to three-dimensional attached flow. In the design processes such flows hence must be understood much better than it is usually the case today. The present book gives the basic principles and the theoretical foundations and thus helps to understand the major aspects of three-dimensional attached viscous flows.

Emphasis is put on three-dimensional attached viscous flows and not on three-dimensional boundary layers. This wider scope is necessary in view of the theoretical and practical problems to be mastered in practice. In design work today the major computation work is made with Navier-Stokes, respectively Reynolds-averaged Navier-Stokes equations. Boundary-layer methods fill some niches.

Boundary-layer theory and viscous-interaction theory—the former regarding weak interaction, the latter strong interaction between the viscous flow and its external inviscid flow—however permit much insight into the phenomena of three-dimensional attached viscous flow. Of course, some of the phenomena of interest can only be understood with the help of discussions of the Navier-Stokes equations.

In the present book thus the properties of three-dimensional viscous flow past realistic shapes of finite extension are at the center of attention. The flight speed range is that of civil air transport vehicles, however also the flow past other flight vehicles up to hypersonic ones are considered. Therefore the results and the findings hold for the entire continuum-flow domain.

Weak and strong interactions as well as the locality principle find due attention. The possibilities to influence three-dimensional flows are considered, including the thermal surface effects which can be of importance for the design and both the numerical and the ground-facility simulation. The governing equations are given and discussed, also with regard to their characteristic properties and the wall-compatibility conditions, thus allowing many statements about the general flow properties.

Important too is the topic of connections and interactions of viscous and inviscid flow, because in aerodynamics the concept of inviscid flow, even of potential flow is widely used. The displacement effect of the attached viscous flow, which is exerted on the inviscid flow, is a phenomenon of very large practical interest. This effect usually is seen only regarding the boundary-layer displacement thickness. Here, it is discussed in view of almost all of its implications in aerodynamic design.

The understanding of three-dimensional flow can be fostered by the consideration of the flow topology. The basic concepts are presented and especially properties of attachment points and lines, and, to a certain extent also of separation points and lines, are studied in detail.

Quasi-one-dimensional and quasi-two-dimensional flows are of large practical interest, appearing for instance in the concept of the infinite swept wing and its extension, the locally infinite swept wing. These and other particular flow cases are considered in detail, too.

Turbulence is considered in view of three-dimensional flow and in particular also laminar-turbulent transition in order to support the approach to transition-sensitive flow problems. Examples, discussed in detail, are given finally in order to illustrate the most important of the treated concepts and the phenomena found in attached three-dimensional viscous flows.

The authors of the book are from the aerospace field and were—in teaching, research, and industrial application—deeply involved in phenomenological, mathematical and computational issues of attached three-dimensional viscous flow. They wish to give the student and, in particular, also the practical aerospace engineer the needed knowledge about three-dimensional attached viscous flow.

February 2013

Ernst Heinrich Hirschel
Jean Cousteix
Wilhelm Kordulla

Acknowledgements

The authors are much indebted to several colleagues, who read chapters of the book and provided critical and constructive comments. These colleagues are D. Arnal, B. Aupoix, S. Hein, G. Simeonides and C. Weiland. Their suggestions and input were very important and highly appreciated. Very helpful too was the information and the input received from J. Délery, R. Friedrich, D.I.A. Poll, and D. Schwamborn.

Data and illustrative material were made available for the book or have been permitted to being used by many colleagues. We wish to thank J. Häberle, S. Hein, R. Höld, H.-P. Kreplin, H.U. Meier, F. Monnoyer, Ch. Mundt, S. Riedelbauch, M.A. Schmatz, D. Schwamborn, K.M. Wanie, and C. Weiland. Several of these colleagues were former doctoral students of the first author. An important illustration was prepared for use in the book by M. Frey whom we thank very much for his contribution. Many thanks go to Ms. H. Reger for the drawing of several figures.

Special thanks are due to C.-C. Rossow, director of the DLR-Institute of Aerodynamics and Flow Technology at Braunschweig, Germany, who permitted to prepare and to make available a number of important figures for the book. It is in particular O.P. Brodersen to whom we are very grateful for great and very informative pictures. We are also very grateful for the input coming from Th. Kilian, N. Krimmelbein, and Th. Schwarz. Thanks go to K. Becker of Airbus for a special permission.

Ernst Heinrich Hirschel
Jean Cousteix
Wilhelm Kordulla

Table of Contents

1

Introduction

Three-dimensional viscous flow is not a simple extension of two-dimensional viscous flow. Actually, it is the general appearance in reality. Because the number of these appearances is large, only a limited number of flow cases will be considered in this book. These cases basically are attached viscous compressible or incompressible flows past realistic shapes of flight vehicles, like fuselages, fuselage-like bodies, and large and small aspect-ratio wings.

Attached viscous flow in its generality is described by the Navier-Stokes (NS) equations. It can be considered as boundary-layer flow, if the Reynolds number is large enough and if strong interaction is absent. The boundary-layer (BL) equations describe this flow.

The topic of this book is three-dimensional attached viscous flow rather than three-dimensional boundary-layer flow. This distinction is made for two reasons. The first reason is that several of the interesting phenomena and properties of such flows analytically can only be treated on the level of the Navier-Stokes equations. However, there are enough phenomena and properties left, which analytically can be approached only at the level of the boundary-layer equations. This holds especially if geometrical properties of the body surface come into play.

The second reason is that nowadays in aerodynamic design work computational simulations more and more are made on the basis of the NS, in particular the Reynolds-Averaged Navier-Stokes (RANS), equations. Of course, in early vehicle definition phases, in optimization schemes and so on, inviscid methods in combination with boundary-layer methods are used. The latter then usually are not exact three-dimensional methods. However, exact three-dimensional boundary-layer methods still have their niche of application, for instance, if laminar flow control is studied. Since about two decades, exact three-dimensional boundary-layer methods are no more at the center of attention of research and development.

The need for the distinction, but also the possibilities, which the different equation levels offer, lead to the concept of the boundary layer as phenomenological model of weakly interacting attached viscous flow. This is discussed in Section 1.1.

The book deals with attached viscous flow on realistic body shapes of finite extent. Therefore one must face the fact that at the end of a body the

E.H. Hirschel, J. Cousteix, and W. Kordulla, *Three-Dimensional Attached Viscous Flow*,
DOI: 10.1007/978-3-642-41378-0_1,

flow leaves the surface, or better, separates from the surface. Kinematically active and inactive vorticity enters the wake [1], see also Section 10.2 of the present book. If a NS/RANS method is applied, this in principle is taken correctly into account. But the question arises, whether and to what extent boundary-layer considerations and solutions are permissible, if the wake is not or only indirectly taken into account. This is the topic of Section 1.2.

In Section 1.3 a short survey of the development of the field of three-dimensional attached viscous flow is given. In Section 1.4 the presentation of scope and content of the book closes the introduction.

1.1 Phenomenological Models of Attached Viscous Flow

The mathematical model of attached viscous flow are the Navier-Stokes equations together with the continuity equation and the energy equation. As we will see in Chapter 3, the discussion of this system of equations yields a host of information about viscous flow. However, when dealing with three-dimensional flow, additional information can be gained from the discussion of the boundary-layer equations, Chapter 4, which are a subset of the NS equations. Moreover, a NS/RANS solution contains all the information about the flow phenomena under consideration, but as was often seen, only BL solutions, Appendix A.3.2, and there especially integral solutions, give a direct and uncomplicated access to the understanding of phenomena.

If the characteristic Reynolds number[1] of the flow is large enough, attached viscous flow is of the boundary-layer type. Boundary-layer type means that the viscous layer obeys the boundary-layer criteria, Sub-Section 4.1.2: the boundary layer is thin[2] and in direction normal to the body surface the (static) pressure in it is nearly constant (flat body surface) or centrifugal-force induced not constant (general curved body surface). These two factors indicate "weak interaction" (Section 1.2), i.e., the (hypothetical) inviscid flow past the body does not differ or differs very little from the real flow. This holds in particular for the surface pressure. If the above is true, the boundary layer can be considered as phenomenological model of attached viscous flow, Fig. 1.1.

Indicated on the upper level of this figure is that the Navier-Stokes equations describe attached viscous flow. Laminar–turbulent transition and turbulence phenomena in principle can be described with these equation (direct numerical simulation, DNS). DNS today is a potent research tool, practical applications of DNS, however, are far in the future. It is not yet seen, how realistic surface and free-stream properties can be described, and also the

[1] For the derivation of the Reynolds number and its discussion see Section 3.4.

[2] The thickness of the viscous layer is inversely proportional to some power of the Reynolds number.

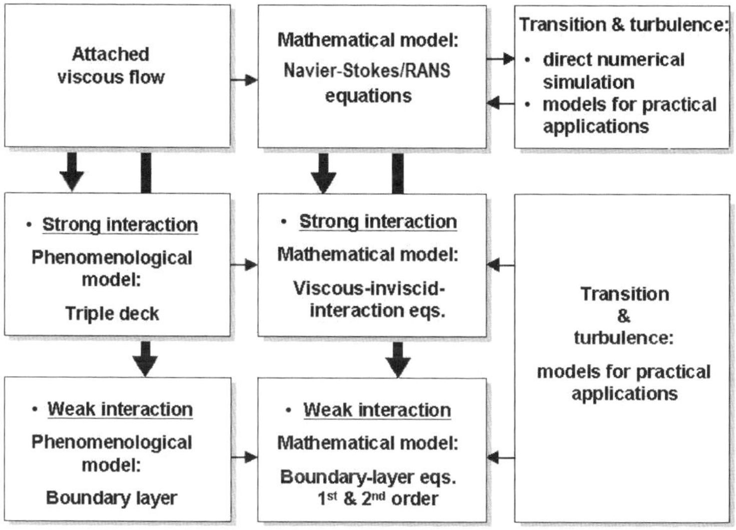

Fig. 1.1. Phenomenological and mathematical models of attached viscous flow shown in descending order of complexity. The continuity equation and the energy equation are not explicitly mentioned.

computational effort for DNS is much too heavy—and that for many years to come. Therefore transition criteria and models of different description levels are in use, as well as statistical turbulence models, Section 9.6, in the frame of the RANS equations. These may be combined in zonal solutions for separated flow with, for instance, large eddy simulation (LES) approaches of different kinds. The description of realistic surface and free-stream properties (receptivity problem) is still away from the level which is actually needed [2].

The concept of boundary-layer flow has been extended to some classes of flow with strong interaction. They encompass separation regions of small extent as well as flow-off separation at sharp edges, e.g., trailing edges. The concept of "strong coupling" between viscous and inviscid flow has led to very efficient methods, however, mostly for two-dimensional flow, Chapter 6. On the second level of Fig. 1.1 hence "strong interaction"—between boundary-layer flow and external inviscid flow, see next section—is placed. This is a model consideration with the triple deck as phenomenological model, Sub-Section 6.2.3. The mathematical model are the viscous-inviscid interaction equations.

On the third level finally the weak interaction between boundary-layer flow and external inviscid flow is indicated (see next section). This is the kind of attached viscous flow which predominantly is considered in this book. We regard the boundary layer as phenomenological model of weakly interacting viscous flow. A boundary layer is given, if the boundary-layer criteria are fulfilled, Sub-Section 4.1.2. Two-dimensional boundary layers are described by one equation, the boundary-layer equation, which is a momentum transport equation. Three-dimensional boundary layers are described by two momentum boundary-layer equations.

At high Reynolds numbers the boundary layer is thin, and the pressure field of the external inviscid flow is "impressed" across its entire thickness. This means, as mentioned, that at a flat body surface for large Reynolds numbers the pressure in the boundary layer becomes constant in direction normal to the surface (zero pressure gradient), or at curved body surfaces, that the pressure gradient is induced by centrifugal forces, Sub-Section 4.1.2.[3] The boundary-layer equations hence cannot describe pressure fluctuations in the flow, a kind of DNS is not possible with them. Therefore in boundary-layer methods transition and turbulence phenomena always must be treated on the basis of criteria, respectively models.

We note that boundary-layer theory is a well developed topic. It permits to obtain a large amount of application-oriented knowledge, although this concerns mainly two-dimensional boundary layers, and less three-dimensional boundary layers, see, e.g., [3, 4]. For the latter see, e.g., [5]. Of course, a Navier-Stokes theory exists, too, but not—in view of practical problems—to the same extent as the boundary-layer theory.

Modelling of transition and turbulence can be considered as flow-physics modelling, if a coherent approach to laminar-turbulent transition *and* turbulence in view of real-life receptivity conditions, see, e.g., [6, 2], is taken, Section 9.4. Depending on the kind of application, chemical reactions, low-density effects etc. have to be included [7].

1.2 Three Kinds of Interaction and the Locality Principle

If the boundary layer is to serve as phenomenological model of attached viscous flow, and further, if a boundary-layer method is employed for the simulation of this flow, one has to ask, under what conditions this is permissible.

In the frame of the boundary-layer concept, the flow past a body is considered to be composed of the external inviscid flow field and the boundary layer. Apart from the basic fact that the inviscid flow transmits its pressure field to the boundary layer, two kinds of interaction between them are usually distinguished: weak and strong interaction.[4] Classical boundary-layer theory treats mainly two-dimensional flow at infinite or semi-infinite bodies, and if

[3] The situation is different in certain mathematical models which describe strong interaction. In the triple-deck theory, for instance, the boundary layer is still a separate entity, however, with the pressure field of the inviscid flow not simply impressed on it, Section 6.2.

[4] In the latter case the pressure field results from the interaction of the boundary layer with the inviscid flow—"viscous-inviscid interaction".

three-dimensional flow is considered, this usually also is not done for bodies of finite dimensions [3]. Because our topic is attached three-dimensional flow past bodies of finite extent, we add a third kind of interaction, which we call global interaction.

The three kinds of interaction are characterized now:

- Weak interaction: The presence of attached viscous flow virtually changes the contour of the body (displacement effect of the boundary layer[5]). At large Reynolds numbers this effect is small. One speaks of weak interaction between the viscous flow—the boundary layer—and the (hypothetical) inviscid flow past the body, if the latter is only weakly, even negligibly, affected by the presence of the boundary layer.

  In boundary-layer computations the virtual thickening of the body can be taken into account—regarding the external inviscid flow—by adding the displacement thickness to the body contour, or, more elegantly, by introducing on the body surface an equivalent inviscid source distribution, e.g., the transpiration velocity, Chapter 5. An extended discussion of weak interaction effects, in particular of the displacement effect in view of flight-vehicle aerodynamics, is given in Section 6.1.
- Strong interaction: Separation, either flow-off separation at sharp trailing edges of airfoils or wings (also at sharp leading edges of delta wings) or squeeze-off separation—the classical separation[6]—or other kinds of separation, leads to a strong interaction, i.e., inviscid flow and viscous flow can no more be treated independently of each other, Chapter 6. This holds locally and downstream of the separation location. The separation locations in two or in three dimensions usually can be determined to a sufficient approximation by boundary-layer computations. The separation process itself—strong interaction—cannot be described in the frame of classical boundary-layer theory, it cannot be computed by means of boundary-layer methods alone. The same holds for flow at low Reynolds numbers, where strong interaction can occur—not necessarily in presence of separation.
- Global interaction: Attached viscous flow on realistic body shapes of finite extent separates from the body surface basically either by flow-off or by squeeze-off separation. Regardless of how and where this happens, kinematically active and inactive vorticity leaves the surface and then is present in the wake flow. At lifting large-aspect-ratio wings, for instance, this wake is initially a vortex sheet which soon rolls up to a pair of trailing vortices, the wing vortices[7]. The kinematically active vorticity in this wake causes

[5] In two-dimensional boundary layers the displacement in general is positive (exceptions are cold-wall cases, Sub-Section 2.3.3), i.e., the displacement thickness $\delta_1 > 0$ enlarges the virtual volume of the body. In three-dimensional boundary layers δ_1 can also be negative, Chapter 5.

[6] For a generalized definition of separation see Sub-Section 7.1.4.

[7] In the literature they are sometimes called tip vortices. This is wrong. Tip vortices are a different phenomenon, see the discussion in Section 10.3.

a global interaction, which results in the well understood downwash and induced drag of the lifting wing. The upstream changes of the flow field, however, usually are rather small.

This is in contrast to wings with large leading-edge sweep, i.e., delta wings. For a certain combination of sweep angle and angle of attack, a pair of lee-side vortices, usually accompanied by secondary and even higher-order vortex phenomena, is present, Section 10.4. At the leeward side of the wing, strong interaction happens, the flow field there is altered completely. At the windward side, however, the flow field generally is not changed much.

This discussion of the three kinds of interaction is supported by a number of observations from experimental and theoretical/numerical flow field investigations. The observations point to the fact that to a certain extent a 'locality principle' exists. The concept of the locality principle was put forward in [8] and later independently in [9].

– Locality principle: The principle means that a local change in body shape, or the separation of flow—with or without kinematically active and inactive vorticity in the wake—changes the flow only locally or downstream of that region, respectively the separation region. Upstream of that region, the changes generally are small.[8] This holds also for subsonic flows, although mathematically their characteristic propagation properties are of elliptic nature, such that always a global interaction occurs.

  On the NS/RANS modelling level all three interaction kinds are described correctly, problems may arise regarding turbulence modelling (RANS) in the very separation region, and if the wake flow is unsteady. Due to the mathematically parabolic character of the boundary-layer equations, the BL modelling level, however, implicitly assumes semi-infinite or infinite bodies. Therefore on the BL modelling level one must ask, whether and how the interaction must be taken into account, either for phenomenological considerations or for simulation purposes. Of course on that level only the attached part of the flow can be described.

  The locality principle is observed to hold in general, except in cases, where massive separation occurs,[9] or where a structurally unstable topology of the velocity field is present, Section 7.2.3. If the principle holds, it is possible to study attached viscous flow phenomena at bodies of finite length on the boundary-layer level without taking into account the wake flow.

[8] This contradicts van Dykes statement that a wake exerts a first-order influence even in the flow upstream [10]. Of course, the integral forces and moments, which the flow exerts on the body, are affected by separation.

[9] See in this regard the flow past the helicopter fuselage discussed in Section 10.1. Despite the massive separation at the rear of the fuselage, which was not modelled in the panel method, the locality principle obviously holds.

However, if computations on the BL level are performed, if possible the global interaction should be taken into account. In the case of large-aspect-ratio lifting wings, for instance, this is done in most methods automatically with the computation of the inviscid flow. In linear (potential) methods, today usually panel methods, an explicitly imposed Kutta condition at the trailing edge serves this purpose. In Euler methods, it is the implicit Kutta condition, which is present at sharp trailing edges. In all these cases, kinematically active vorticity in one or the other form is present in the wing's wake [1].

1.3 Short Survey of the Development of the Field

Viscous flow, as we consider it here, is described by the Navier-Stokes equations. These were established, [11], in the first half of the 1800s (Claude Louis Navier, 1823, Siméon Denis Poisson, 1831, Adhémar Jean-Claude Barré de Saint-Venant, 1843, and George Gabriel Stokes, 1845). The question whether the fluid sticks to a body surface (no-slip wall boundary condition) still was not settled in the second half of the 1800s. A slip of the flow along the surface was considered to be possible [12].

The question was answered with Ludwig Prandtl's paper entitled "Über Flüssigkeitsbewegung bei sehr kleiner Reibung" ("On Fluid Motion with Very Small Friction") presented at the Third International Congress of Mathematicians in August 1904 in Heidelberg [13]. Prandtl did show that viscous effects appear mainly in a thin layer adjacent to the body surface—the boundary layer. In a few lines, he laid the foundations of the boundary-layer theory, wrote the boundary-layer equations, the boundary conditions—in fact the no-slip condition at the body surface—and suggested a possible numerical technique to solve the problem. These very inspired ideas have opened the way for understanding the motion of fluids at high Reynolds numbers.

It took a long time before Prandtl's concept was accepted. In the first decade after the presentation of his paper, only seven papers on boundary layers from five authors, all at Göttingen, were published. In the second decade six papers were published but the interest in boundary layers had spread out of Göttingen. After this period of digestion of a very new concept many studies have been devoted to this subject and it is impossible to list all the papers which have been written. Today, even if the tendency is to treat any flow problem with the Navier-Stokes equations, an aerodynamicist must have a minimum of knowledge in this field which must be considered as one of the key fields of aerodynamics.

In the first half of the 1900s predominantly problems of two-dimensional boundary layers were treated [11]. This holds also for turbulent boundary layers and for laminar-turbulent transition, see, e.g., [11, 3]. In view of three-dimensional attached viscous flow past realistic flight vehicle configurations, as treated in this book, it appears that three-dimensional boundary layers became of interest only after 1945, see also [14]. Then the German work on the swept wing and other means to shift the transonic drag divergence to higher sub-sonic Mach numbers became known.[10]

The aerodynamic and flight-mechanical problems of the swept wing instigated research and development work on three-dimensional boundary layers notably in Great Britain and in the USA.[11] In Great Britain especially two phenomena were studied early on [17]. The first phenomenon was the stalling behavior of swept wings. The three-dimensionality of the boundary layer leads to a particular kind of separation. This finally was controlled by boundary-layer fences and vortex generators. The second phenomenon was laminar-turbulent transition. It was discovered during flight tests that all but the smallest swept-wing aircraft had fully turbulent boundary-layer flow. This led to performance problems—the thrust levels of the turbo engines were still low, the fuel consumption was high. Cross-flow instability and leading-edge contamination, Section 9.3, were discovered as new transition phenomena occurring in addition to the classical Tollmien-Schlichting transition path which was well known from two-dimensional flow. Both in Great Britain and in the USA these phenomena were of concern in view of the attempts to enforce laminar flow past swept wings [18].

Instability and transition criteria then and still today are correlated with certain boundary-layer properties. This makes necessary methods for the computation of three-dimensional boundary layers. In 1945 Prandtl considered the quasi-two-dimensional flow past a swept cylinder [19]. He did show that the momentum equation for the flow in direction normal to the cylinder axis together with the continuity equation is—in a strict sense only for the laminar case—decoupled from the momentum equation for the flow parallel to the cylinder axis. The "independence principle" of Prandtl is the underlying principle[12] of the many computation methods—initially on the basis of integral relations—for the boundary-layer flow past infinite swept wings in the 1950s and 1960s, see, e.g., [14, 11].

The first finite-difference method for the solution of the three-dimensional boundary-layer equations appears to be that of G.S. Raetz, documented in 1957 [21]. (M.G. Hall notes in [22] that applications of Raetz's method are not known.) The development of the method certainly is to be seen in relation to the work on laminar flow control (LFC) initiated by J. Northrop during 1949. In 1962 finally this work at Northrop Corporation officially resulted in the X-21A program [23]. The X-21A had a leading-edge sweep of 35°, and

---

10 Besides the swept wing it was the supercritical airfoil, called at that time "Schnellflug-Profil" (fast-flight airfoil) and the area rule [15, 16].

11 It was indeed the swept wing with its design problems which in the aeronautical field triggered the work on three-dimensional boundary layers. For this reason we concentrate the following discussion somewhat on the swept wing topic.

12 In [20] the "principle of prevalence" was introduced. It assumes that the streamwise flow can be decoupled from the cross flow, if the cross flow is small (small cross-flow hypothesis), Appendix A.2.2.

laminar flow control was performed with the help of suction through slots distributed along given per cent chord lines in the wing surface. Results from the program were reported for instance in [24].

At the end of the 1960s several 3-D finite-difference methods were operational, see for instance [25], but not yet for realistic wing or fuselage shapes. Infinite swept wing solutions for compressible laminar flow appeared in the first half of the 1970s. The computational effort was still large for finite-difference codes—at that time the computer power was still small—and consequently in the second half of the 1970s integral methods were developed which made it possible to treat also turbulent three-dimensional boundary layers. At the beginning of the 1980s finally the intricacies of handling the geometry of realistic wing and fuselage shapes for three-dimensional boundary-layer computations were overcome, see, e.g., [26].

At the IUTAM symposium on Three-Dimensional Turbulent Boundary Layers 1982 in Berlin, Germany, predominantly experimental work was reported (17 papers) [27]. The USA and Japan with five papers each had the largest contingents. Only 10 papers dealt with theoretical and computational work. Apart from flow past simple shapes, boundary layers over ship hulls, road vehicles, and in turbo-machines were treated. Only one paper was devoted to an airplane topic, viz. fuselage flow.

Five presentations were given by specialists from France, Germany, the Netherlands, the United Kingdom, and the USA during a AGARD Fluid Dynamics Panel discussion on three-dimensional boundary layers on May 24 1984 in Brussels [28]. In his summary remarks on the panel discussion R.E. Whitehead, USA, mentions that turbulence modelling, already unsatisfactory for two-dimensional flow is even less suitable for three-dimensional flow. This was in particular true in cases of separated flow. (One has to note that the discussed computation results were based on the solution of the boundary-layer equations.) Flow separation cases exposed that viscous/inviscid interaction approaches have deficiencies, though inverse techniques appeared promising. Finally the lack of accurate, extensive experimental data was mentioned.[13] Insight into flow mechanisms and validation of computational results was thus hampered. The topic of laminar-turbulent transition was not mentioned by Whitehead, although R. Michel from France gave details of the state of the art. Also here the picture did not look too good.

The second author of the present book, J. Cousteix, acted in 1986 as the director of a AGARD/VKI Special Course in Rhode-St-Genèse, Belgium. The topic of the course was the computation of three-dimensional boundary

[13] Still, Whitehead notes implicitly the high quality data cited in the proceedings of the discussion [28]. B. van den Berg and A. Elsenaar for instance studied experimentally in the first half of the 1970s the three-dimensional incompressible turbulent boundary layer under infinite swept wing conditions (cited in [29]), and H.U. Meier and H.-P. Kreplin in the second half of the 1970s the three-dimensional incompressible laminar and turbulent flow past an inclined prolate spheroid (cited in [30]).

layers including separation [31]. All individual lectures treated topics and material based on the three-dimensional boundary-layer equations, two lectures treated the three-dimensional separation problem in the form of viscous-inviscid interaction approaches.

In 1988 then the third author of this book, W. Kordulla, organized an international workshop on the numerical simulation of compressible viscous-flow aerodynamics under the motto "towards the validation of viscous flow codes" [32]. "Viscous flow codes" now predominantly meant RANS codes. The shift from the boundary-layer equations to the Reynolds-averaged Navier-Stokes equations for the computation of attached and separated three-dimensional viscous flow became possible because of two developments.

The first was the general algorithm development, the second the increase of computer power, for both see, e.g., [2]. The microprocessor emerged in the early 1970s. At the time of the international workshop (1988) the chip performance was still small, but was beginning to rise strongly, Fig. 1.2.

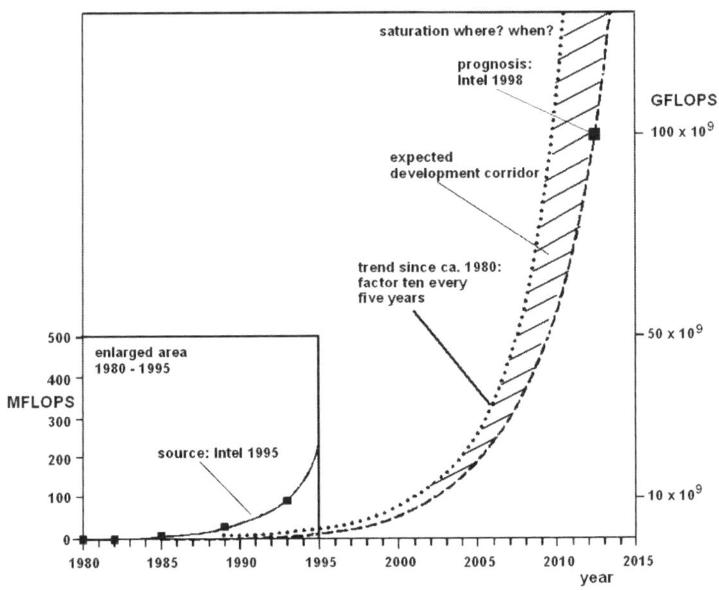

Fig. 1.2. Development of computer chip performance since the 1980s [6].

The rise of computer chip performance, together with new computer architectures, which had appeared in the mid 1980s, then led to an enormous increase of computer power, Fig. 1.3. In combination with very large and fast storage devices this made practical applications possible, and on the long run made numerical methods interesting also for industry.

What actually motivated—for both research and practical applications—the shift from boundary-layer methods towards Navier-Stokes and RANS methods? It were the phenomena associated with stream-wise and transverse

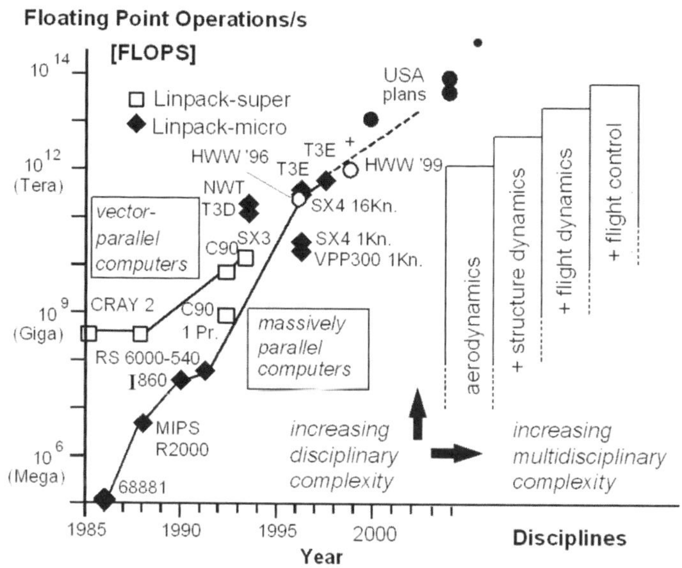

Fig. 1.3. The impact of new computer architectures since the second half of the 1980s and—schematically—the possible increase of disciplinary and multidisciplinary complexity [6].

surface curvature, the influence of vorticity of the external inviscid flow etc., and finally, of very large practical importance, separation phenomena. All these can be handled more easily in practical applications with single-domain methods rather than with two- or three-domain methods, like coupled Euler or potential equation/boundary-layer methods.

The separation phenomenon actually was the problem which led L. Prandtl to the development of the boundary-layer concept [13]. When later solutions of the boundary-layer equation became available, they showed that—in the presence of adverse pressure gradients—the skin friction decreases and eventually vanishes [33]. This phenomenon led to a host of investigations and finally to viscous/inviscid interaction and inverse methods, see the overview in Chapter 6. These methods have been developed for two-dimensional laminar flow first. For three-dimensional flow, equivalent results have been obtained, at least when the cross-wise length scale is not much shorter than the stream-wise length scale. In particular we note that interactive boundary-layer methods are available due to the work of J.C. Le Balleur et al. in France, see, e.g., [34].

Surface curvature, external-flow vorticity and related phenomena can be handled in the frame of second-order boundary-layer theory. M. van Dyke proposed this on the basis of a sound mathematical theory, viz. the formalism of matched asymptotic expansions [10]. Second-order boundary-layer methods found wide application especially for hypersonic flow computations.

A noteworthy example in Europe is the second-order boundary-layer method SOBOL, developed by F. Monnoyer in the 1980s [35, 36]. At MBB, later Dasa, Ottobrunn/München, Germany, it found application in the European (ESA) manned space plane HERMES project in the late 1980s/early 1990s. SOBOL was extended by Ch. Mundt to include equilibrium and non-equilibrium high-temperature real-gas effects. It was then employed, coupled to a corresponding Euler code by M. Pfitzner, for the determination of thermal loads on the vehicle at the windward (lower) side including the forward stagnation point region [37, 38]. The numerical simulations were probably the most complex boundary-layer computation cases ever performed: large flight speed and altitude, thick three-dimensional boundary layers, laminar flow or turbulent flow with arbitrary transition location, no-slip wall condition, longitudinally and transversally curved vehicle surface, entropy-layer swallowing, thermal radiation cooling of the surface (radiation-adiabatic wall [7]), thermo-chemical equilibrium or non-equilibrium gas model, finite-rate catalytic, fully catalytic or non-catalytic vehicle surface.

The path towards the Navier-Stokes/RANS methods was not straight forward. For blunt body re-entry problems viscous shock-layer methods, introduced in 1970 by R.T. Davis [39], found wide use. Thin-layer approximations—the name coined by B. Baldwin and H. Lomax [40]—applicable in all speed domains, were developed, as well as parabolized Navier-Stokes equations, see the review [41].

Zonal solutions, i.e. three-domain approaches, couple in the weak interaction domain the Euler equations with the, by necessity second-order boundary-layer equations, together with the equivalent inviscid wall source distribution. In the strong interaction domain finally the Navier-Stokes/RANS equations are necessarily employed. This works well for realistic two- and three-dimensional, laminar and turbulent flow cases, see, e.g., [42, 43]. However, if one uses an automatic search to find the boundary between the weak and the strong interaction domain, the computational effort becomes so large that the zonal approach has no advantage compared to a single-domain Navier-Stokes solution.

In the second half of the 1990s Navier-Stokes/RANS methods were in wide use at universities, research establishments and industry. About twenty major European methods were discussed in the year 2002 in Vol. 38 of Progress in Aerospace Sciences [44]. In the same volume, a review was published on the computational fluid dynamics (CFD) capabilities to predict high lift [45]. One year later, in 2003, F.T. Johnson et al. reported on thirty years of development and application of CFD at Boeing commercial airplanes [46], demonstrating the implementation of CFD in the design of cost-effective and high-performance commercial transport aircraft. In 2009 C.-C. Rossow and L. Cambier gave an overview of European Numerical Aerodynamics

Simulation Systems [47].[14] Each of the six major national European aerospace research establishments[15] develops, supplies and maintains now such a system. Because of the special demands on numerical aerodynamics simulation, commercial software, see, e.g., [48], so far is not much used, low speed applications partly being an exception.

Flow-physics modelling, Chapter 9 of the present book, concerns prediction and modelling of laminar-turbulent transition and modelling of turbulence. It is certainly fair to say that, of course with notable exceptions, research so far always concentrated more on the phenomena and problems of two-dimensional flow than on those of three-dimensional flow. This observation still holds today.

It was already noted above that in the attached viscous flow past wings with swept leading edge besides the two-dimensional Tollmien-Schlichting instability/transition path towards turbulence, cross-flow instability and leading-edge contamination are phenomena of interest.

With Prandtl's independence principle in mind, it was originally argued that laminar-turbulent transition would be unaffected by the sweep, see the short historical review in [49]. Flight tests at the Royal Aircraft Establishment in 1951 and 1952, however, yielded another picture [50]. Obviously leading-edge sweep led to a transition mechanism which was not the classical two-dimensional one. The concept of 'cross-flow instability' in three-dimensional boundary layers became established.

Of course the stability/instability behavior of the boundary layer along the leading edge, i.e. the attachment-line flow, must be considered, too. The question remains, whether a connection between a possible attachment-line instability and the cross-flow instability exists. It appears that this has most completely and positively been answered only in 1999 by F.P. Bertolotti [51].

Attachment line or leading-edge contamination is a phenomenon not connected to these instabilities. It is due to the turbulence of the flow coming from the fuselage/wing junction or from local disturbances at the leading edge (surface irregularities, insect cadavers) which then travels along the leading edge. The result is eventually a fully turbulent flow along the leading edge and the whole wing.

The phenomenon was discovered independently at Northrop Corporation in the USA and at Handley Page Limited in Great Britain during the early 1960s attempts to produce fully laminar wing flow by surface suction. Both W. Pfenninger from Northrop, [52], and M. Gaster from the College of Aeronautics of the Cranfield Institute of Technology, Great Britain, [53], appear to share the credit for identifying it. Industrial work on laminar flow control ceased with the respective programs in the mid 1960s, but leading-edge

[14] In [2] also overviews of algorithms and code developments in Germany (W. Haase and E.H. Hirschel), Japan (K. Fujii and N. Satofuka), Russia (Yu.I. Shokin), and the USA (B. van Leer) can be found.

[15] ONERA (France), DLR (Germany), CIRA (Italy), NLR (The Netherlands), FOI (Sweden), and ARA (United Kingdom).

contamination today is still a major challenge for skin-friction reduction on swept wings.

Several so-called energy crises during the 1970s, as well as the increasing importance of environmental compatibility, triggered new interest in flow-physics modelling for three-dimensional flow, see, e.g., [54, 55]. Laminar flow control (LFC) on swept wings and swept vertical stabilizers again was a major topic: natural laminar flow (NLF), created by proper airfoil shaping, and hybrid laminar flow (HLF), combining shaping and suction. At the same time turbulent skin-friction drag reduction by, for instance, riblets, large-eddy break-up (LEBU) devices, and so on, became a big topic, too. Around the end of the 1990s the interest declined again.

The flight at subsonic, but supercritical Mach numbers (transonic flight) became the preferred commercial flight mode in the second half of the 1900s. The related aerodynamic vehicle shape definition and the data-set generation was done for a long time and still is mostly done with tests in transonic wind tunnels which properly simulate the Mach number, but not the Reynolds number. Already in the second half of the 1960s the need for high Reynolds number transonic tunnels was a topic of AGARD. The quest for wind tunnels with true Mach number/Reynolds number simulation finally led to the so-called cryogenic wind tunnel concept.

In the USA the decision was made in 1978 to proceed at the NASA Langley Laboratory with the National Transonic Facility (NTF) [56]. In Europe cryogenic wind tunnel activities at ONERA/CERT, Toulouse and somewhat later at DLR Köln helped to pave the way towards the European Transonic Wind Tunnel (ETW), a common tunnel of the four participating countries France, Germany, The Netherlands and the United Kingdom [57, 58]. Tunnels of this kind permit a true simulation of the attached viscous flow including laminar-turbulent transition, boundary-layer displacement effects, and, if present, shock-wave/boundary-layer interaction and flow separation.

When it came to the numerical simulation of three-dimensional attached viscous flow with boundary-layer and RANS methods, the latter also simulating separation, two-dimensional statistical turbulence models were used. Of course it was known that in three-dimensional flow the turbulence is non-isotropic, that is, the vector of the shear stress is in general not parallel to the vector of the mean velocity gradient. Non-isotropic eddy viscosity models, as for instance proposed by J.C. Rotta, [59], hence should be used. In general this did not happen. It seems not to be a problem as long as the flow is not strongly three-dimensional.

When computer power became a less limiting factor, and after some algorithmic difficulties were overcome, Reynolds-stress models came into use, which in principle take into account non-isotropy of turbulence, see, e.g., [60, 61]. Turbulent flow separation still remains to be a major issue. Since about one decade zonal approaches are studied which couple RANS methods in the attached flow regime with Large-Eddy Simulation (LES) in the separation regime, see, e.g., [62]. Maybe hybrid RANS-LES methods, Section

9.6, are the ultimate industrial methods for the simulation of flow fields past realistic vehicle configurations.

However, flow-physics modelling for numerical simulation tools in general must be advanced further [2]. Non-empirical transition prediction methods, perhaps on the basis of non-local and non-linear approaches, are needed. They require receptivity models, which are also needed for turbulence modelling approaches of all kinds. Realistic operational free-stream fluctuations and noise, as well as realistic vehicle surface irregularities (roughness, holes, joints of all kinds, steps, discontinuities et cetera) affect laminar-turbulent transition as well as turbulent transport phenomena (skin friction, thermal loads).

Three-dimensional attached viscous flow and the related flow-physics phenomena still pose large research challenges. Numerical simulation tools on the other hand have a potential for both research and industrial applications which today is only beginning to be exploited.

1.4 Scope and Content of the Book

The book gives an introduction to three-dimensional attached viscous flow past realistic flight vehicle configurations. The speed domain of the vehicles ranges from low subsonic to hypersonic speeds. The flow is continuum flow, it may be laminar or turbulent, incompressible or compressible. Considered in general, the flow is steady up to its primary separation from the body surface.

The goal is to develop an understanding, i.e. knowledge of the phenomena present in the flow and their relevance for vehicle design and operation. This understanding is important already for the student, but especially for the design engineer. On the one hand, it is needed in view of the quantification of the phenomena, either with experimental or with computational means. This regards the prediction of properties and performance characteristics in a vehicle's aerodynamic shape definition process. On the other hand, it regards also problem diagnostics if, for example, it is discovered in the design verification phase that the performance goals of the vehicle are not met or if particular problems are encountered. This then may make necessary flow manipulation, either passive, by shape changes, or active, by suction or the like. For taking such decisions a deep understanding of the involved phenomena is necessary.

The following Chapter 2 provides a general characterization of attached viscous flow past realistic configurations, with stagnation points, attachment and detachment lines, symmetry lines et cetera, which in later chapters will be treated in detail. Further, the different used nomenclatures and coordinate systems are presented, and the matter of velocity profiles is examined. Basic considerations are made in Cartesian coordinates. Other coordinate systems are introduced, the most general ones being the surface-oriented non-orthogonal curvilinear (locally monoclinic) coordinates. In some of the following chapters these go together with tensorial concepts. The reader not familiar with them should not be scared. They are used only if necessary and all effort is made to foster the understanding of the derivations. Finally considered are factors — apart from the vehicle shape and the pressure field — which influence flow three-dimensionality. These are surface suction and normal injection (blowing), as well as the for high-speed applications relevant thermal surface effects,[16] the latter in conjunction with the effects of surface properties on attached viscous flows.

Chapter 3 is devoted to the presentation and discussion of the equations of motion, i.e., the Navier-Stokes equations (for laminar flow) — together with the continuity and an energy equation — in Cartesian coordinates. The chapter begins with a consideration of material and transport properties for air in the temperature interval 50 K $\lessapprox T \lessapprox$ 1,500 K. If van der Waals or high-temperature real-gas effects need consideration, the reader is referred to the pertinent literature. After the discussion of the equations of motion — perfect gas is assumed throughout —, initial and boundary conditions, similarity parameters, and boundary-layer thicknesses are discussed. Finally particularities of the equations of motion for turbulent flow are addressed. The equations of motion in general coordinates are given in Appendix A.1.

In Chapter 4 the first-order boundary-layer equations for laminar flow in Cartesian coordinates are derived. It follows the derivation for turbulent flow. After that the characteristic properties of attached viscous flow as well as the wall compatibility conditions are discussed. This is done here and not in Chapter 3, because the boundary-layer as phenomenological model of attached viscous flow permits a more convenient approach. The boundary-layer equations in non-orthogonal curvilinear coordinates, the small cross-flow formulation and the boundary-layer equations in contravariant formulation are given in Appendix A.2.

Properties of three-dimensional boundary layers, especially the integral relations like the displacement thickness et cetera, are not in any case simple extensions of those of two-dimensional boundary layers. In Chapter 5 the displacement thickness, the equivalent inviscid source distribution and other relations are discussed in Cartesian coordinates. In non-orthogonal curvilinear coordinates some of them are given in Appendix A.2.

When employing the boundary layer as phenomenological model of attached viscous flow, the so called higher-order effects — from the view of boundary-layer theory — may come into play. These are discussed as connections and interactions of viscous flow and inviscid flow in Chapter 6. Boundary-layer displacement effects are considered, higher-order effects are classified, viscous-inviscid interaction and the related methods are treated, examples are given. Information about the second-order boundary-layer equations in contravariant formulation is given in Appendix A.2.5.

---

[16] Thermal surface effects are the influence of the wall temperature and/or the temperature gradient in the gas at the wall on the flow [7].

A topic usually not of interest in two-dimensional boundary-layer theory is flow topology. In this book on three-dimensional attached viscous flow, however, the topology of skin-friction lines is of particular interest, but also that of the inviscid surface flow which is used, when the boundary-layer model is applied. In Chapter 7 the basic approach to flow topology is developed. Singular points, for instance attachment/stagnation points and lines, and separation points and lines are discussed. Topological rules are given. Whereas separation is very simply defined in two-dimensional flow, it is shown that the situation in three dimensions is totally different.

Chapter 8 is devoted to the discussion of quasi-one-dimensional and quasi-two-dimensional attached viscous flow. This concerns the stagnation point, symmetry lines and infinite-swept-wing approximations. The flows are described with the help of first-order boundary-layer formulations.

Laminar–turbulent transition and turbulence in three-dimensional attached viscous flow are the topic of Chapter 9. The most important phenomena for practical applications, criteria and models, together with their shortcomings and limitations, as well as transition and turbulence control possibilities are sketched, and references to special literature are given.

In Chapter 10 concluding illustrating examples, obtained with NS/RANS and BL simulations are presented and analyzed. Fuselage and wing flows are considered in view of some of the theoretical concepts discussed in the foregoing chapters.

Solutions of the problems which are posed at the ends of Chapters 2 to 9 are given in Chapter 11.

Appendix A presents the equations of motion in general coordinates. These are the Navier-Stokes/RANS equations, the boundary-layer equations in non-orthogonal curvilinear coordinates, the small cross-flow equations and the boundary-layer equations in contravariant formulation. Also given are the boundary-layer integral parameters in the respective curvilinear coordinates. Regarding the formulation of higher-order boundary-layer equations only references are provided.

A note on computation methods including the problem of grid generation for three-dimensional attached viscous flow is given in Appendix A.3.1. The methods are mainly discrete numerical methods, mainly from the aerospace sector, for the solution of the NS/RANS and the BL equations. A number of—in one or the other sense—exact solutions is available for two-dimensional flow. Regarding three-dimensional flow, similarity solutions are available only for some special cases. They can be quite useful for accuracy checks of, for instance, numerical solution schemes for the NS and the BL equations.

In Appendix B useful approximate relations for boundary-layer properties are provided. The latter, given in reference-temperature formulation, are valid for two-dimensional laminar and turbulent flow. They permit quick and sufficiently accurate estimations of important properties also of general attached viscous flow, if that is not too strongly three-dimensional. In particular they allow a qualitative understanding of Mach number and wall-temperature effects. The equations are useful in design work and for checking results of experimental as well as computational simulations.

Appendix C gives for two configuration classes the metric properties, which are used in the boundary-layer equations in surface-oriented locally monoclinic coordinates, as well as transformation and other needed laws and relations.

The book closes with constants, air properties, atmosphere data, units and conversions in Appendix D, symbols, abbreviations and acronyms in Appendix E, and, following the acknowledgement of copyright permissions, the author and the subject index.

References

1. Hirschel, E.H.: Vortex Flows: Some General Properties, and Modelling, Configurational and Manipulation Aspects. AIAA-Paper 96-2514 (1996)
2. Hirschel, E.H., Krause, E. (eds.): 100 Volumes of 'Notes on Numerical Fluid Mechanics'. Notes on Numerical Fluid Mechanics and Multidisciplinary Design, vol. 100. Springer, Heidelberg (2009)
3. Schlichting, H., Gersten, K.: Boundary Layer Theory, 8th edn. Springer, Heidelberg (2000)
4. Cebeci, T., Cousteix, J.: Modeling and Computation of Boundary-Layer Flows, 2nd edn. Horizons Publ., Springer, Long Beach, Heidelberg (2005)
5. Hirschel, E.H., Kordulla, W.: Shear Flow in Surface-Oriented Coordinates. NNFM, vol. 4. Vieweg, Braunschweig Wiesbaden (1981)
6. Hirschel, E.H.: Present and Future Aerodynamic Process Technologies at Dasa Military Aircraft. Viewgraphs presented at the ERCOFTAC Industrial Technology Topic Meeting in Florence, Italy. Dasa-MT63-AERO-MT-1018, Ottobrunn, Germany (October 26, 1999)
7. Hirschel, E.H.: Basics of Aerothermodynamics. Progress in Astronautics and Aeronautics, vol. 204. AIAA, Springer, Reston, Heidelberg (2004)
8. Hirschel, E.H.: On the Creation of Vorticity and Entropy in the Solution of the Euler Equations for Lifting Wings. MBB-LKE122-Aero-MT-716, Ottobrunn, Germany (1985)
9. Dallmann, U., Herberg, T., Gebing, H., Su, W.-H., Zhang, H.-Q.: Flow-Field Diagnostics: Topological Flow Changes and Spatio-Temporal Flow Structure. AIAA Paper 95-0791 (1995)
10. Van Dyke, M.: Perturbation Methods in Fluid Mechanics. Academic Press, New York (1964)
11. Tani, I.: History of Boundary-Layer Theory. Annual Review of Fluid Mechanics 9, 87–111 (1977)
12. Goldstein, S.: Fluid Mechanics in the First Half of the Century. Annual Review of Fluid Mechanics 1, 1–28 (1969)
13. Prandtl, L.: Über Flüssigkeitsbewegung bei sehr kleiner Reibung. In: Proceedings 3rd Intern. Math. Congr., Heidelberg, pp. 484–491 (1904)
14. Eichelbrenner, E.A.: Three-Dimensional Boundary Layers. Annual Review of Fluid Mechanics 5, 339–360 (1973)
15. Hirschel, E.H., Prem, H., Madelung, G. (eds.): Aeronautical Research in Germany—from Lilienthal until Today. Springer, Heidelberg (2004)

16. Meier, H.U. (ed.): German Development of the Swept Wing—1935-1945. Library of Flight, AIAA, Reston (2010)
17. Poll, D.I.A.: Personal communication (2010)
18. Lachmann, G.V. (ed.): Boundary Layer and Flow Control: Its Principles and Application, vol. 2. Pergamon Press (1961)
19. Prandtl, L.: Über Reibungsschichten bei dreidimensionalen Strömungen. Festschrift zum 60. Geburtstage von A. Betz, Göttingen, Germany, pp. 134–141 (1945)
20. Eichelbrenner, E.A., Oudart, A.: Méthode de calcul de la couche limite tridimensionnelle. Application à un corps fuselé incliné sur le vent. O.N.E.R.A. Publication 76 (1955)
21. Raetz, G.S.: A Method of Calculating Three-Dimensional Laminar Boundary Layers of Steady Compressible Flows. Northrop Aircraft, Inc., Rep. No. NAI-58-73, BLC-144 (1957)
22. Hall, M.G.: A Numerical Method for Calculating Steady Three-Dimensional Laminar Boundary Layers. Royal Aircraft Establishment, Techn. Rep. 67145 (1967)
23. Miller, J.: The X-Planes X-1 to X-45. Midland Publishing, Hinckley (2001)
24. N.N.: Recent Developments in Boundary-Layer Research. AGARDograph 97, Part IV (1965)
25. Krause, E., Hirschel, E.H., Bothmann, T.: Die numerische Integration der Bewegungsgleichungen dreidimensionaler laminarer kompressibler Grenzschichten. DGLR-Fachbuchreihe, Band 3, Braunschweig, Germany, 03-1–03-49 (1968)
26. Hirschel, E.H.: Boundary-Layer Coordinates on General Wings and Bodies. Zeitschrift für Flugwissenschaften und Weltraumforschung, ZFW 6, 194–202 (1982)
27. Fernholz, H.H., Krause, E. (eds.): Three-Dimensional Turbulent Boundary Layers. Proc. IUTAM Symp. Springer, Heidelberg (1982)
28. N.N.: Three-Dimensional Boundary Layers. Rep. AGARD FDP Round Table Discussion, Brussels, Belgium, May 24, 1984. AGARD-R-719 (1985)
29. Van den Berg, B.: Three-Dimensional Boundary-Layer Research at NLR. AGARD-R-719, 4-1–4-17 (1985)
30. Hornung, H.: Three-Dimensional Boundary Layers—A Report on Work in Germany. AGARD-R-719, 3-1–3-22 (1985)
31. N.N.: Computation of Three-Dimensional Boundary Layers Including Separation. AGARD/VKI Special Course, Rhode-St-Genése, Belgium, April 14-18, 1986. AGARD-R-741 (1987)
32. Kordulla, W.: Numerical Simulation of the Transonic DFVLR-F5 Wing Experiment. In: Proc. Int. Workshop "Numerical Simulation of Compressible Viscous-Flow Aerodynamics", Göttingen, Germany, September 30-October 2, 1987. NNFM, vol. 22. Vieweg, Braunschweig Wiesbaden (1988)
33. Goldstein, S.: Concerning some Solutions of the Boundary Layer Equations in Hydrodynamics. In: Proc. Camb. Phil. Soc. XXVI, Part I, pp. 1–30 (1930)
34. Le Balleur, J.C., Girodroux-Lavigne, P.: Calculation of Fully Three-Dimensional Separated Flows with an Unsteady Viscous-Inviscid Interaction Method. In: Proc. 5th Int. Symp. on Numerical and Physical Aspects of Aerodynamical Flows. California State University, Long Beach CA; also T.P. ONERA no 1992-1 (1992)
35. Monnoyer, F.: The Effect of Surface Curvature on Three-Dimensional, Laminar Boundary-Layers. Doctoral thesis, Université libre de Bruxelles, Belgium (1985)

36. Monnoyer, F.: Calculation of Three-Dimensional Viscous Flow on General Configurations Using Second-Order Boundary-Layer Theory. ZFW 14, 95–108 (1990)
37. Monnoyer, F., Mundt, C., Pfitzner, M.: Calculation of the Hypersonic Viscous Flow Past Reentry Vehicles with an Euler/Boundary-Layer Coupling Method. AIAA-Paper 90-0417 (1990)
38. Mundt, C., Monnoyer, F., Höld, R.: Computational Simulation of the Aerothermodynamic Characteristics for the Reentry of HERMES. AIAA-Paper 93-5069 (1993)
39. Davis, R.T.: Numerical Solution of the Hypersonic Viscous Shock-Layer Equations. AIAA J. 8(5), 843–851 (1970)
40. Baldwin, B., Lomax, H.: Thin-Layer Approximation and Algebraic Model for Separated Turbulent Flows. AIAA-Paper 78-257 (1978)
41. Rubin, S.G., Tannehill, J.C.: Parabolized/Reduced Navier-Stokes Computational Techniques. Annual Review of Fluid Mechanics 24, 117–144 (1992)
42. Wanie, K.M., Schmatz, M.A., Monnoyer, F.: A Close Coupling Procedure for Zonal Solutions of the Navier-Stokes, Euler and Boundary-Layer Equations. ZFW 11, 347–359 (1987)
43. Wanie, K.M., Hirschel, E.H., Schmatz, M.A.: Analysis of Numerical Solutions for Three-Dimensional Lifting Wing Flow. ZFW 15, 107–118 (1991)
44. Vos, J.B., Rizzi, A., Darracq, D., Hirschel, E.H.: Navier-Stokes Solvers in European Aircraft Design. Progress in Aerospace Sciences 38, 601–697 (2002)
45. Rumsey, C.L., Ying, S.X.: Prediction of High Lift: Review of Present CFD Capability. Progress in Aerospace Sciences 38, 145–180 (2002)
46. Johnson, F.T., Tinoco, E.N., Yu, N.Y.: Thirty years of Development and Application of CFD at Boeing Commercial Airplanes, Seattle. AIAA-Paper 2003-3439 (2003)
47. Rossow, C.-C., Cambier, L.: European numerical aerodynamics simulation systems. In: Hirschel, E.H., Krause, E. (eds.) 100 Volumes of 'Notes on Numerical Fluid Mechanics'. NNFM, vol. 100, pp. 189–208. Springer, Heidelberg (2009)
48. Boysan, H.F., Choudhury, D., Engelman, M.S.: Commercial CFD in the Service of Industry: The first 25 years. In: Hirschel, E.H., Krause, E. (eds.) 100 Volumes of 'Notes on Numerical Fluid Mechanics'. NNFM, vol. 100, pp. 451–461. Springer, Heidelberg (2009)
49. Poll, D.I.A.: Some Aspects of the Flow near a Swept Attachment Line with Particular Reference to Boundary Layer Transition. Doctoral thesis, Cranfield, U. K., CoA Report 7805/L (1978)
50. Gray, W.E.: The Effect of Wing Sweep on Laminar Flow. R.A.E. TM 255 (ACR 14,929), and The Nature of the Boundary Layer Flow at the Nose of a Swept Wing. R.A.E. TM 256 (ACR 15,021) (1952)
51. Bertolotti, F.P.: On the Connection between Cross-Flow Vortices and Attachment-Line Instabilities. In: Fasel, H.F., Saric, W.S. (eds.) Laminar-Turbulent Transition. Proc. IUTAM Symposium on Laminar-Turbulent Transition, Sedona, AZ, USA, pp. 625–630. Springer, Heidelberg (2000)
52. Pfenninger, W.: Flow Phenomena at the Leading Edge of Swept Wings. Recent Developments in Boundary Layer Research - Part IV, AGARDograph 97 (1965)
53. Gaster, M.: On the Flow along Swept Leading Edges. The Aeronautical Quarterly XVIII, 165–184 (1967)
54. N.N.: Fluid Dynamics of Three-Dimensional Turbulent Shear Flows and Transition. In: Proc. AGARD Symposium, Çeşme, Turkey, Oktober 3-6, 1988, pp. 3–6. AGARD-CP-438 (1989)

55. N.N.: Advances in Laminar-Turbulent Transition Modelling. NATO Research and Technology Organisation (RTO), RTO-EN-AVT-151 (2008) ISBN 978-92-837-0900-6
56. Polhamus, E.C., Kilgore, R.A., Adcock, J.B., Ray, E.J.: The Langley Cryogenic High Reynolds Number Wind-Tunnel Program. Astronautics and Aeronautics 12(10) (1974)
57. Van der Bliek, J.A.: ETW, a European Resource for the World of Aeronautics. The History of ETW in the Context of European Aeronautical Research and Development Cooperation. ETW, Köln-Porz, Germany (1996)
58. Green, J., Quest, J.: A Short History of the European Transonic Wind Tunnel ETW. Progress in Aerospace Sciences 47, 319–368 (2011)
59. Rotta, J.C.: A Family of Turbulence Models for Three-Dimensional Thin Shear Layers. In: Proc. Symp. on Turbulent Shear Flows, Pennsylvania State University, University Park, Pa., USA, April 18-20, pp. 10.27–10.34 (1977)
60. Pope, S.B.: Turbulent Flows. Cambridge University Press, Cambridge (2000)
61. Wilcox, D.C.: Turbulence Modelling for CFD, 3rd edn. DCW Industries, La Cañada (2000)
62. Fu, S., Haase, W., Peng, S.-H., Schwamborn, D. (eds.): Progress in Hybrid RANS-LES Modelling. NNFM, vol. 117. Springer, Heidelberg (2012)

2

Properties of Three-Dimensional Attached Viscous Flow

In this chapter first three-dimensional attached viscous flow is characterized, having in mind, in particular, flow past fuselage/wing configurations with the forward stagnation point, attachment and detachment lines, symmetry lines et cetera, which in later chapters will be treated in detail. Boundary-layer (BL) concepts serve to understand details. Then the coordinate systems and nomenclatures, which are used in this book, are presented. Basic considerations usually are made in Cartesian coordinate systems. External streamline-oriented coordinate systems are a kind of natural coordinates for three-dimensional boundary layers, whereas surface-oriented non-orthogonal curvilinear (locally monoclinic) coordinate systems are the most general ones. Again details can be found in later chapters. Finally some means of flow control are discussed especially in view of their potential influence on flow three-dimensionality. Suction and surface-normal injection into the flow (blowing), as well as the consequences of thermal surface effects are considered, partly with the help of results of numerical investigations. A few problems, posed at the end of the chapter, will help the reader to develop a feeling for the quantities of flow properties.

2.1 Characterization of the Flow

The flow past a body exhibits, beginning, for instance, at the forward stagnation point of a fuselage, a thin layer close to the body surface, where viscous effects play a role. They are due to the fact that in the continuum regime the fluid sticks to the surface which is called the no-slip wall boundary condition.[1] We speak about attached viscous flow. Outside of this layer the flow is considered to be inviscid, i.e. viscous forces can be neglected there. Of course the inviscid flow field behind a flight vehicle, and at large angle of attack also above it, contains vortex sheets and vortices, which are viscous phenomena. At transonic and supersonic speeds, also shock waves may be present, which interact with the viscous flow. Here we consider basically only attached viscous flow without vortex or shock-wave interactions.

[1] In hypersonic flight at high altitudes or in vacuum machinery the flow may stick not fully to the surface, and we have slip-flow boundary conditions.

E.H. Hirschel, J. Cousteix, and W. Kordulla, *Three-Dimensional Attached Viscous Flow*,
DOI: 10.1007/978-3-642-41378-0_2,

If the body under consideration is an airplanes fuselage, see, e.g., Section 10.3, the flow downstream of the stagnation point first encounters a blunt-cone like surface and then a more or less cylinder-like surface portion. The flow—at zero and small angles of attack—generally is only weakly three-dimensional. At the end of the fuselage, which in general will have a boattailing in order to permit a high angle of attack of the airplane during take-off or landing, the flow, partly considerably three-dimensional, leaves—separates from—the surface. The Mangler effect, Section 8.7, plays a role both at the front part and, as reverse effect, at the rear part of the fuselage.

If the fuselage is at an angle of attack, at the lower symmetry line an attachment line exists. Both the attached viscous flow and the inviscid flow diverge from this line and stream up the fuselage sides, i.e. the flow is three-dimensional. At the upper symmetry line then a detachment line exists, Sub-Section 7.1.5, where the inviscid flow leaves the body, whereas the boundary layer only thickens. Both holds only, if the angle of attack is small, otherwise the flow will separate at the sides of the fuselage.

At a large-aspect-ratio wing, Section 10.3, the flow coming from the forward part of the fuselage bifurcates at the wing root and partly becomes—at the wing's leading edge—the attachment line flow. The attachment line in general is slightly curved. From the attachment line the flow diverges and streams over the upper (suction-) and the lower (pressure-) side of the wing. If the wing is swept[2], the flow at the leading edge is highly three-dimensional. Downstream of the leading edge, depending on the sweep angle of the wing and on the character of the static pressure distribution in downstream direction—different on the upper and the lower side—, the flow is only slightly three-dimensional on large portions of the wing's surface.[3] Similar flow situations are found at the elevator assembly, at the vertical stabilizer and also at small-aspect-ratio wings.

We consider now the steady flow past a fuselage or a wing. We put at some suitable place at the configuration an imaginary straight stem normal to the surface. The streamlines crossing this stem are forming a stream surface. The flow speed along this stream surface is changing continuously and usually slowly as long as no embedded shock waves are present. In direction normal to the surface the speed changes also only slowly. Near to the body surface, however, in the viscous layer—the boundary layer—the speed drops very fast to zero at the body surface (no-slip wall boundary condition). At large Reynolds numbers, this layer is very thin. Above this layer, viscous forces are negligible small, and we have the already mentioned external inviscid flow.

---

[2] Transonic transport jet airplanes have leading-edge sweep angles of up to approximately $\varphi_{LE} = 40°$, see, e.g., [1].

[3] This is a characteristic of rooftop and super-critical transonic airfoils which have rather flat pressure distributions in chord direction in contrast to so-called peaky airfoils which have a suction peak on the upper side close to the airfoil's nose.

The thickness of the viscous layer[4] needs a special consideration, because it is not clearly defined. A practical definition of the thickness would be—coming from the body surface—the location of vanishing vorticity: $|\omega|_{vl} \leqq \epsilon$ [4]. In the frame of boundary-layer theory the outer edge of a laminar boundary layer lies at infinity [5]. The outer edge of a turbulent boundary layer in reality has a rugged unsteady pattern. In the frame of statistical turbulence theory (RANS model) it is defined as smooth time-averaged edge, see, e.g., [6]. For practical purposes one usually defines the boundary-layer thickness as the location, where the surface-tangential velocity of the boundary-layer flow has reached 99 per cent of the velocity of the "external" inviscid flow [5].

The viscous layer is indeed very thin. Its thickness is inversely proportional to some power of the Reynolds number. In order to get a feeling for that, we have made a simple estimation, assuming two-dimensional incompressible boundary-layer flow past a flat surface and using relations given in Appendix B.[5] Table 2.1 gives the estimated boundary-layer thicknesses δ and displacement thicknesses δ_1 at different down-stream locations x. For fully laminar flow the thicknesses δ are in the cm domain, even at the 50 m location. For fully turbulent flow we find larger thicknesses, but still in the cm domain. The displacement thicknesses δ_1, see the discussion of weak interaction in Section 1.2, are very small, too, such that the boundary-layer approach is valid.

Now we return to the discussion of the stream surface. If it is not skewed downstream of the imaginary stem, neither in the "inviscid", nor in the "boundary-layer" flow part, we have got two-dimensional flow throughout. Fig. 2.1 a) shows the boundary-layer part of the stream surface.

To be exact, we have two-dimensional "mean" flow. If the viscous layer is turbulent, we find there—beginning already in the laminar-turbulent transition zone—small-scale three-dimensional fluctuations and hence also unsteadiness. This holds for the inviscid flow part, too, where generally atmospheric free-stream turbulence is present[6]. These fluctuations, however, are disregarded in the following discussion.

---

[4] The classical viscous layer or boundary layer is a phenomenon connected to momentum transport. A general consideration shows that transport of (thermal) energy and of (molecular) mass gives rise to related phenomena, viz. the thermal and the mass-concentration boundary layer. All three layers, although of different thickness, can co-exist in flow fields past high-velocity flight vehicles, in propulsion systems, in devices of process engineering, et cetera. The reader interested in this regard is referred to, e.g., [2, 3].

[5] The reader is asked to solve Problem 2.1 to obtain own experience.

[6] In a wind tunnel free-stream turbulence is a special phenomenon, which receives particular attention, because it influences laminar-turbulent transition [8], for transonic and supersonic flow see [9, 10].

Table 2.1. Estimated boundary-layer thicknesses δ and displacement thicknesses δ_1 at different down-stream locations x on a flat surface. Incompressible flow, unit Reynolds number $Re^u_\infty = 6{,}847{,}400\ \mathrm{m}^{-1}$ ($v_\infty = 100\ \mathrm{m\,s}^{-1}$, $\rho = 1.225\ \mathrm{kg\,m}^{-3}$, $\mu_\infty = 1.789 \cdot 10^{-5}\ \mathrm{N\,s\,m}^{-2}$), the boundary-layer flow is assumed to be either fully laminar or fully turbulent.

x [m]	δ_{lam} [cm]	$\delta_{1,lam}$ [cm]	δ_{turb} [cm]	$\delta_{1,turb}$ [cm]
1	0.19	0.066	1.59	0.20
5	0.43	0.15	5.76	0.72
10	0.60	0.21	10.03	1.25
50	1.35	0.46	36.33	4.55

If the stream surface is skewed, the (mean) flow is three-dimensional, and that much stronger in the attached viscous layer, the boundary-layer, Fig. 2.1 b), than in the inviscid part.

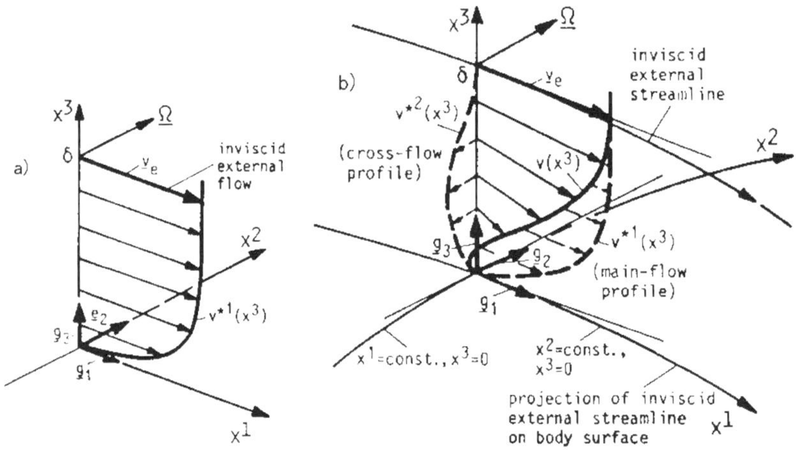

Fig. 2.1. Wall-near viscous part of the stream surface, schematics of boundary-layer profiles [7]: a) two-dimensional flow (not skewed), b) three-dimensional flow (skewed). External inviscid streamline-oriented coordinate systems: x^1 and x^2 are the surface-tangential coordinates with $v^{*1}(x^3)$ and $v^{*2}(x^3)$ the related stream-wise (main-flow) and cross-flow viscous-flow profiles, x^3 is the surface-normal coordinate. The resulting velocity profile in the 3-D case is $v(x^3)$. $\underline{\Omega}$ is the vector of vorticity content of the boundary layer [7].

To understand this better, we consider it in the frame of the BL picture, Fig. 2.1. There the inviscid flow is considered only as "external inviscid flow", i.e., as flow at the edge of the boundary layer. This means in the two-dimensional BL case, that the external inviscid flow is one-dimensional, and in the three-dimensional BL case, that it is two-dimensional. In both cases the static pressure of the external inviscid flow is impressed on the boundary layer with zero or nearly zero gradient in direction normal to the surface, as long as the flow is in the weak interaction regime.

With that we see the difference between two-dimensional and three-dimensional attached viscous flow. In the two-dimensional case only a stream-wise pressure gradient exists in the external inviscid flow: pressure-gradient vector and inviscid external flow vector are coincident. The pressure gradient is either negative (accelerated flow), zero (the classical flat plate case) or positive (decelerated flow, leading finally to separation).

In the three-dimensional case pressure-gradient vector and inviscid external flow vector are not coincident. Besides the stream-wise, or main-flow[7] pressure gradient, also a cross-flow pressure gradient exists. The main-flow external pressure gradient acts like in two-dimensional flow, whereas the cross-flow pressure gradient leads to the lateral deflection (curving) of the streamlines of the external inviscid flow. In the viscous layer then the cross-flow pressure gradient acts on a deflected flow with decreasing speed in direction towards the body surface. The result is that the streamlines in the viscous layer are much more curved than the streamlines of the external inviscid flow.

This can easily be understood by a consideration of circular flow, with a look at the centrifugal term only, see, e.g., [2]:

$$\rho \frac{v_\theta^2}{r} = \frac{\partial p}{\partial r}. \tag{2.1}$$

With the external pressure gradient $\partial p/\partial r$ acting throughout the viscous layer, the inscribed streamline radius $r(z)$ must reduce towards the surface, because the circular velocity term $v_\theta^2(z)$ reduces towards the surface. At the surface, for $z = 0$ and $v_\theta = 0$ (no-slip wall boundary condition), this however does not lead to $r = 0$. The more one approaches the surface, the more viscous forces play a role, which are neglected completely in eq. (2.1). Except for some singular points, the curvature radius of skin-friction lines is always non-zero.

This explains why the pressure field with its main-flow and cross-flow gradients, as impressed on the viscous layer, leads to the skewing of the stream surface, which is stronger there than in the inviscid part of the flow

[7] The main-flow and the cross-flow direction are related to the external-streamline coordinates, Fig. 2.1 and Section 2.2.

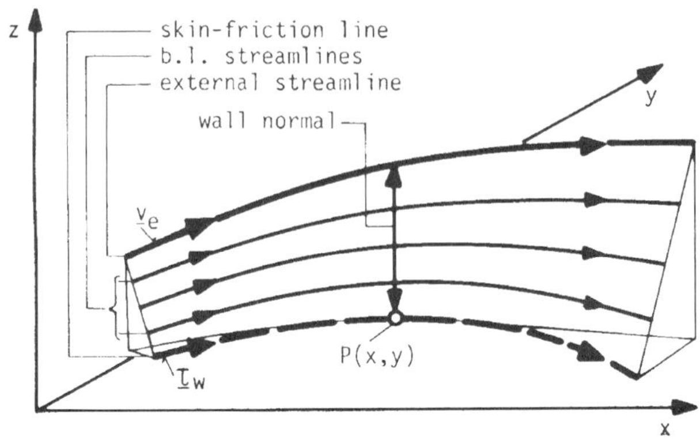

Fig. 2.2. Schematic of stream-surface skewing in three-dimensional attached viscous flow [13]. Cartesian coordinate system, x and y are the surface coordinates, z is the coordinate normal to the surface. The wall normal imaginary stem (not indicated as such) is located at $P(x, y)$ on the surface. Its extension into the external inviscid part of the flow field is not indicated.

field.[8] In Fig. 2.2 this is schematically illustrated. The matter of skewing of the stream surface will be of interest again in Section 4.4, where the characteristic properties of attached viscous flow are considered.

We draw the following conclusions [13]:

1. Any boundary-layer streamline including the skin-friction line is curved in the same sense but stronger than the inviscid external streamline.[9]
2. The skin-friction lines, for instance from an oil-flow picture, do not have the same direction as the external streamlines, if the boundary layer is three-dimensional.
3. Any deceleration in main-flow direction (point of inflection appears in the stream-wise or main-flow profile, Fig. 4.3) leads to a strong deflection in cross-flow direction (three-dimensional separation, wing trailing-edge flow).

---

[8] This three-dimensionality of the attached viscous flow field is a pressure-field driven three-dimensionality. The cross-flow in this case can be considered as secondary flow. This would be Prandtl's secondary flow of the first kind. Other kinds of three-dimensionality exist, for instance that driven by rotation of an axisymmetric body, see, e.g., [5]. Three-dimensionality driven by anisotropic turbulence is Prandtl's secondary flow of the second kind [11], see also, e.g., [12]. It appears in corner flow, curved pipes, and in curved channels. Other kinds of three-dimensionality exist, for instance due to flow unsteadiness (secondary flow of the third kind) and of course the above mentioned small-scale three-dimensionality in transitional and turbulent flow, including vortices of cross-flow instability, and Görtler vortices, Chapter 9.

[9] If the inviscid external streamline has a point of inflection, the situation is different, see below.

Besides the variation of the direction of the streamlines along the normal to the wall—the stream-surface skewing—also the divergence of the streamlines, for instance downstream of the stagnation point of a fuselage, see page 24 above, is considered as characteristic for attached three-dimensional viscous flow. This effect can be understood from an axially symmetric flow on a body of revolution whose axis is aligned with the free-stream. In this case, no stream-surface skewing occurs but flow divergence is observed. Consequently the boundary-layer has a tendency to get thinner, compared to a two-dimensional flow. This happens because the section of the body becomes larger and individual stream tubes are stretched in width which implies the thinning of the boundary layer. This so-called Mangler effect is treated in Section 8.7.

2.2 Coordinate Systems and Velocity Profiles

Coordinate systems have two aspects in the frame of this book. The first one is the discussion of the equations of motion and the analysis of phenomena, both basic ones and those connected to real configurations. For this discussion mostly the BL view is taken, and boundary-layer coordinates are employed. The other is the application of numerical —NS/RANS and also BL[10]—methods for flow simulations on such configurations. This aspect is discussed in Appendix A.3.

When making use of the boundary-layer hypothesis, we employ in this book Cartesian, streamline-oriented, and general surface-oriented non-orthogonal curvilinear coordinates. In all of these coordinate systems one line of the coordinates is always chosen to be normal to the wall. In Fig. 2.2 it is the wall normal—the imaginary stem—in $P(x, y)$. The other two lines of coordinates are always constructed from two families of lines defined on the body surface.

In general, it is not possible to construct a triply orthogonal axis system in this way. However, in first-order boundary-layer studies it is possible to introduce simplifying hypotheses. The metric coefficient along the normal to the wall is taken as unity, see below, and the metric coefficients along the other two axes do not depend on the distance to the wall. This means that the construction of a coordinate system reduces to the definition of the two families of lines lying on the body surface (locally monoclinic coordinates, see below). Many choices are possible, we consider three of them.[11]

[10] Three-dimensional boundary-layer methods usually are space-marching methods. These must obey spatially the Courant-Friedrichs-Lewy (CFL) condition, see, e.g., [14]. That makes a special orientation of the surface coordinates, as well as special discretization approaches necessary, see Appendix A.3.2.

[11] The authors of this book regret that it was not practical to use only one coordinate system and only one kind of axis nomenclature throughout. The reader therefore is asked to have an alert eye on the coordinate systems and nomenclatures used in a given derivation, discussion or numerical example.

2.2.1 Cartesian Coordinates

Most convenient to use are Cartesian coordinate systems.[12] They are usually right-handed systems, with the x-coordinate pointing in main-flow direction, the y-coordinate in lateral direction, whereas the z-coordinate is normal to the surface.[13] The related velocity components are u, v, w. The profiles of the surface-tangential velocity components u and v can have any form, depending on the evolution of the flow field, which depends on the pressure field of the external inviscid flow.

In the literature on two-dimensional inviscid or viscous/boundary-layer flow past flat surfaces the stream-wise coordinate and hence the coordinate along the surface customarily is the x-coordinate and that normal to the surface the y-coordinate, with u and v the respective velocity components. Often also the equations of three-dimensional flow are given in this manner. Then the z-coordinate is the lateral coordinate, u and w are the surface-tangential velocity components. This has the advantage that he reader can stay with the familiar 2-D notation for the direction normal to the surface.

For this reason we also use this convention in Chapter 4, where the three-dimensional boundary-layer equations are derived and discussed, and also in some other chapters and sections. However, we also use other conventions and notations, for instance in many of the figures given in this book. Of course, we point out for every figure the actually used coordinate system and notation.

2.2.2 External Inviscid Streamline-Oriented Coordinates

These coordinates are in a sense "natural" coordinates for three-dimensional boundary layers. In such a system, the stream-wise flow locally is contained in a plane normal to the wall and tangential to the external streamline. The streamline of the inviscid flow at the boundary-layer edge is projected onto the body surface, Figs. 2.1 and 2.3. The cross-flow direction is locally contained in a plane normal to the wall and normal to the external streamline. The advantage of this streamline coordinate system is that the stream-wise velocity profile looks like a two-dimensional one and the value of the cross-flow velocity is zero at the wall and at the edge of the boundary layer.

It can be shown that, if an external inviscid streamline is a surface geodesic, Appendix A.2.3, a possible solution is the so-called zero cross-flow solution. When the flow is not too far away from the geodesic condition, the three-dimensionality is small and the stream-wise flow is prevalent. The so-called "principe de prévalence" introduced by Eichelbrenner and Oudart,

---

[12] These coordinates are more generally called orthonormal Cartesian coordinates.

[13] This resembles the convention mostly used in airplane aerodynamics. There x is the coordinate lying in the airplane's axis, pointing backwards, y is the coordinate pointing outwards in the direction of the right wing, and z points upwards.

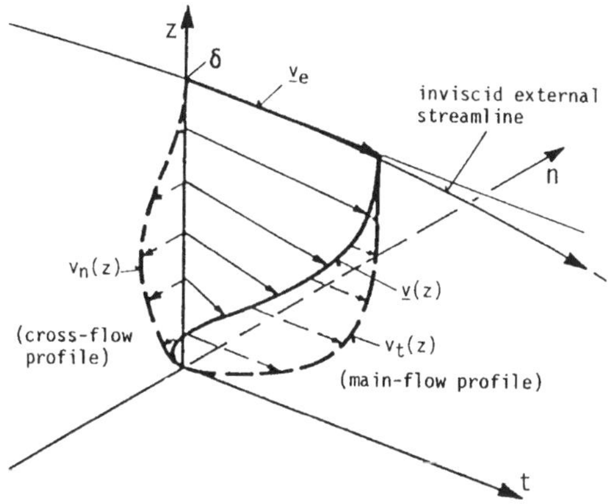

Fig. 2.3. Right-handed external inviscid streamline-oriented coordinate system on a flat surface [13]. The surface coordinates t and n locally are orthogonal to each other, the surface-normal coordinate is z. $v_t(z)$ is the main-flow or stream-wise flow profile, $v_n(z)$ the cross-flow profile. The resulting velocity profile is $\underline{v}(z)$. At $z = \delta$ we have $v_{t_e} = v_e$.

describes such cases [15]. In laminar flow, the continuity equation and the stream-wise momentum equation—in which the cross-flow is neglected—form the same system of equations as the axially symmetric boundary-layer equations, Appendix A.2.2.

We introduce the external inviscid stream-line oriented coordinates as right-handed coordinates, too, Fig. 2.3: the t-coordinate and the n-coordinate lie in the surface, the z-coordinate lies normal to it. The t-coordinate is oriented locally tangential at the surface-projection of the external inviscid streamline. The stream-wise velocity profile in the figure, the main-flow profile $v_t(z)$, resembles that of a two-dimensional boundary layer at zero or positive pressure gradient, Fig. 4.3.[14]

The n-coordinate is oriented normal to the surface-projected main-flow streamline. As we will see below, the cross-flow profile $v_n(z)$ can have very different shapes. In this case it points in negative n-direction due to the indicated curvature of the external inviscid streamline. $v_n(z)$ is zero at both the surface, $z = 0$, and the external inviscid streamline, the boundary-layer edge, $z = \delta$.

Because the cross-flow profile meets tangentially the inviscid flow profile at the boundary-layer edge, Section 4.5, it has a point of inflection in its upper part. The resulting total flow profile $\underline{v}(z)$ is skewed, Fig. 2.3.

In Fig. 2.3 the cross-flow profile points in negative n-direction, because the external inviscid streamline has a negative curvature (the second derivative is negative). If the external streamline would have a positive curvature,

---

[14] See also Fig. 2.1.

the cross-flow profile would point in positive n-direction. However, in reality a "history" effect can exist. If the external inviscid streamline has a point of inflection, the cross-flow profile will swing over, but not immediately. It therefore initially may have an s-like, Fig. 2.4 b), or even more complicated profile.

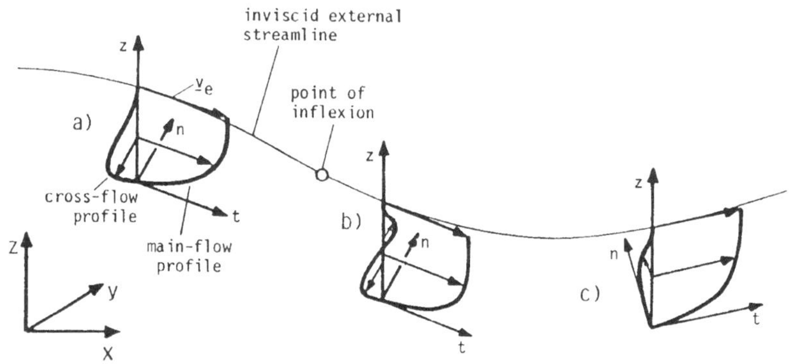

Fig. 2.4. Cross-flow profile dependence on the curvature of the external inviscid streamline (schematically) [13]. Curvature of the external inviscid streamline: a) negative, b) zero, c) positive. Coordinate system like in Fig. 2.3.

2.2.3 Surface-Oriented Non-Orthogonal Curvilinear Coordinates

These coordinates are more general than the external inviscid stream-line oriented coordinates. They basically take into account the shape of the body surface under consideration. Because we have realistic flight vehicle configurations in the background of our discussions, we consider in Appendix C two canonical coordinate systems: fuselage cross-section coordinates and percent-line wing coordinates.

Here first we introduce a change of notation. Fig. 2.5 a) shows the classical notation of surface coordinates, for convenience only the percent-line coordinates of flat wing shapes.[15] The shape is embedded in the orthogonal x, y reference coordinate system. The coordinate in chord direction is the x-coordinate, which sometimes is also called ξ-coordinate. In span direction, however along the percent-lines, we have the z- or η-coordinate. The coordinate normal to the surface—not indicated—may be called n- or ζ-coordinate.

In Fig. 2.5 b) we change to the index notation. The orthogonal reference coordinate system is now denoted as $x^{1'}, x^{2'}$ system. The span-wise (chord percent-line) coordinates—the x^2-coordinates—are called $x^1 =$ const. coordinates, whereas the chord-wise (span percent-line) coordinates—the x^1-coordinates—are called $x^2 =$ const. coordinates. The coordinate normal to the surface is the x^3-coordinate. Note that the span-wise coordinate x^2 now

[15] "Percent-line" here holds both in chord and in span direction.

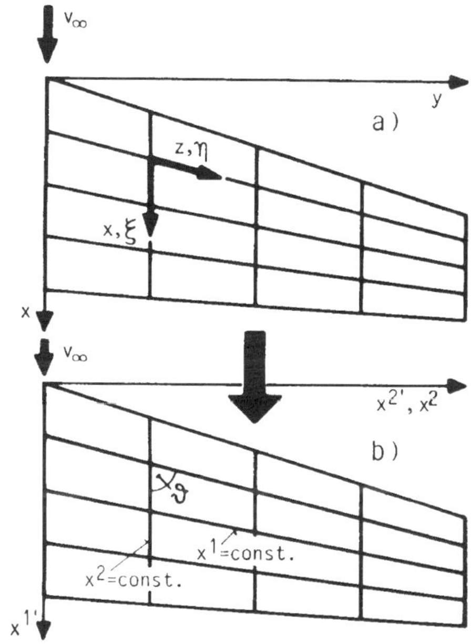

Fig. 2.5. Flat-wing percent-line coordinates [13]: a) classical notation, b) index notation. The surface-normal coordinates are not indicated.

is counted along the $x^{2'}$-coordinate of the Cartesian reference coordinate system.

This all is generalized in Fig. 2.6, where a surface element is embedded in the orthogonal $x^{j'}$ reference system ($j' = 1,2,3$). The general boundary-layer coordinates (x^i-system, $i = 1,2,3$) are defined on the surface. The x^1-coordinates, i.e. the lines x^2 = const. and the x^2-coordinates, i.e. the lines x^1 = const. lie on the surface. The x^3-coordinate is always rectilinear and locally normal to the surface—therefore we speak also of "surface-oriented locally monoclinic coordinates". Both x^1 and x^2, or x^α, $\alpha = 1,2$—called Gaussian parameters—in general have no length properties. Both parameters are not necessarily counted along the coordinate lines.

In Fig. 2.6 the Cartesian reference coordinate system has the base vectors $\underline{e}_{j'}$, $j' = 1,2,3$, which are unit vectors. The coordinate base on the surface element is called a covariant base, see, e.g., [16]. The covariant base vectors are $\underline{a}_1$ and $\underline{a}_2$, for details see Appendix C. The third base vector $\underline{a}_3$ is a unit vector which points in x^3-direction, Appendix C.1.

The surface-tangential velocity components again can have any form, depending on the orientation of the flow field relative to the surface coordinate system. Stream-wise and cross-flow profiles, for instance, are found only in external streamline-oriented coordinates.

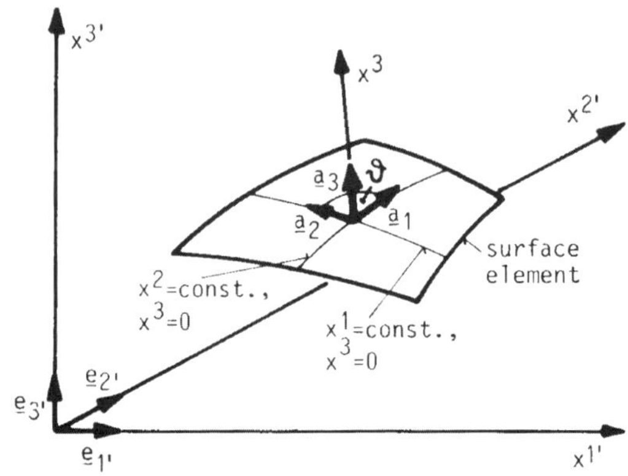

Fig. 2.6. Surface element in general surface-oriented locally monoclinic non-orthogonal curvilinear coordinates [13]. The orthogonal reference system has the axes $x^{1'}$, $x^{2'}$, $x^{3'}$. The curvilinear surface-oriented coordinates are—on the surface of the element—the x^1-coordinate (x^2 = const., $x^3 = 0$) and the x^2-coordinate (x^1 = const., $x^3 = 0$). The rectilinear surface-normal coordinate is the x^3-coordinate.

Surface-oriented non-orthogonal curvilinear (locally monoclinic) coordinates are very appropriate if phenomena with surface-normal characteristics in accordant formulation are to be described. In our case it holds for the boundary-layer equations, their discussion and solution. The appropriate formulation is given in Appendix A.2.4. For NS/RANS solutions other approaches are taken, Appendix A.1.

Surface-oriented curvilinear coordinates, however, have restrictions, if flow past concave surface portions is to be treated. This is due to the fact that the coordinate normal to the wall plays a particular role. Fig. 2.7 shows such a situation.

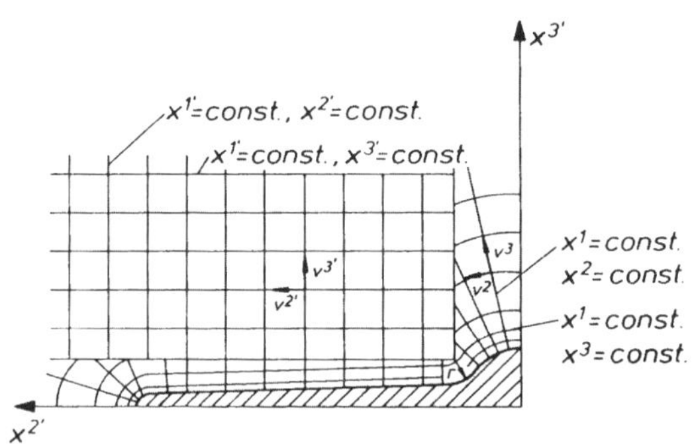

Fig. 2.7. Schematic of surface-oriented curvilinear coordinates at a concave surface portion with embedded Cartesian coordinates [4]. Coordinate convention like in Fig. 2.6.

No problem exists, if the thickness of the viscous layer is small compared to the radius r of the concave surface. Otherwise, special measures must be taken. In the worst case the problem is not treatable with these coordinates.

In Appendix C we give the necessary details for the creation and application of surface-oriented non-orthogonal curvilinear coordinates for two canonical shapes, viz. fuselage- and wing-like configurations.

2.3 Influencing Attached Viscous Flow and Flow Three-Dimensionality

Assume a flight vehicle with given configuration and attitude at steady level flight in a calm atmosphere with given speed v_∞ at altitude H. Hence the shape of the elastic vehicle is known as well as free-stream temperature $T_\infty(H)$ and density $\rho_\infty(H)$. If further the boundary-layer receptivity scenario (atmospheric fluctuations, vehicle surface properties, noise, vibrations, Section 9.4) is known, location and shape of the laminar-turbulent transition zone are known, too. In this case the flow past the vehicle—the attached viscous flow field being of particular interest for us—macroscopically is fully determined. It can only be influenced through the vehicle's surface, and that in several ways.

In this section we discuss these ways in a quantitative manner, concentrating on three-dimensional flow. The results in principle hold for two-dimensional flow, too. When dealing with the wall compatibility conditions in Section 4.5, we will also obtain analytical results. They are valid for two-dimensional flow, but certainly can be extended to not too strongly three-dimensional flow.

In the following sub-section we look at active flow manipulation by suction, in the next one by surface-normal injection (blowing) and in the third one at the—usually—passive influence of thermal surface effects. We are not looking at the consequences which these manipulations have on vehicle design; we only wish to illustrate how viscous flow and in particular its three-dimensionality is influenced by them.

Only in passing is considered the influence which surface properties—roughness, waviness, steps, gaps and the like—can have on the flow development. They influence laminar-turbulent transition and, especially in turbulent flow, wall-shear stress and heat transfer in the gas at the wall. We come back to these topics in Section 9.4.

In any case we have the requirement that the said manipulations do not disturb the boundary-layer character of the attached viscous flow. This means, an introduced disturbance must be sufficiently small. We note further that in some cases we have knowledge from two-dimensional boundary-layer flow only. However, if a flow under consideration is not too strongly three-dimensional, we can carry over the knowledge from two-dimensional flow.

The application background is briefly mentioned, see, e.g., [17]–[19]: suction is applied for the control of laminar-turbulent transition, shock-wave/boundary-layer interaction and flow separation, surface-normal injection (blowing) for surface cooling.

Thermal surface effects are a special case. They are of utmost importance in the high-speed flight domain, in particular for hypersonic flight. This holds for entry/re-entry and especially for airbreathing airplane-like vehicles [3]. In the lower flight regimes, they are of importance, too. This regards, for instance, boundary-layer stability—in view of laminar-turbulent transition (a topic particularly of interest in cryo wind-tunnel technology)—, flush heat exchangers located at a wing's or a fuselage's surface, propulsion systems of all kinds and process engineering in general. Even skin-friction control via surface heating is an interesting option.

In any case we consider each flow manipulation item separately. In reality some of these manipulations may be present simultaneously which will lead to overall (combined) effects on the flow field.

2.3.1 Surface Suction

Suction influences attached viscous flow by removing low stream-wise momentum flow. This leads to a fuller tangential velocity profile and a thinner boundary layer. In three-dimensional flow the fuller profile then is better able to balance the cross-flow pressure gradient. Hence suction reduces flow three-dimensionality. This is in direct analogy to the phenomenon that turbulent two-dimensional flow can negotiate a larger adverse pressure gradient than laminar flow before separating.

We illustrate the suction effect with results from a numerical simulation, [20]. Consider the three-dimensional incompressible laminar boundary-layer flow over a flat surface, Fig. 2.8.

The external inviscid flow has the property $u_e = 1$, $v_e = x$,[16] such that the static pressure reduces linearly with y, Problem 2.3. The external inviscid streamlines are a host of parabolas $y = 0.5\, x^2 + C$, where C is an arbitrary constant. Suction is applied in a square region in the domain $0.13 \leqq x \leqq 0.4$ and $1.13 \leqq y \leqq 1.4$. The stretched (with $\sqrt{Re_L}$) suction velocity $w_{surface} = -1.0$ is ramped up linearly from the edges of the square ($w_{surface} = 0$) over a distance $\triangle x = 0.06$.

Fig. 2.8 shows that the skin-friction lines crossing the suction area indeed are deflected towards the external inviscid streamlines. The skin-friction line, which begins at $y = 1.08$, is the most affected one, because it fully crosses the suction area.

[16] This is one of the few two-dimensional inviscid flow fields, for which a similar solution of the resulting three-dimensional boundary layer exists [21]. See also Appendix A.3.3.

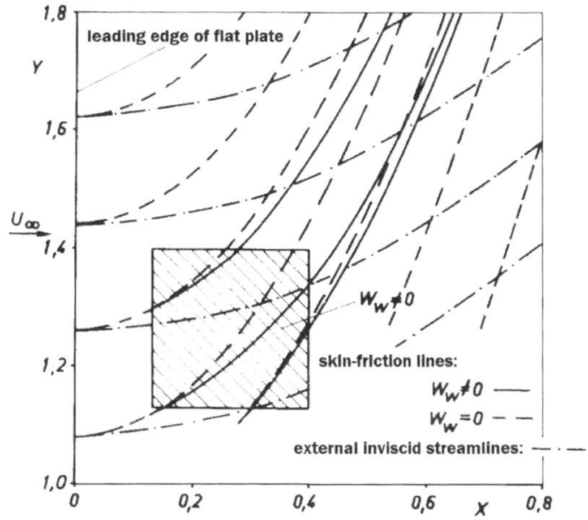

Fig. 2.8. Suction in a three-dimensional incompressible laminar boundary layer [20]. The square hatched area is the suction region. The surface is flat, x and y are the Cartesian surface coordinates, u and v are the surface-tangential velocity components. The surface-normal velocity component is w, with $w_w \equiv w_{z=0}$.

Suction can act as a virtual boundary-layer fence without the side effects which go together with a real solid fence. In [20] it was shown that it is possible to force the skin-friction lines completely to follow the external inviscid streamlines. This was done with a suitable suction distribution.

The effect of such strong suction on the velocity profiles of the boundary layer is illustrated in Fig. 2.9.

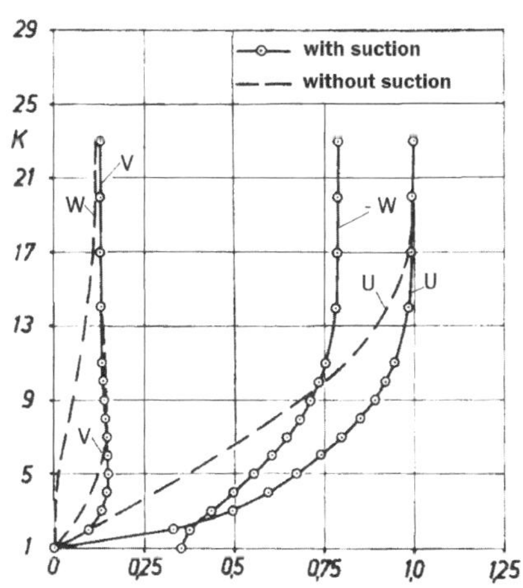

Fig. 2.9. Suction as virtual boundary-layer fence: velocity profiles $u(K)$ and $v(K)$ at $x = 0.125$ in the suction area, Fig. 2.10 [20]. The normal velocity component $w(K)$ is normalized. K is the parameter of the distance z normal to the wall. Coordinate and velocity component convention like in Fig. 2.8.

The location of the profiles is in the suction area, Fig. 2.10, at $x = 0.125$. In this case the normalized suction velocity at the wall is $w_w = -0.35$. The normal velocity component at the boundary-layer edge is $w = -0.78$, compared to $w = 0.12$ for the case without suction. The u-profile (x-direction) is much fuller in the case with suction than without. The v-profile (y-direction) is not so much changed. From the u-profile we see clearly that the boundary layer thickness is reduced. Note that the velocity gradient in direction normal to the surface at the wall is enlarged for both u and v. This means a larger skin-friction for the case with suction which is typical for such kind of flow and easy to understand.

In Fig. 2.10 the effect of suction as virtual boundary-layer fence is shown. The skin-friction lines are indeed forced to follow more or less the direction of the external inviscid streamlines.

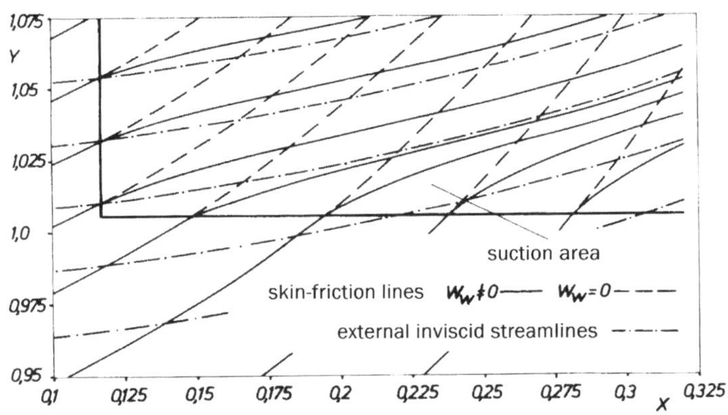

Fig. 2.10. Suction as virtual boundary-layer fence: pattern of external inviscid streamlines and of skin-friction lines without and with suction [20]. Coordinate and velocity component convention like in Fig. 2.8.

This demonstrates well the possibility to influence three-dimensionality of attached viscous flow. It shows also that, for instance, on swept wings laminar flow control with suction will influence to a degree the three-dimensionality of the flow. Further, laminar flow manipulated in this way has a larger skin friction than the non-manipulated flow. However, this skin friction is still smaller than the one, which one would have if the flow had become turbulent.

2.3.2 Surface-Normal Injection (Blowing)

Surface-normal injection into the boundary layer—blowing—adds flow with normal momentum, and effectively reduces the momentum of the tangential flow components as shown in Fig. 2.11 [22]. The undisturbed flow there is the same as above, where the suction effect was studied. The skin-friction

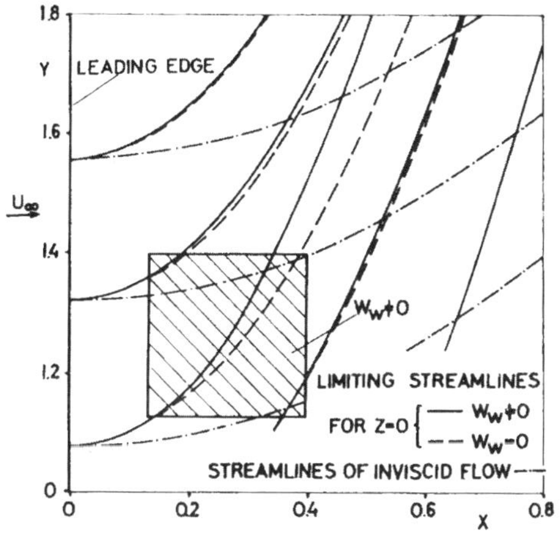

Fig. 2.11. Deflection of skin-friction lines (limiting streamlines) due to normal injection (blowing) into a three-dimensional boundary layer [22]. The square hatched area is the blowing region. Coordinate and velocity component convention like in Fig. 2.8.

lines crossing the hatched blowing are more strongly curved in the case with blowing, compared to that without blowing.

Fig. 2.12 shows the velocity profiles at a location in the blowing area which is the hatched square in Fig. 2.11.

In the case with blowing—again being ramped up in the same fashion as described for the suction case—the normalized normal velocity at the wall is $w_w = 0.2$.

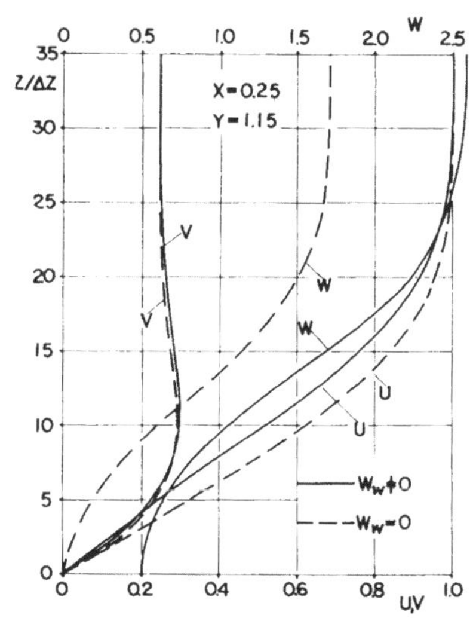

Fig. 2.12. Velocity profiles in the blowing area [22]. Coordinate and velocity component convention like in Fig. 2.8.

The normal velocity component at the boundary-layer edge is $w = 2.6$, compared to $w = 1.7$ for the case without blowing. The u-profile (x-direction) now is less full than in the case without blowing. Moreover, with blowing it has a point of inflection at $z/\triangle z \approx 8$. See in this respect Section 4.5. The v-profile (y-direction) is not much changed. The boundary-layer thickness has increased. Note that the velocity gradient in direction normal to the surface is reduced for both u and v. This means a smaller skin-friction for the case with blowing which is typical for this kind of flow and also easy to understand.

2.3.3 Thermal Surface Effects

Thermal surface effects are due to both the temperature in the gas at the wall and the temperature gradient in the gas at the wall normal to it [3]. In the continuum regime the gas temperature at the wall and wall temperature are the same: $T_{gw} = T_w$. This does not hold for the temperature gradient. As long as heat transport by thermal radiation q_{rad} towards and away from the wall is not present, and tangential heat conduction is negligible, only the heat fluxes in the wall q_w and in the gas in direction normal to the wall q_{gw} are equal [23]. The only exception in this case is the adiabatic wall, where both gas and wall temperature are the same, as well as the heat fluxes which are zero.

The classical consideration of boundary layers adds to the topic of compressibility simply the topic of heating. If one is concerned with high-speed flight, with cryo wind-tunnel technology, with propulsion systems etc., a more detailed view is necessary, i.e. to work only with the Stanton-number concept is not sufficient. In [3] instead of the latter the concept of the 'thermal state of the surface' was introduced as well as the concept of 'thermal surface effects'.

The thermal state of the surface is defined by *both* the wall temperature T_w and the wall-normal temperature gradient in the gas at the wall $(\partial T/\partial n)|_{gw}$. The thermal state of the surface governs the thermal loads on the wall, as well as the thermal surface effects, Fig. 2.13.

Two basic kinds of thermal surface effects are distinguished [3]: (1) viscous and (2) thermo-chemical effects. For us of primary interest are viscous thermal surface effects.[17] These encompass, for instance, the increase of the boundary-layer thickness, the displacement thickness etc. with increasing wall temperature, the lowering (!) of skin friction with increasing wall temperature, for turbulent flow stronger than for laminar flow, the increase of separation disposition with increasing wall temperature, and the stabilization/destabilization of the boundary layer depending on the heat flux in the gas at the wall and the wall temperature. Influenced too are wall heat flux, shock wave/boundary-layer interaction, and hypersonic viscous

[17] An approximative qualitative discussion is possible with the help of the generalized reference-temperature relations given in Appendix B.3.

interaction. Thermo-chemical thermal surface effects concern surface catalycity and transport properties at and near the vehicle surface.

For a more detailed discussion of the implications of the thermal state of the surface and thermal surface effects the reader is referred to [3] and to [23, 24].[18]

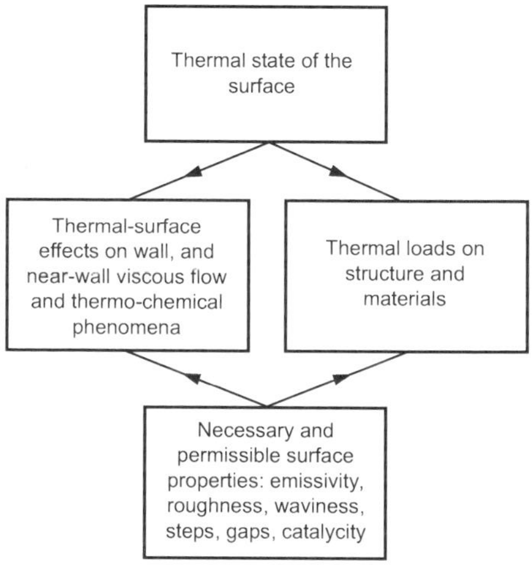

Fig. 2.13. The thermal state of a (hypersonic) vehicle surface and its different aero-thermal design implications [3].

Indicated in Fig. 2.13 are also surface properties, which are of deciding importance for a number of flow phenomena. In view of the attached viscous flow treated in this book we note only that surface roughness, for instance, is an important parameter in laminar-turbulent transition, and, once the flow is turbulent, it enlarges skin friction and wall heat transfer, Sub-Section 9.4.1.

We consider now some viscous thermal surface effects:

- **Temperature and density distribution in the direction normal to the wall.** The influence of the gas temperature gradient at and normal to the wall on the wall-normal gradient of the tangential velocity is similar to that of the stream-wise pressure gradient or of suction and normal injection. This tells us the discussion of the wall-compatibility conditions for two-dimensional flow, Section 4.5.

  A positive temperature gradient—heat is transported out of the boundary layer into the wall (cold wall, cooling of the boundary layer)—acts like a favorable pressure gradient or like suction: the velocity profile becomes fuller. A negative temperature gradient—heat is transported out of the wall into the boundary layer (hot wall, heating of the boundary layer)—acts like

[18] The for many design issues important thermal surface effects generally are not or only marginally treated in the bulk of text books on boundary-layers.

an adverse pressure gradient or like surface-normal blowing: the velocity profile gets an inflection-point, see Table 4.1 and Fig. 4.3 in Section 4.5, page 93 f..

This all holds only for air and gases in general, but not for liquids. The reason is that for gases the viscosity increases with temperature, whereas it decreases in liquids.

These results for two-dimensional flow also hold for three-dimensional flow, as long as the three-dimensionality is not too strong. This means that a positive temperature gradient will act like wall suction, decreasing three-dimensionality, and a negative one like normal injection, increasing three-dimensionality.

Regarding the density, we recall that in attached viscous flow the gradient of the static pressure normal to the wall is small and in the large Reynolds number limit of flat-plate flow even zero. This means that in direction normal to the wall the pressure is equal—or nearly equal—to the pressure at the outer edge of the boundary layer, i.e., to that of the external inviscid flow:

$$p \approx p_e. \tag{2.2}$$

Consequently, with the equation of state, Sub-Section 3.1.1, we have in the boundary layer

$$\rho T = \rho_e T_e = constant, \tag{2.3}$$

and hence we obtain the proportionality

$$\rho \propto \frac{1}{T}. \tag{2.4}$$

This means that a hot wall leads to a small density at and above the wall. For a cold wall then we find a large density. In this case the average tangential momentum flux $< \rho u^2 >$ is larger than in the hot-wall case. In analogy to the effect of the stream-wise pressure gradient we again can hence expect—for compressible flow—that a cold wall has the tendency to reduce and a hot wall to increase flow three-dimensionality, see below.

We demonstrate with a computational example the influence of the wall temperature on the laminar boundary layer past a flat plate [25]. With the help of a numerical solution of the Navier-Stokes equations the influence of the wall temperature on the boundary layer was studied, Fig. 2.14.

We find five effects in the figure. (1) With the larger wall temperature the average density $< \rho >$ in the boundary layer decreases. (2) If the same mass flow is present, the boundary layer thickness δ increases as well as the displacement thickness δ_1. (3) The velocity gradient $\partial u/\partial z$ at the wall is reduced and—although the viscosity μ at the wall is enlarged—(4) the wall shear stress τ_w is reduced (see below). (5) Because of the smaller density the average momentum flux $< \rho u^2 >$ is reduced, hence the boundary layer

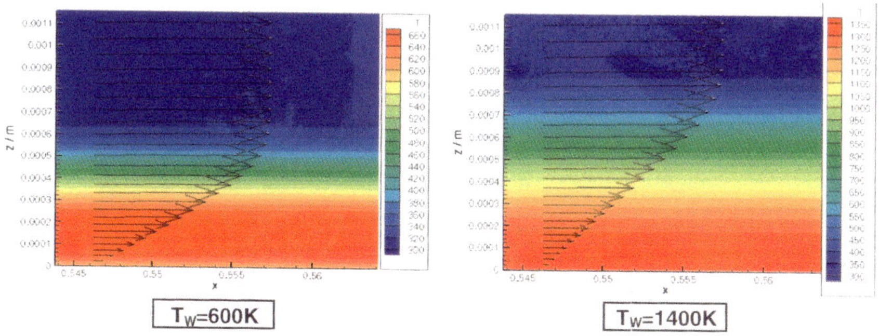

Fig. 2.14. Tangential velocity profiles at a flat plate with a wall temperature (left) of T_w = 600 K and (right) of T_w = 1,400 K (note the different color bar scales) [25]. Laminar flow, $M_\infty = 4$, $Re_{ref} = 2.036 \cdot 10^7$, $L_{ref} = 0.467$ m.

at the hot wall is more at separation risk, see the HOPPER example on page 46.

– **Dependence of boundary-layer thicknesses on the wall temperature.** The reference temperature approach, proposed by several authors in the late 1940s and in the 1950s, see, e.g., [26, 27], permits to take into account in a simple approximate way the influences of the wall temperature, of the temperature of the external inviscid flow, and of the recovery temperature—hence the flight Mach number—on the boundary-layer properties. The reader can find generalized reference-temperature relations for two-dimensional laminar and turbulent flow in Appendix B.3 of this book. For the purpose of our discussion of special interest are the boundary-layer thicknesses δ_{lam} and δ_{turb}, and the displacement thicknesses $\delta_{1,lam}$ and $\delta_{1,turb}$. Regarding the displacement thickness in three-dimensional flow see Chapter 5. The characteristic boundary-layer thickness Δ_c regarding wall shear stress and wall heat flux for laminar flow is the boundary-layer thickness δ_{lam}, and for turbulent flow the thickness of the viscous sub-layer δ_{vs}, Appendix B.3.1.

  We consider flat-plate flow and employ the generalized reference-temperature formulations given in Appendix B.3.1. For the viscosity coefficient the power-law formulation given in Sub-Section 3.1.2 is used, with $\omega = 1$ for $T \lessapprox 200$ K, and $\omega = 0.65$ for $T \gtrapprox 200$ K.

  Being interested in the dependence of the characteristic thicknesses Δ_c as well as of the boundary-layer thicknesses δ and the displacement thicknesses δ_1 on the reference temperature ratio T^*/T_∞, we obtain the following proportionalities (note that they are the same for the thicknesses δ and the displacement thicknesses δ_1):

$$\Delta_{c,lam}(= \delta_{lam}) \propto \left(\frac{T^*}{T_\infty}\right)^{0.5\,(1+\omega)}, \quad \Delta_{c,turb}(= \delta_{vs}) \propto \left(\frac{T^*}{T_\infty}\right)^{0.8\,(1+\omega)}, \tag{2.5}$$

and

$$\delta_{lam} \propto \left(\frac{T^*}{T_\infty}\right)^{0.5\,(1+\omega)}, \quad \delta_{1,lam} \propto \left(\frac{T^*}{T_\infty}\right)^{0.5\,(1+\omega)}, \tag{2.6}$$

$$\delta_{turb} \propto \left(\frac{T^*}{T_\infty}\right)^{0.2\,(1+\omega)}, \quad \delta_{1,turb} \propto \left(\frac{T^*}{T_\infty}\right)^{0.2\,(1+\omega)}. \tag{2.7}$$

The result is: the larger the wall temperature T_w and with it the reference temperature T^*, the larger are the thicknesses. The influence of T_w is weakest on δ_{turb} and $\delta_{1,turb}$ and strongest on δ_{vs}. Assumed is $T^* > T_\infty$ and moderate temperature differences. These results are for two-dimensional flow, but they hold also for not too strongly three-dimensional flow.

In three-dimensional attached viscous flow the displacement thickness δ_1 can become negative. This may happen in regions with strong flow divergence, for instance, at attachment lines. For an example see Section 10.1. In two-dimensional flow δ_1 can become negative, too, if the wall is strongly cooled. Then, according to eq. (2.4), the density becomes large at the wall and above it. (The above proportionalities do not hold for this case.)

We illustrate this with an example of rocket nozzle boundary-layer flow [28]. The figure is by courtesy of M. Frey. The turbulent flow with high-temperature real-gas modelling was computed with the method described in [29]. The flow, Fig. 2.15, is characterized by the following data at the boundary-layer edge: $M_e = 2.23$, $u_e = 3{,}119.51$ m s^{-1}, $\rho_e = 0.1881$ kg m^{-3}, $T_e = 2{,}418.24$ K. The edge unit Reynolds number, with $\mu_e = 7.27 \cdot 10^{-5}$ Pa s, is $Re^u_e = \rho_e\, u_e/\mu_e = 8.071 \cdot 10^6$ m^{-1}.

The highly cooled nozzle wall, $T_w = 510$ K, leads to a high density at the wall, almost five times larger (not shown in the figure) than that at the boundary-layer edge, and to a tangential mass flux ρu at maximum almost twelve per cent larger than that at the edge, Fig. 2.15. The resulting displacement thickness, eq. (5.2), is negative, with $\delta_1 = -0.27$ m.

In closing this item, we consider body surface properties, in particular surface roughness. In laminar flow domains surface roughness influences laminar-turbulent transition. In turbulent flow domains it influences, if effective, strongly both wall shear stress and wall heat flux. For turbulent flow, for instance, with a hotter surface and therefore a thicker viscous sub-layer a larger surface roughness can be tolerated.

How much flow three-dimensionality is influenced by an effective roughness is not known. If roughness is effective, and the wall shear stress increases, this would mean also an increase of the wall-near gradient of the tangential velocity profiles; then one can presume an effect of surface roughness similar to that of suction.

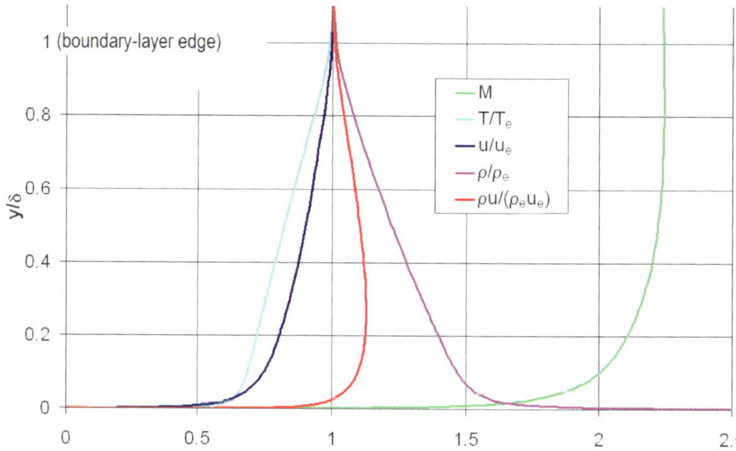

Fig. 2.15. Rocket-nozzle flow downstream of the nozzle throat [28]: Profiles of Mach number M, temperature T/T_e, velocity u/u_e, density ρ/ρ_e, and mass-flow $\rho u/(\rho_e u_e)$ in the boundary layer. The boundary-layer edge values are given in the text.

– **Dependence of skin friction on the wall temperature.** The skin friction depends on the wall temperature, too. This can be demonstrated for flat-plate flow also with the generalized reference-temperature formulations, Appendix B.3. A simpler consideration is the following one. We assume that we can approximate the skin-friction relation by

$$\tau_w = \mu \frac{\partial u}{\partial y}|_w \approx \mu \frac{u_e}{\triangle}. \tag{2.8}$$

For laminar flow we take as the characteristic thickness $\triangle = \delta_{lam}$, with the proportionality eq. (2.7), and for turbulent flow $\triangle = \delta_{vs}$, with the proportionality eq. (2.6).

With the above used power-law formulation of the viscosity coefficient we arrive at

$$\tau_{w,lam} \propto \left(\frac{T^*}{T_\infty}\right)^{-0.5\,(1-\omega)}, \quad \tau_{w,turb} \propto \left(\frac{T^*}{T_\infty}\right)^{-0.8\,(1-0.25\omega)}. \tag{2.9}$$

The result is: the larger the wall temperature T_w and with it the reference temperature T^*, the smaller is the skin friction. The reason is that for both laminar and turbulent flow the characteristic thickness—being in the denominator of eq. (2.8)—rises stronger than the viscosity in the numerator[19]. The reduction of skin friction is appreciably stronger for turbulent flow than for laminar flow.

[19] Note that the viscosity at the wall is the same for laminar and turbulent flow.

We illustrate the influence of the wall temperature on the skin friction with numerical simulation data of the flow past the windward side of the re-entry vehicle HOPPER, Fig. 2.16 [25].[20]

In this case we observe three thermal surface effects. For the case with larger wall temperature (right side of the figure) we see (1) a distinctly smaller skin-friction coefficient than for the cold temperature case (left side).

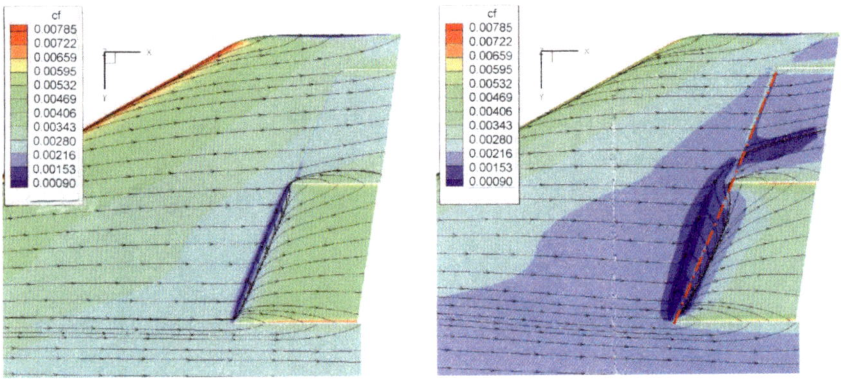

Fig. 2.16. Computed influence of the wall temperature on the skin-friction at the left lower-side aft part of the HOPPER configuration [25]. $M_\infty = 3.2$, $Re_{L,\infty} = 2.31 \cdot 10^7$, $L = 50.2$ m, turbulent flow, $\alpha = 15°$, deflection of the inboard wing flap: $\eta_{iwf} = +20°$ (downward). Left part: radiation-adiabatic wall temperature $T_w = T_{ra} \approx 500$ K (surface emissivity coefficient $\varepsilon = 0.8$), right part: $T_w \approx 1{,}600$ K.

(2) The patterns of the skin-friction lines show that the flow three-dimensionality is enlarged in the hotter case, however, only slightly. (3) a considerably larger separation zone appears around the hinge line of the flap. Flow three-dimensionality is enlarged on the flap, flap efficiency is reduced [25].

This reflects the influence of the wall temperature on the tangential velocity profile of the boundary layer, see Fig. 2.14. The wall-near momentum flux is reduced and—if an adverse pressure gradient is present—the separation disposition is enlarged.

This is the result for a winged re-entry vehicle, where the skin-friction drag reduction is not so important. The reduction is important for any high-speed airbreathing flight vehicle, [24], and also for elements of propulsion systems.

[20] For a detailed discussion of this case see [23].

We learn, too, that a flow-control measure may lead to a lower drag, but that this may have as a side effect the degradation of the efficiency of a trim or control surface. If one locally manipulates the flow on a part of a flight vehicle, one always has to have a look also at the whole vehicle.

2.4 Problems

Problem 2.1. In Table 2.1 boundary-layer thicknesses are given for a flat plate case. Compute for that case with the relations given in Appendix B the thicknesses also at the locations $x = 20$ m, $x = 40$ m. Verify the thicknesses given in Table 2.1.

Problem 2.2. In view of Fig. 2.7: what is the limitation of curvilinear monoclinic coordinate systems for realistic configurations?

Problem 2.3. Determine for the external inviscid flow field shown in Fig. 2.8 a) the pressure gradients $\partial p/\partial x$ and $\partial p/\partial y$. Give b) the function $p = p(x, y)$ and c) the streamline function $y = f(x)$. Assume $\rho = 1$.

Problem 2.4. We consider the flow in a supersonic wind tunnel. The tunnel is fed by atmospheric air at a pressure of 10^5 Pa and at a temperature of 300 K. Downstream of the nozzle throat the flow is supersonic in the inviscid region. The air is considered as perfect gas.

A flat plate with a very small thickness without incidence is placed in the region of uniform flow where the Mach number is equal to $M_e = 2$.

1. Give the value of the stagnation or total temperature T_t and of the static temperature T_e at the edge of the boundary layer on the plate.
2. Calculate the wall temperature of the plate when the thermal equilibrium is achieved. Consider the cases of a laminar and of a turbulent boundary layer.
3. What happens in a nozzle in which the Mach number would be $M_e = 5$?

Problem 2.5. We consider the flow past a blunt-nosed body in supersonic flight with $M_\infty = 2$ at the altitude $H = 10$ km. It is assumed that the body is axisymmetric and its incidence is zero. The boundary-layer flow is laminar. The reference axis-system moves with the body.

1. Calculate the total temperature T_{t_∞} of the free stream in the reference axis-system.
2. Is it permitted to assume perfect gas?
3. Calculate the wall temperature at the forward stagnation point of the body.
4. It is assumed that the wall is adiabatic. Calculate the wall temperature at a point where the Mach number at the edge of the turbulent boundary layer is $M_e = 3$.
5. Are the results restricted to axisymmetric bodies?

Problem 2.6. Consider an airplane flying with $M_\infty = 0.8$ at $H = 10$ km. Approximate the fuselage as circular cylinder with length $l = 10$ m and diameter $d = 1$ m. Assume the boundary-layer thickness to be small compared to the fuselage radius and estimate, see below, several boundary-layer properties at the fuselage with the relations for two-dimensional flat-plate flow and the reference temperature, Appendix B.

Assume perfect gas, fully turbulent flow, a Prandtl number $Pr = 0.74$. The wall temperature T_w is: case a) the recovery temperature T_r, and case b) a temperature 50 K higher. Use the power-law approximation for the viscosity coefficient with $\omega = 0.65$.

1. Compute for the two cases at the mid-location $L/2 = x = 5$ m the boundary-layer thickness δ, the displacement thickness δ_1, the thickness of the viscous sublayer δ_{vs}, and the wall shear stress τ_w.
2. Compute δ_1 also with the alternate formulation eq. (B.16).
3. Compare and discuss the outcome of the two cases.

Problem 2.7. Estimate the skin-friction drag D_{visc} acting on the fuselage for the two cases of Problem 2.6. Compare and discuss the outcome of the two cases.

Problem 2.8. The tangential velocity profile in the case $T_w = 1{,}400$ K in Fig. 2.14 has a weak point of inflection near to the wall. What is the reason for that?

Problem 2.9. Why does a turbulent boundary-layer flow separate later than a laminar one assuming the same positive pressure gradient?

Problem 2.10. How is the thermal state of a surface defined. What does it govern?

Problem 2.11. List the major viscous thermal surface effects.

References

1. Raymer, D.P.: Aircraft Design: A Conceptual Approach, 4th edn., Reston, Va. AIAA Education Series (2006)
2. Bird, R.B., Stewart, W.E., Lightfoot, E.N.: Transport Phenomena, 2nd edn. John Wiley & Sons, New York (2002)
3. Hirschel, E.H.: Basics of Aerothermodynamics. Progress in Astronautics and Aeronautics, AIAA, Reston, Va., vol. 204. Springer, Heidelberg (2004)
4. Hirschel, E.H., Kordulla, W.: Shear Flow in Surface-Oriented Coordinates. NNFM, vol. 4. Vieweg, Braunschweig Wiesbaden (1981)
5. Schlichting, H., Gersten, K.: Boundary Layer Theory, 8th edn. Springer, Heidelberg (1999)
6. Wilcox, D.C.: Turbulence Modelling for CFD, 3rd edn. DCW Industries, La Cañada (2000)

7. Eberle, A., Rizzi, A., Hirschel, E.H.: Numerical Solutions of the Euler Equations for Steady Flow Problems. NNFM, vol. 34. Vieweg, Braunschweig Wiesbaden (1992)
8. Saric, W.S., Reshotko, E.: Review of Flow Quality Issues in Wind Tunnel Testing. AIAA Paper 98-2613 (1998)
9. Pate, S.R.: Effects of Wind Tunnel Disturbances on Boundary-Layer Transition With Emphasis on Radiated Noise: a Review. AIAA Paper 80-0431 (1980)
10. Dougherty, N.S., Fisher, D.F.: Boundary-Layer Transition on a 10°-Cone: Wind Tunnel/Flight Data Correlation. AIAA Paper 80-0154 (1980)
11. Prandtl, L.: Führer durch die Strömungslehre. Fiedr. Vieweg & Sohn, Braunschweig (1942)
12. Townsend, A.A.: Turbulence. In: Streeter, V.L. (ed.) Handbook of Fluid Dynamics, pp. 10-1–10-33. McGraw-Hill, New York (1961)
13. Hirschel, E.H.: Evaluation of Results of Boundary-Layer Calculations with Regard to Design Aerodynamics. AGARD R-741, 5-1–5-29 (1986)
14. Hirsch, C.: Numerical Computation of Internal and External Flow, 2nd edn. Fundamentals of Computational Fluid Dynamics, vol. 1. Elsevier, Amsterdam (2007)
15. Eichelbrenner, E.A., Oudart, A.: Méthode de calcul de la couche limite tridimensionnelle. Application à un corps fusel'e incliné sur le vent. O.N.E.R.A. Publication 76 (1955)
16. Aris, R.: Vectors, Tensors, and the Basic Equations of Fluid Mechancis. Prentice Hall, Englewood Cliffs (1962), unabridged Dover republication (1989)
17. Thiede, P. (ed.): Aerodynamic Drag Reduction Technologies. Proc. of the CEAS/DragNet European Drag Reduction Conference, Potsdam, Germany, June 19-21. NNFM, vol. 76. Springer, Heidelberg (2001)
18. Stanewsky, E., Délery, J., Fulker, J., de Matteis, P. (eds.): Drag Reduction by Shock and Boundary Layer Control. Results of the Project EUROSHOCK II, Supported by the European Union, 1996-1999. NNFM, vol. 80. Springer, Heidelberg (2002)
19. King, R. (ed.): Active Flow Control II. Papers contributed to the Conference "Active Flow Control II 2010", Berlin, Germany, May 26-28. NNFM, vol. 108. Springer, Heidelberg (2010)
20. Krause, E., Hirschel, E.H., Bothmann, T.: Die numerische Integration der Bewegungsgleichungen dreidimensionaler laminarer kompressibler Grenzschichten, DGLR-Fachbuchreihe, Band 3, Braunschweig, Germany, 03-1–03-49 (1968)
21. Yohner, P.L., Hansen, A.G.: Some Numerical Solutions of Similarity Equations for Three-Dimensional Incompressible Boundary-Layer Flows. NACA TN 4370 (1958)
22. Krause, E., Hirschel, E.H., Bothmann, T.: Normal Injection in a Three-Dimensional Laminar Boundary Layer. AIAA J. 7(2), 367–369 (1969)
23. Hirschel, E.H., Weiland, C.: Selected Aerothermodynamic Design Problems of Hypersonic Flight Vehicles, Reston, Va. Progress in Astronautics and Aeronautics, AIAA, vol. 229. Springer, Heidelberg (2009)
24. Hirschel, E.H., Weiland, C.: Design of Hypersonic Flight Vehicles: Some Lessons from the Past and Future Challenges. CEAS Space J. 1(1), 3–22 (2011)
25. Häberle, J.: Einfluss heisser Oberflächen auf aerothermodynamische Flugeigenschaften von HOPPER/PHOENIX (Influence of Hot Surfaces on Aerothermodynamic Flight Properties of HOPPER/PHOENIX). Diploma Thesis, Institut für Aerodynamik und Gasdynamik, Universität Stuttgart, Germany (2004)

26. Rubesin, M.W., Johnson, H.A.: A Critical Review of Skin Friction and Heat Transfer Solutions of the Laminar Boundary Layer of a Flat Plate. Trans. ASME 71, 385–388 (1949)
27. Eckert, E.R.G.: Engineering Relations of Friction and Heat Transfer to Surfaces in High-Velocity Flow. J. Aeronautical Sciences 22(8), 585–587 (1955)
28. Frey, M.: Personal communication (2011)
29. Nickerson, G.R., Dang, L.D., Coats, D.E.: Two Dimensional Reference Computer Program. NAS 8-35931, MSFC, Huntsville Ala. (1985)

3

Equations of Motion

In this chapter the equations of motion for three-dimensional attached viscous flow are discussed. Most of the material given is valid for any type of flow, because it concerns the Navier-Stokes equations. Assumed is Newtonian fluid, thermally perfect gas, and steady flow. At the beginning of the chapter the material and transport properties of air are presented. Then the equations of motion—called summarily Navier-Stokes (NS) equations[1]—are considered for the transported entities mass, momentum, and energy. The basic considerations are made in Cartesian coordinates for laminar flow. Initial and boundary conditions as well as similarity parameters and boundary-layer thicknesses are given. After that the particularities of the equations of motion for turbulent flow—the Reynolds-averaged Navier-Stokes (RANS) equations—are discussed. The time-dependent NS/RANS equations in general coordinates are given in Appendix A.1.

3.1 Material and Transport Properties of Air

We give the material and transport properties of air for moderate temperatures and pressures as is the custom in fluid mechanics and aerodynamics, see, e.g., [1]. We assume absence of van der Waals effects, but permit thermally perfect (though calorically imperfect) gas, i.e. equilibrium vibration excitation. In view of the simulation problems, which the practitioner usually meets, we provide the data of air for the temperature range 50 K $\lessapprox T \lessapprox$ 1,500 K.[2] We note that below approximately 400 K air can be considered as calorically and thermally perfect gas.

3.1.1 Equation of State and Specific Heat at Constant Pressure

The equation of state relates the static pressure p with the density ρ and the static temperature T:

[1] To be precise, the Navier-Stokes equations only describe the transport of momentum. However, the term is often used in a wider context to include the continuity equation and the energy equation.

[2] At the low temperature side of this interval possible condensation and liquefaction effects must be regarded.

E.H. Hirschel, J. Cousteix, and W. Kordulla, *Three-Dimensional Attached Viscous Flow*,
DOI: 10.1007/978-3-642-41378-0_3,

$$p = \rho RT, \tag{3.1}$$

where R is the specific gas constant.[3]

For our purposes it suffices generally to consider air as a non-dissociated binary gas consisting of molecular nitrogen N_2 and molecular oxygen O_2.[4] We assume thermodynamic equilibrium. The specific heat at constant pressure, [3], then can be determined from

$$c_p = 3.5R + \omega_{O_2} c_{v_{vibr_{O_2}}} + \omega_{N_2} c_{v_{vibr_{N_2}}} \tag{3.2}$$

with the specific heats of the vibration energy at constant volume:

$$c_{v_{vibr_{O_2}}} = R \left(\frac{\Theta_{vibr_{O_2}}}{T} \right)^2 \frac{e^{\Theta_{vibr_{O_2}}/T}}{(e^{\Theta_{vibr_{O_2}}/T} - 1)^2} \tag{3.3}$$

and

$$c_{v_{vibr_{N_2}}} = R \left(\frac{\Theta_{vibr_{N_2}}}{T} \right)^2 \frac{e^{\Theta_{vibr_{N_2}}/T}}{(e^{\Theta_{vibr_{N_2}}/T} - 1)^2}. \tag{3.4}$$

The mass fractions of air in the low-temperature range are $\omega_{O_2} = 0.26216$, and $\omega_{N_2} = 0.73784$, see, e.g., [4]. The characteristic vibration temperatures Θ_{vibr} of N_2 and O_2 are given in Appendix D.1.

Fig. 3.1 shows in terms of the ratio of specific heats γ that non-negligible vibration excitation sets in already at around $T = 400$ K. It depends on the case under consideration and on the needed accuracy above what temperature air can not be treated anymore as calorically and then thermally perfect gas.

3.1.2 Transport Properties

The molecular transport of the two entities momentum and heat basically obeys similar laws, in which the entities are linearly proportional to the gradients of flow velocity and temperature. The coefficients of the respective transport relations are viscosity and thermal conductivity, see, e. g., [1]. The fluids, which can be described in this way, are called "Newtonian fluids".

Because attached viscous flow—as considered in this book—is present as laminar and turbulent flow, we have besides the molecular transport also

[3] For the value of R and other constants of air as well as the units and their conversions from the SI system into US units see Appendix D. Data of the Earth standard atmosphere can be found in [2], we give some data in Appendix D.2. If a high accuracy is needed, the actual atmospheric or simulation data must be used.

[4] Again, if a high accuracy is needed, the spurious gases must be regarded, which always in small amounts are present in air, such as argon (Ar), carbon dioxide (CO_2) et cetera [1, 3].

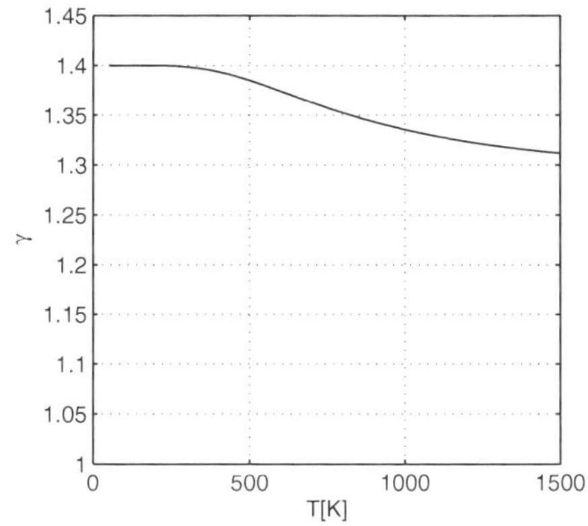

Fig. 3.1. The ratio of specific heats $\gamma = c_p/c_v$ as function of the temperature T in the interval 50 K $\leqq T \leqq$ 1,500 K.

turbulent transport of momentum and energy.[5] Turbulence is not a property of the fluid. The—apparent—turbulent transport is described by terms which are added to the laminar (molecular) transport terms of the Navier-Stokes equations and the energy equation.

In the following the molecular transport properties of air are given. The matter of apparent turbulent transport properties is treated in Section 3.5, see also Section 9.6.

The transport properties viscosity μ and thermal conductivity k of a gas, in our case air, are basically functions only of the temperature. We give relations with different degrees of accuracy for the temperature range 50 K $\lessapprox T \lessapprox$ 1,500 K, i.e., for not or, at the higher temperatures, only weakly dissociated air. Some emphasis is put on simple power-law approximations. They are useful for quick estimates and for qualitative considerations.

Viscosity The viscosity of pure monatomic, but also of polyatomic gases, in this case air, can be determined in the frame of the Chapman-Enskog theory, see, e.g., [1]. For our purposes the Sutherland equation is sufficient:

$$\mu_{Suth} = 1.458 \cdot 10^{-6} \frac{T^{1.5}}{T + 110.4}. \tag{3.5}$$

With the temperature T given in [K] the dimension of μ is [kg m^{-1}s^{-1}].

A simple power-law approximation is $\mu = c_\mu T^{\omega_\mu}$. We find for the temperature range $T \lessapprox$ 200 K the approximation (with the constant $c_{\mu 1}$ determined at T = 97 K [4]):

[5] In flow with thermodynamic non-equilibrium, and in general in mixing flows also turbulent transport of mass is present besides the molecular transport of mass—mass diffusion, see, e.g., [4].

$$\mu_1 = c_{\mu 1} T^{\omega_{\mu 1}} = 0.702 \cdot 10^{-7} T, \tag{3.6}$$

and for $T \gtrsim 200$ K (with the constant $c_{\mu 2}$ determined at $T = 407.4$ K [4]):

$$\mu_2 = c_{\mu 2} T^{\omega_{\mu 2}} = 0.04644 \cdot 10^{-5} T^{0.65}. \tag{3.7}$$

In Fig. 3.2 we compare results of the three above relations and the exact Chapman-Enskog relation in the temperature range up to $T = 1{,}500$ K, with the understanding, that a more detailed consideration might be necessary due to possible dissociation around $T \approx 1{,}500$ K.

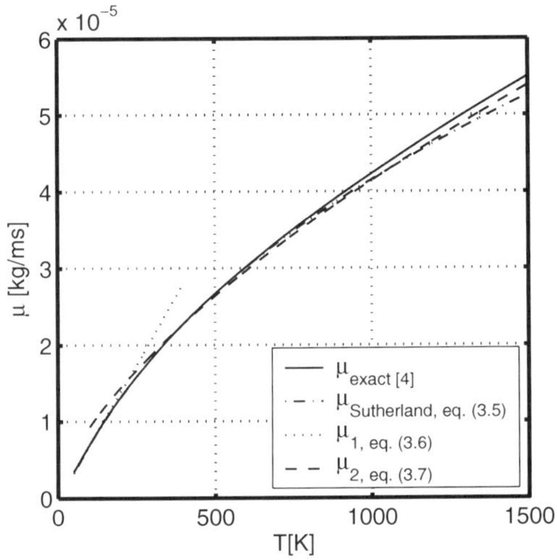

Fig. 3.2. Different approximations of the viscosity μ of air as function of the temperature T [4].

The Figure shows that the data from the Sutherland relation compare well with the exact data except for the large temperatures, where they are noticeably smaller. The power-law relation for $T \lesssim 200$ K fails above $T = 200$ K. The second power-law relation gives good data for $T \gtrsim 300$ K. For $T = 200$ K the error is less than nine per cent. At high temperatures the exact data are better approximated by this relation than by the Sutherland relation.

Thermal Conductivity. The thermal conductivity of pure monatomic gases can be determined in the frame of the Chapman-Enskog theory, but not that of polyatomic gases [1]. An approximate relation, which takes into account the exchanges of rotational as well as vibrational energy of polyatomic gases, is the semi-empirical Eucken formula [1], where c_p is the specific heat at constant pressure:

$$k = \left(c_p + \frac{5}{4} R \right) \mu. \tag{3.8}$$

The dimension of k is $[\mathrm{W\,m^{-1}K^{-1}}]$. The monatomic case is included, if for the specific heat $c_p = 2.5\,R$ is taken.

An important dimensionless quantity in heat transfer is the Prandtl number Pr (Section 3.4, page 64):

$$Pr = \frac{\mu c_p}{k}.$$

With eq. (3.8) the following relation can be derived for it:

$$Pr = \frac{\mu c_p}{k} = \frac{c_p}{c_p + 1.25R} = \frac{4\gamma}{9\gamma - 5}, \tag{3.9}$$

which is a good approximation for both monatomic and polyatomic gases [3]. $\gamma = c_p/c_v$ is the ratio of the specific heats.

For temperatures up to 1,500 K, an approximate relation due to C.F. Hansen—similar to Sutherland's equation for the viscosity of air—can be used [6]:

$$k_{Han} = 1.993 \cdot 10^{-3} \frac{T^{1.5}}{T + 112.0}. \tag{3.10}$$

A simple power-law approximation can also be formulated for the thermal conductivity: $k = c_k T^{\omega_k}$. For the temperature range $T \lessapprox 200$ K the approximation reads (with the constant c_{k1} determined at $T = 100$ K [4]):

$$k_1 = c_{k1} T^{\omega_{k1}} = 9.572 \cdot 10^{-5}\,T, \tag{3.11}$$

and for $T \gtrapprox 200$ K (with the constant c_{k2} determined at $T = 300$ K [4]):

$$k_2 = c_{k2} T^{\omega_{k2}} = 34.957 \cdot 10^{-5}\,T^{0.75}. \tag{3.12}$$

In Fig. 3.3 we compare the results of the four above relations in the temperature range up to $T = 1{,}500$ K, again with the understanding that a more detailed consideration due to possible dissociation around $T \approx 1{,}500$ K might be necessary. The data computed with eq. (3.8) were obtained for non-dissociated air with vibration excitation effects on the specific heats determined with eq. (3.2).

The figure shows that the data from the Hansen relation initially compare well with the Eucken data. For temperatures above approximately 600 K they are noticeably smaller. The power-law relation for $T \lessapprox 200$ K fails for $T \gtrapprox 200$ K. The second power-law relation gives good data for $T \gtrapprox 200$ K.

It should be noted that non-negligible vibration excitation sets in already at around $T = 400$ K, see above. This is reflected in the behavior of c_p/R, Fig. 3.4, where also the Prandtl number is given.

To obtain, as it is often done, the thermal conductivity simply from eq. (3.9) with a constant Prandtl number would introduce errors above $T \approx 400$ K

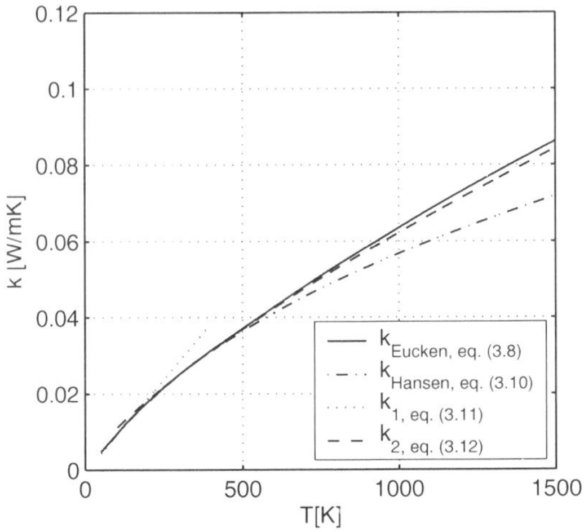

Fig. 3.3. Thermal conductivity k, different approximations, as function of the temperature T [4].

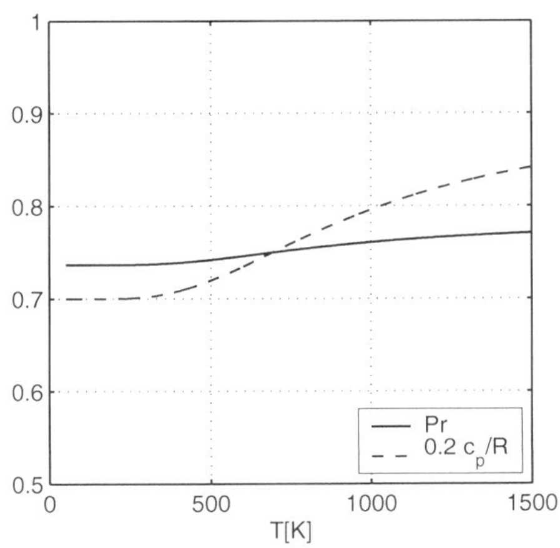

Fig. 3.4. Prandtl number Pr and specific heat at constant pressure c_p of air as function of the temperature T [4].

which, however, are not large. The Prandtl number of air in the considered temperature range is well between 0.73 and 0.77.[6] Generally we find $Pr < 1$ in a large temperature and pressure range [6, 7].

[6] In the literature values for the Prandtl number of air at ambient temperatures are given as low as $Pr = 0.72$, compared to $Pr = 0.737$ in Fig. 3.4. A gas-kinetic theory value of $Pr = 0.74$ for $T = 273.2\ K$, compared to an observed value of $Pr = 0.73$, is quoted in [1].

3.2 Equations of Motion for Steady Laminar Flow in Cartesian Coordinates

The equations of motion of fluid flow describe the transport of mass, momentum and energy. We give these equations here for steady compressible flow, assuming continuum flow (no-slip wall boundary condition), Newtonian fluid, no mass and energy sources, and thermally perfect gas. Cartesian coordinates are employed, and the frame of reference is Galilean, see, e.g., [8]. For the detailed derivation of these equations see, e.g., [1]. We choose the x- and the y-coordinate to be parallel to a (flat) body surface, and the z-coordinate normal to it, with $z = 0$ denoting the wall.

3.2.1 Transport of Mass: The Continuity Equation

Mass is a scalar entity. Its transport is described by the continuity equation. For three-dimensional flow it reads:

$$\frac{\partial \rho u}{\partial x} + \frac{\partial \rho v}{\partial y} + \frac{\partial \rho w}{\partial z} = 0. \tag{3.13}$$

The equation represents the gain of mass in the unit control volume by convective transport.

3.2.2 Transport of Momentum: The Navier-Stokes Equations

Momentum is a vectorial entity. Its transport is described by the Navier-Stokes equations, with one equation for each coordinate direction. For three-dimensional flow we write first the Cauchy equations without body forces in non-conservative form:[7]

x-direction

$$\rho u \frac{\partial u}{\partial x} + \rho v \frac{\partial u}{\partial y} + \rho w \frac{\partial u}{\partial z} = -\frac{\partial p}{\partial x} + \left(\frac{\partial \tau_{xx}}{\partial x} + \frac{\partial \tau_{yx}}{\partial y} + \frac{\partial \tau_{zx}}{\partial z} \right), \tag{3.14}$$

y-direction

$$\rho u \frac{\partial v}{\partial x} + \rho v \frac{\partial v}{\partial y} + \rho w \frac{\partial v}{\partial z} = -\frac{\partial p}{\partial y} + \left(\frac{\partial \tau_{xy}}{\partial x} + \frac{\partial \tau_{yy}}{\partial y} + \frac{\partial \tau_{zy}}{\partial z} \right), \tag{3.15}$$

[7] Note that we write the shear-stress terms without the negative sign which sometimes is found in the literature. We follow the notation used in [9].

z-direction

$$\rho u \frac{\partial w}{\partial x} + \rho v \frac{\partial w}{\partial y} + \rho w \frac{\partial w}{\partial z} = -\frac{\partial p}{\partial z} + \left(\frac{\partial \tau_{xz}}{\partial x} + \frac{\partial \tau_{yz}}{\partial y} + \frac{\partial \tau_{zz}}{\partial z} \right). \tag{3.16}$$

The terms on the left-hand side of each equation represent the gain of momentum in the unit control volume by convective transport. On the right-hand side we have first the pressure force acting on the unit control volume and then the viscous forces.

With the modelling of the viscous stresses by assuming a Newtonian fluid, Sub-Section 3.1.2, we arrive from the Cauchy equations at the Navier-Stokes equations.

The components of the viscous stress tensor $\underline{\underline{\tau}}$, [1], in eqs. (3.14) to (3.16), with the bulk viscosity κ neglected[8], then read:

$$\tau_{xx} = \mu \left[2\frac{\partial u}{\partial x} - \frac{2}{3} \left(\frac{\partial u}{\partial x} + \frac{\partial v}{\partial y} + \frac{\partial w}{\partial z} \right) \right], \tag{3.17}$$

$$\tau_{yy} = \mu \left[2\frac{\partial v}{\partial y} - \frac{2}{3} \left(\frac{\partial u}{\partial x} + \frac{\partial v}{\partial y} + \frac{\partial w}{\partial z} \right) \right], \tag{3.18}$$

$$\tau_{zz} = \mu \left[2\frac{\partial w}{\partial z} - \frac{2}{3} \left(\frac{\partial u}{\partial x} + \frac{\partial v}{\partial y} + \frac{\partial w}{\partial z} \right) \right], \tag{3.19}$$

$$\tau_{xy} = \tau_{yx} = \mu \left(\frac{\partial u}{\partial y} + \frac{\partial v}{\partial x} \right), \tag{3.20}$$

$$\tau_{yz} = \tau_{zy} = \mu \left(\frac{\partial v}{\partial z} + \frac{\partial w}{\partial y} \right), \tag{3.21}$$

$$\tau_{zx} = \tau_{xz} = \mu \left(\frac{\partial w}{\partial x} + \frac{\partial u}{\partial z} \right). \tag{3.22}$$

By adding to eq. (3.14) the continuity equation (3.13) multiplied by u we find the so-called conservative formulation. This form of the governing equations needs to be applied in discrete numerical computation methods in order to appropriately capture shock waves.

For the momentum transport, e.g., in x-direction, we obtain the respective Navier-Stokes equation

$$\frac{\partial}{\partial x} \left(\rho u^2 + p - \tau_{xx} \right) + \frac{\partial}{\partial y} \left(\rho v u - \tau_{xy} \right) + \frac{\partial}{\partial z} \left(\rho w u - \tau_{xz} \right) = 0. \tag{3.23}$$

For the conservative form of the y- and of the z-component of the Navier-Stokes equations see Problem 3.3.

[8] This is permitted because its influence is very small except if very strong compression or—in diatomic or polyatomic gases—rotational non-equilibrium occurs, see, e.g., [3].

3.2.3 Transport of Energy: The Energy Equation

Energy is a scalar entity. Energy transport is described by means of the energy equation which expresses the first principle of thermodynamics. Many formulations are possible, see, e.g., [1]. We choose for our considerations the enthalpy-transport form of the convective operator:

$$\begin{aligned} \rho u \frac{\partial h}{\partial x} + \rho v \frac{\partial h}{\partial y} + \rho w \frac{\partial h}{\partial z} = &- \left[\frac{\partial q_x}{\partial x} + \frac{\partial q_y}{\partial y} + \frac{\partial q_z}{\partial z} \right] + \\ &+ u \frac{\partial p}{\partial x} + v \frac{\partial p}{\partial y} + w \frac{\partial p}{\partial z} + \left[\tau_{xx} \frac{\partial u}{\partial x} + \tau_{yy} \frac{\partial v}{\partial y} + \tau_{zz} \frac{\partial w}{\partial z} \right] + \\ &+ \left[\tau_{xy} \left(\frac{\partial u}{\partial y} + \frac{\partial v}{\partial x} \right) + \tau_{xz} \left(\frac{\partial u}{\partial z} + \frac{\partial w}{\partial x} \right) + \tau_{yz} \left(\frac{\partial v}{\partial z} + \frac{\partial w}{\partial y} \right) \right]. \end{aligned} \tag{3.24}$$

The terms on the left-hand side represent the gain of energy in the unit control volume by convective transport. The first three terms on the right-hand side represent the molecular transport of energy into the unit volume (heat conduction), the next three terms the work on the fluid by pressure forces (compression work), and the terms in the subsequent square brackets the work by viscous forces (dissipation work).

The components of the heat-conduction vector $\underline{q}$, the heat fluxes, read with k being the thermal conductivity:

$$q_x = -k \frac{\partial T}{\partial x}, \quad q_y = -k \frac{\partial T}{\partial y}, \quad q_z = -k \frac{\partial T}{\partial z}. \tag{3.25}$$

The enthalpy of air—considered as thermally perfect gas—is defined by:

$$h = \int c_p \, dT. \tag{3.26}$$

For thermally and calorically perfect gas—air at temperatures below approximately 400 K, see also Sub-Section 3.1.1—c_p = const. and the left-hand side of eq. (3.24) can be written as

$$c_p \left(\rho u \frac{\partial T}{\partial x} + \rho v \frac{\partial T}{\partial y} + \rho w \frac{\partial T}{\partial z} \right) = ... \tag{3.27}$$

3.3 Initial and Boundary Conditions

The equations of motions are a system of coupled non-linear partial differential equations with first- and second-order spatial derivatives.

Second-order derivatives of u, v, w, and T appear in eqs. (3.14) to (3.16), and (3.24). This means that for this quadruplet in general two boundary conditions per spatial direction are needed. In the direction normal to the surface of our flat body—the z-direction—we can provide at the surface for $z = 0$, the no-slip wall boundary condition for a non-permeable wall:

$$z = 0 : u = 0,\ v = 0,\ w = 0. \tag{3.28}$$

In the case of surface-normal suction ($w_{wall} < 0$) or injection ($w_{wall} > 0$) the boundary conditions read, with w_{wall} either being constant on the surface or a function of x and/or y:

$$z = 0 : u = 0,\ v = 0,\ w = w_{wall} = \text{const. or } w = w_{wall}(x, y). \tag{3.29}$$

For the thermal boundary condition of at $z = 0$ three basic types of wall-boundary conditions are possible, which may also be functions of x and/or y:[9]

- prescribed wall temperature: T_w = const. or $T_w(x, y)$,
- adiabatic wall:[10] wall-normal heat flux $q_{gw} = 0 \rightarrow \partial T/\partial z|_w = 0$,
- prescribed wall-normal heat flux: q_{gw} = const. or $f(x, y) \rightarrow \partial T/\partial z|_w$ = const. or $h(x, y)$.

A special case is given with high-speed flight. Here external vehicle surfaces are radiation cooled [4]. The radiation-adiabatic wall temperature T_{ra} in general is a good approximation of the real wall temperature in the presence of radiation cooling. We have in that case $q_{gw} \neq 0$, the heat flux in the wall is $q_w = 0$, the radiation heat flux is $q_{rad} \neq 0$. The wall temperature T_w $= T_{ra}$ as well as q_{gw} and q_{rad} are functions of the location on the vehicle surface, see also [5]. In the case of radiation cooling in any case $q_w \neq q_{gw}$.

For external flow problems, the other quadruplet of the boundary conditions is defined in principle at infinity away from the body (far-field or external boundary conditions). For internal flows, e.g., inlet flows, diffuser-duct flows, et cetera, boundary conditions are to be formulated in an appropriate way.

The boundary conditions for the other two directions of our flat body surface are inflow and outflow conditions which we do not list here.[11]

The first-order derivatives in the convective operators of the different equations demand also boundary conditions in z-direction. These are prescribed for $z = 0$ like above.

---

[9] Instead of the term 'wall heat flux q_w' we use the more general term 'heat flux in the gas at the wall q_{gw}'. This is done in order to distinguish between the heat flux in the gas at the wall and that in the wall, which we call q_w. Of course, in many cases these are equal. A typical exception is given, if the wall is radiation cooled.

[10] For the adiabatic or recovery (wall) temperature see Appendix B.2.

[11] Note, however, that in particular for internal flow problems the appropriate formulation of these boundary conditions can be difficult.

The continuity equation has first-order derivatives of the density ρ. It is not possible to prescribe the density or a normal density gradient at a body surface. The same holds for the pressure. Boundary conditions for the spatial first-order derivatives of the pressure also cannot be prescribed. We note, however, that in Navier-Stokes/RANS codes often $\partial p/\partial z = 0$ at $z = 0$ is employed as a kind of wall boundary condition which is derived from first-order boundary-layer theory. This condition may constrain the solution to a certain degree.

A possibility would be to use the wall-compatibility condition, which results for $z = 0$ from eq. (3.16):

$$\frac{\partial p}{\partial z}|_{z=0} = \left(\frac{\partial \tau_{xz}}{\partial x} + \frac{\partial \tau_{yz}}{\partial y} + \frac{\partial \tau_{zz}}{\partial z}\right)|_{z=0}. \tag{3.30}$$

The use of this compatibility condition is hampered by the fact that second-order derivatives—some of them, however, being zero anyway—have to be computed with one-sided difference formulae.

Regarding both the surface-normal derivative of the density and the pressure, an alternative approach was proposed in [10]. It makes use of the usually employed asymptotic transient solutions of the governing equations. From the continuity equation, now including the temporal term (see Appendix A.1), we get for $z = 0$

$$\frac{\partial \rho}{\partial t}|_{z=0} = -\rho \frac{\partial w}{\partial z}|_{z=0}, \tag{3.31}$$

or, for a perfect gas and for a prescribed heat flux ($\partial T/\partial t \neq 0$ in the transient phase):

$$\frac{\partial p}{\partial t}|_{z=0} = p \left(\frac{1}{T}\frac{\partial T}{\partial t} - \frac{\partial w}{\partial z}\right)|_{z=0}. \tag{3.32}$$

With this compatibility condition instead of eq. (3.30) only one first-order spatial derivative has to be computed. For the steady state—$t \to \infty$—each term reduces to the familiar steady state condition

$$\frac{\partial \rho}{\partial t}, \frac{\partial p}{\partial t}, \frac{\partial T}{\partial t} \to 0 : \frac{\partial w}{\partial z}|_{z=0} \to 0. \tag{3.33}$$

3.4 Similarity Parameters, Boundary-Layer Thicknesses

We define the basic similarity parameters and boundary-layer thicknesses which concern the Navier-Stokes equations and the energy equation. For other boundary-layer thicknesses, for instance the displacement thickness and other integral parameters in three dimensions see Chapter 5, and for the thickness of the viscous sublayer Appendix B.

Mach Number M. In the Navier-Stokes equations we compare the leading convective x-momentum flux term and the pressure term of eq. (3.14):

$$\rho u \frac{\partial u}{\partial x} + ... \approx -\frac{\partial p}{\partial x} + ..., \tag{3.34}$$

and find from this, using characteristic values

$$\frac{\rho u^2}{p} = \frac{\rho u^2}{\rho R T} = \gamma M^2. \tag{3.35}$$

The Mach number M is defined by:

$$M = \frac{u}{a}, \tag{3.36}$$

with a being the speed of sound:

$$a = \left(\frac{\partial p}{\partial \rho}\right)_s = \sqrt{\gamma R T}, \tag{3.37}$$

with the subscript s indicating constant entropy.

The Mach number is the ratio 'characteristic speed' to 'speed of sound'. Its magnitude characterizes compressibility effects in fluid flow. Here we employ it in order to distinguish two flow types:

- $M \rightarrow 0$: Compressibility effects can be neglected, we speak of incompressible flow. Note that in a strict sense incompressible flow, $M = 0$, would imply—because u is finite—that the temperature T is infinitely large. Actually in this case eq. (3.37) would loose its meaning: fluid mechanics is decoupled from thermodynamics.
- $M > 0$: Compressible flow, compressibility effects may have to be taken into account. At which value of the Mach number this happens, depends on the flight parameters, configurational peculiarities, and on the flow phenomenon under consideration.

Reynolds Number Re. Noting that the momentum flux is a vector entity, we compare now in a schematic way in the Navier-Stokes equations the convective and the molecular x-momentum flux in x-direction in the first large bracket of eq. (3.23). After introducing the simple proportionality $\tau_{xx} \sim \mu(u/L)$—which here does not anticipate the presence of a boundary layer—we obtain

$$\frac{\rho u^2}{\tau_{xx}} \sim \frac{\rho u^2}{\mu(u/L)} = \frac{\rho u L}{\mu} = Re, \tag{3.38}$$

and find in this way the Reynolds number Re, which is the ratio 'convective transport of momentum' to 'molecular transport of momentum' or 'inertial forces' to 'viscous forces'.

The Reynolds number is the similarity parameter characterizing viscous phenomena. The following limiting cases of Re can be distinguished:

- $Re \to 0$: The molecular transport of momentum is much larger than the convective transport, the flow is the "creeping" flow, see, e.g., [1], [11]: the convective transport of momentum can be neglected.
- $Re \to \infty$: The convective transport of momentum is much larger than the molecular transport, the flow can be considered as inviscid, i.e. molecular transport can be neglected. The governing equations reduce to the Euler equations, i.e. eqs. (3.14) to (3.16) without the molecular transport terms. If the flow is also irrotational, the Euler equations can be further reduced to the potential equation.
- $Re = O(1)$: The molecular transport of momentum has the same order of magnitude as the convective transport, the flow is viscous, it is, for instance, boundary-layer, or in general, shear-layer flow.[12]

Péclet Number *Pe*. Noting that the energy flux is a scalar entity we compare now in the energy equation, in the same way as we did for the momentum flux, the convective and the molecular heat flux in x-direction.

First we add to the left-hand side of eq. (3.24) the continuity equation times the enthalpy to find the conservative form of convective and molecular transport:

$$\frac{\partial}{\partial x}\left(\rho u h - k\frac{\partial T}{\partial x}\right) + \frac{\partial}{\partial y}\left(\rho v h - k\frac{\partial T}{\partial y}\right) + \frac{\partial}{\partial z}\left(\rho w h - k\frac{\partial T}{\partial z}\right) = \cdots. \tag{3.39}$$

Assuming perfect gas with $h = c_p T$, and again not anticipating a thermal boundary layer, we then compare the convective and the conductive heat transport in x-direction after introduction of the simple proportionality $\partial T/\partial x \propto T/L$:[13]

$$\frac{\rho u c_p T}{k(\partial T/\partial x)} \propto \frac{\rho u c_p T}{k(T/L)} = \frac{\rho u c_p L}{k} = \frac{\mu c_p}{k}\frac{\rho u L}{\mu} = Pr\ Re = Pe, \tag{3.40}$$

[12] Note that in boundary-layer theory the boundary-layer equations are found for $Re \to \infty$, however, only after the "boundary-layer stretching" has been introduced, Section 4.2.

[13] Pr is the Prandtl number, see below.

and find in this way the Péclet number:

$$Pe = \frac{\rho u c_p L}{k}. \tag{3.41}$$

The Péclet number can be interpreted as the ratio 'convective transport of heat' to 'molecular transport of heat'. Of interest are the limiting cases of of the Péclet number Pe (compare with the limiting cases of the Reynolds number Re):

- $Pe \to 0$: the molecular transport of heat is much larger than the convective transport.
- $Pe \to \infty$: the convective transport of heat is much larger than the molecular transport.
- $Pe = O(1)$: the molecular transport of heat has the same order of magnitude as the convective transport.

Prandtl Number *Pr*. The Prandtl number is found by division of the Péclet number by the Reynolds number

$$Pr = \frac{Pe}{Re} = \frac{\mu\, c_p}{k}. \tag{3.42}$$

The Prandtl number Pr can be written with the kinematic viscosity $\nu = \mu/\rho$:

$$Pr = \frac{\mu/\rho}{k/\rho\, c_p} = \frac{\nu}{\alpha}, \tag{3.43}$$

where

$$\alpha = \frac{k}{\rho\, c_p} \tag{3.44}$$

is the thermal diffusivity, see, e.g., [12], which is a property of the conducting material. The Prandtl number Pr hence can be interpreted as the ratio 'kinematic viscosity' to 'thermal diffusivity'. It is a measure for the capacity of the fluid to diffuse momentum compared to its capacity to diffuse heat. For its meaning in the context of this book see below for the thickness of the thermal boundary layer.

Eckert Number *E*. If we non-dimensionalize the energy equation, eq. (3.24) with proper reference data (p is here non-dimensionalized with ρu^2), we find:

$$\begin{aligned} \rho u c_p \frac{\partial T}{\partial x} + \cdots = {} & \frac{1}{RePr}\left(\frac{\partial}{\partial x}(k\frac{\partial T}{\partial x}) + \cdots\right) + E\left(u\frac{\partial p}{\partial x} + \cdots\right) + \\ & + \frac{E}{Re}\left[2\mu\frac{\partial^2 u}{(\partial x)^2} + \cdots\right]. \end{aligned} \tag{3.45}$$

All entities in this equation are dimensionless. The new parameter is the Eckert number E:

$$\frac{u(\partial p/\partial x)}{\rho u c_p(\partial T/\partial x)} \propto \frac{u\rho u^2}{\rho u c_p T} = \frac{u^2}{c_p T} = (\gamma - 1)M^2 = E, \tag{3.46}$$

with the Mach number defined by eq. (3.36). The Eckert number can be interpreted as the ratio 'kinetic energy' to 'thermal energy' of the flow.

For $E \to 0$, respectively $M \to 0$, we find the incompressible case, in which of course a finite energy transport by both convection and conduction can happen, but where compression work is not done on the fluid, and also dissipation work does not occur. For $E = 0$ actually fluid mechanics and thermodynamics are decoupled.

Wall Temperature Ratio T_w/T_{ref}. The Π or Pi theorem, see, e. g., [13], permits to perform dimensional analysis in a rigorous way. It yields, besides the basic similarity parameters discussed above, for the problems of viscous flow the ratio of wall temperature to free-stream temperature

$$\frac{T_w}{T_\infty}$$

as a similarity parameter [14].

A more general form is given in [11]:

$$\frac{T_w - T_{ref}}{T_{ref}}.$$

This usually ignored similarity parameter is of importance if thermal surface effects, Sub-Section 2.3.3, are present in the flow under consideration.

Boundary-Layer Thickness δ. In the Navier-Stokes equations the convective transport of x-momentum in x-direction ρu^2 is compared now for $Re = O(1)$ with the molecular transport of x-momentum in the direction normal to the wall, in z-direction, τ_{zx}, anticipating a boundary layer with the (asymptotic) thickness δ.[14] We do this with eq. (3.14), and neglecting the second term of τ_{zx} in eq. (3.20):

$$\rho u \frac{\partial u}{\partial x} + \cdots \approx \cdots + \frac{\partial}{\partial z}(\mu \frac{\partial u}{\partial z}) + \cdots. \tag{3.47}$$

Again we introduce in a schematic way characteristic values and find:

$$\rho u \frac{u}{L} \propto \frac{\mu u}{\delta^2}. \tag{3.48}$$

[14] This boundary layer is also called flow or dynamic boundary layer in order to distinguish it from the thermal boundary layer.

After rearrangement we obtain for the boundary-layer thickness δ, which in laminar flow is the characteristic thickness in wall-normal direction[15]

$$\frac{\delta}{L} \propto \sqrt{\frac{\mu}{\rho u L}} = \frac{1}{\sqrt{Re_L}}, \tag{3.49}$$

and, using the boundary-layer running length x as characteristic length for the main-flow direction along the surface under consideration:

$$\frac{\delta}{x} \propto \sqrt{\frac{\mu}{\rho u x}} = \frac{1}{\sqrt{Re_x}} = \frac{1}{\sqrt{Re_L}} \sqrt{\frac{L}{x}}. \tag{3.50}$$

This boundary-layer thickness is the thickness of the flow boundary layer $\delta \equiv \delta_{flow}$ [11]. It is the smaller, the larger the Reynolds number is. We will identify below with the same kind of consideration the thermal boundary layer with a thickness, which is different from the flow boundary-layer thickness. For the approximate determination of the boundary-layer thicknesses for laminar and turbulent flow, as well as other characteristic boundary-layer thicknesses, see Appendix B.

Thermal Boundary-Layer Thickness δ_T. We compare now in the energy equation for $Pe = O(1)$ the convective transport of heat in x-direction $\rho u c_p T$ with the molecular transport of heat in z-direction q_z, anticipating a thermal boundary layer with the thickness δ_T. We do this in the differential form given with eq. (3.39):

$$\rho u c_p \frac{\partial T}{\partial x} + \cdots \approx \cdots + \frac{\partial}{\partial z}(k \frac{\partial T}{\partial z}) + \cdots \tag{3.51}$$

Again we introduce in a schematic way characteristic quantities and find after rearrangement:

$$\rho u c_p \frac{T}{L} \propto \frac{kT}{\delta_T^2}. \tag{3.52}$$

From this we find the thickness δ_T of the thermal boundary layer:

$$\frac{\delta_T}{L} \propto \sqrt{\frac{k}{c_p \rho u L}} = \frac{1}{\sqrt{Pe_L}} = \frac{1}{\sqrt{Re_L Pr}}, \tag{3.53}$$

and, using again the boundary-layer running length x as characteristic length:

$$\frac{\delta_T}{x} \propto \sqrt{\frac{k}{c_p \rho u x}} = \frac{1}{\sqrt{Pe_x}} = \frac{1}{\sqrt{Re_x Pr}}. \tag{3.54}$$

---

[15] In turbulent flow the thickness of the viscous sub-layer δ_{vs}, Appendix B, is the characteristic thickness regarding wall shear stress τ_w and heat flux in the gas at the wall q_{gw}.

The thickness of the thermal boundary layer δ_T is related to the thickness of the flow boundary layer $\delta \equiv \delta_{flow}$ by

$$\frac{\delta_T}{\delta} \propto \frac{1}{\sqrt{Pr}}. \tag{3.55}$$

If the thermal boundary layer is thinner than the flow boundary layer ($Pr > 1$) and is located in the linear or nearly linear part of the velocity profile, we obtain—with a correspondingly scaled reference velocity $u\,\delta_T/\delta$—instead of eq. (3.52):

$$\rho(\frac{\delta_T}{\delta}u)c_p\frac{T}{L} \propto \frac{kT}{\delta_T^2}, \tag{3.56}$$

and get, in the same way as above

$$\frac{\delta_T}{\delta} \propto \frac{1}{\sqrt[3]{Pr}}. \tag{3.57}$$

We find now regarding the limiting cases of Pr, see, e.g., [15]:

- $Pr \to 0$: the thermal boundary layer is much thicker than the flow boundary layer, which is typical for the flow of liquid metals.
- $Pr \to \infty$: the flow boundary layer is much thicker than the thermal boundary layer, which is typical for liquids.
- $Pr = O(1)$: the thermal boundary layer has a thickness of the order of that of the flow boundary layer. This is typical for gases, in our case air. However, since in the interesting temperature and density/pressure domain $Pr < 1$, see Section 3.1, the thermal boundary layer is somewhat thicker than the flow boundary layer.[16]

3.5 Equations of Motion for Steady Turbulent Flow

The Navier-Stokes equations, Sub-Section 3.2.2, describe also turbulent flow including laminar-turbulent transition. This was disputed for quite a time, because of the linear relationship between stress tensor and deformation tensor in the equations for Newtonian fluids. Today it is accepted that for the flow which we consider here, the Navier-Stokes equations are valid. The values of fluid properties and dependent variables are then understood as their instantaneous values [9, 16].

However, the scales of the motion in turbulent flow are always very different from the molecular scales (molecular transport) so that the fluid is a continuum for turbulence motions. A direct approach to solve the equations for

[16] This can be important for computation methods. If using a boundary-layer method, one must take into account, if given, the different thicknesses by choosing the computation domain normal to the surface according to the largest thickness.

turbulent flows—direct numericalsimulation (DNS)—is to solve them for appropriate boundary conditions and initial values that include time-dependent quantities. Mean values are needed in most practical cases, so an ensemble of solutions of the time-dependent Navier-Stokes equations (with laminar transport properties) is required.

If a high accuracy of DNS is desired, a large number of points in space and time is necessary which increases rapidly with the Reynolds number [16]. Even for less demanding cases, DNS becomes a difficult and extremely expensive computing problem because the unsteady turbulent motions spread over a wide range of scales.

This does not mean that this approach is useless, on the contrary. Generally, the objective of DNS currently is not to reproduce the flow which occurs at a large Reynolds number. For example, to simulate the behavior of dissipative structures which have very small length scales, it can be done at moderate Reynolds numbers [17]. However, to test the hypothesis of isotropy of small structures, it is necessary to reach very large Reynolds numbers.

If the question is to simulate the flow around a complete airplane and to describe the details of turbulence events, it is illusory to call for DNS. Today, there is no algorithm and no computer for such a task. A way out is large eddy simulation (LES), seȩ, e.g., [18]. In this approach the Navier-Stokes equations are filtered and the largest scales of the flow are calculated. The interaction of the large scales with the smaller scales is modelled. In this way, the computing effort is reduced for a given Reynolds number.[17]

DNS and LES are becoming more and more useful tools in aerodynamics. But today they are not much more than a complement to statistical turbulence models which are closer to satisfying engineering requirements. Instead of solving the Navier-Stokes equations, as in DNS, and performing averages on an ensemble of solutions, the idea is to average the Navier-Stokes equations first and to solve the resulting equations. Due to the averaging process and the non-linearity of the convection terms, additional terms are present in the averaged equations. These terms look like additional stresses and heat transfer contributors, and they are called turbulent or apparent stresses and heat transfers.

Incompressible Flow. In incompressible flow, the averaging process is based on a decomposition of the flow due to O. Reynolds, in which an average flow and a turbulent flow are distinguished. For the sake of generality, the average is taken as an ensemble average, in which a large number of realizations of the same flow is considered. In thought experiments, it is very easy to imagine that the flow is realized a large number of times under the same initial and boundary conditions and that the average flow is an ensemble average over the different realizations.

[17] There exist many variants of this approach which is also called scale-resolving approach, see Section 9.6.

The instantaneous velocity $\tilde{u}$ is decomposed into the average value u and the turbulent fluctuation quantity u':

$$\tilde{u} = u + u'. \tag{3.58}$$

In the same way, for $\tilde{v}$, $\tilde{w}$ and the pressure $\tilde{p}$ we have:

$$\tilde{v} = v + v', \ \tilde{w} = w + w', \ \tilde{p} = p + p'. \tag{3.59}$$

The average velocity component u, for example then is:

$$u = \lim_{N \to \infty} \frac{1}{N} \sum_{i=1}^{N} \tilde{u}_i, \tag{3.60}$$

where $\tilde{u}_i$ is the value of the velocity taken in the ith realization of the flow. Each value of $\tilde{u}_i$ in the different realizations of the flow is taken under the same conditions, i.e. at the same point in space and at equivalent times in the different realizations.

When the continuity and momentum equations are averaged, the form is exactly the same as eq. (3.13) with $\rho =$ const., and eqs. (3.14) to (3.16), but the expressions of the stresses eqs. (3.17) to (3.22) become

$$\tau_{xx} = 2\mu \frac{\partial u}{\partial x} - \rho < u'u' >, \tag{3.61}$$

$$\tau_{yy} = 2\mu \frac{\partial v}{\partial y} - \rho < v'v' >, \tag{3.62}$$

$$\tau_{zz} = 2\mu \frac{\partial w}{\partial z} - \rho < w'w' >, \tag{3.63}$$

$$\tau_{xy} = \tau_{yx} = \mu \left(\frac{\partial u}{\partial y} + \frac{\partial v}{\partial x} \right) \quad \rho < u'v' >, \tag{3.64}$$

$$\tau_{yz} = \tau_{zy} = \mu \left(\frac{\partial v}{\partial z} + \frac{\partial w}{\partial y} \right) - \rho < v'w' >, \tag{3.65}$$

$$\tau_{zx} = \tau_{xz} = \mu \left(\frac{\partial w}{\partial x} + \frac{\partial u}{\partial z} \right) - \rho < w'u' > . \tag{3.66}$$

In these equations, $< u'v' >$, for example, denotes the average value of $u'v'$. Compared to laminar flow, turbulent terms are added to the viscous terms in the expression of the (laminar) stress terms. These turbulent terms are called the Reynolds stresses, which lead to the closure problem of turbulence models, because these terms need to be determined.

Compressible Flow. In compressible flow, the problem is more involved. The average can also be defined as Reynolds's average. For example, the average velocity component of $\tilde{u}$ is the ensemble average $< \tilde{u} >$ of this velocity component. Very often, however, the average velocity is defined as Favre's average, [19], which is a mass-weighted average:

$$u = \frac{< \tilde{\rho}\tilde{u} >}{\rho}, \tag{3.67}$$

where ρ is the ensemble average $< \tilde{\rho} >$ of the density. All components of the average velocity are defined as mass-weighted averages. Mass-weighted averages are also used for temperature, enthalpy, internal energy, and entropy. Ensemble averages are used for density and pressure:

$$\rho =< \tilde{\rho} >, \tag{3.68}$$

$$p =< \tilde{p} > . \tag{3.69}$$

The turbulent fluctuation of any quantity is defined as the difference between the instantaneous value and the mean value (see also eq. (3.58)), for example:

$$\rho' = \tilde{\rho} - \rho, \tag{3.70}$$

$$u' = \tilde{u} - u. \tag{3.71}$$

In one case, the fluctuation is centered, in the other case the fluctuation is not centered:

$$< \rho' >= 0, \tag{3.72}$$

$$< u' >= -\frac{< \rho' u' >}{\rho}. \tag{3.73}$$

The stagnation enthalpy is related to the static enthalpy by:

$$h_{stag} = h + \frac{u^2 + v^2 + w^2}{2} + \frac{1}{2}\frac{< \tilde{\rho}(u'^2 + v'^2 + w'^2) >}{\rho}. \tag{3.74}$$

By taking the average of the continuity and momentum equations, the resulting equations are the same as eqs. (3.13) to (3.16), but the density, the velocity components and the pressure are to be understood as average quantities.

The expressions of the stresses become, with $\tilde{\mu}$ being calculated with the average temperature T:

$$\tau_{xx} = \left\langle \tilde{\mu} \left[2\frac{\partial \tilde{u}}{\partial x} - \frac{2}{3}\left(\frac{\partial \tilde{u}}{\partial x} + \frac{\partial \tilde{v}}{\partial y} + \frac{\partial \tilde{w}}{\partial z} \right) \right] \right\rangle - < \tilde{\rho} u' u' >, \tag{3.75}$$

$$\tau_{yy} = \left\langle \tilde{\mu} \left[2\frac{\partial \tilde{v}}{\partial y} - \frac{2}{3}\left(\frac{\partial \tilde{u}}{\partial x} + \frac{\partial \tilde{v}}{\partial y} + \frac{\partial \tilde{w}}{\partial z} \right) \right] \right\rangle - < \tilde{\rho} v' v' >, \tag{3.76}$$

$$\tau_{zz} = \left\langle \tilde{\mu} \left[2\frac{\partial \tilde{w}}{\partial z} - \frac{2}{3}\left(\frac{\partial \tilde{u}}{\partial x} + \frac{\partial \tilde{v}}{\partial y} + \frac{\partial \tilde{w}}{\partial z} \right) \right] \right\rangle - < \tilde{\rho} w' w' >, \tag{3.77}$$

$$\tau_{xy} = \tau_{yx} = \left\langle \tilde{\mu} \left(\frac{\partial \tilde{u}}{\partial y} + \frac{\partial \tilde{v}}{\partial x} \right) \right\rangle - < \tilde{\rho} u' v' >, \tag{3.78}$$

$$\tau_{yz} = \tau_{zy} = \left\langle \tilde{\mu} \left(\frac{\partial \tilde{v}}{\partial z} + \frac{\partial \tilde{w}}{\partial y} \right) \right\rangle - < \tilde{\rho} v' w' >, \tag{3.79}$$

$$\tau_{zx} = \tau_{xz} = \left\langle \tilde{\mu} \left(\frac{\partial \tilde{w}}{\partial x} + \frac{\partial \tilde{u}}{\partial z} \right) \right\rangle - < \tilde{\rho} w' u' > . \tag{3.80}$$

Generally, with some approximations, the above expressions are written in the usual notation, for example

$$\tau_{xx} = \mu \left[2\frac{\partial u}{\partial x} - \frac{2}{3}\left(\frac{\partial u}{\partial x} + \frac{\partial v}{\partial y} + \frac{\partial w}{\partial z} \right) \right] - < \rho u' u' >, \tag{3.81}$$

$$\tau_{xy} = \tau_{yx} = \mu \left(\frac{\partial u}{\partial y} + \frac{\partial v}{\partial x} \right) - < \rho u' v' > . \tag{3.82}$$

In the energy equation, eq. (3.24), the components of the molecular heat-flux vector become with the enthalpy h:

$$q_x = - < \tilde{k} \frac{\partial \tilde{T}}{\partial x} > + < \tilde{\rho} h' u' >, \tag{3.83}$$

$$q_y = - < \tilde{k} \frac{\partial \tilde{T}}{\partial y} > + < \tilde{\rho} h' v' >, \tag{3.84}$$

$$q_z = - < \tilde{k} \frac{\partial \tilde{T}}{\partial z} > + < \tilde{\rho} h' w' > . \tag{3.85}$$

This can be written in the usual notation, too, for example

$$q_x = -k \frac{\partial T}{\partial x} + < \rho h' u' > . \tag{3.86}$$

With the same type of approximation, the dissipation terms become:

$$\begin{aligned} &\tau_{xx}\frac{\partial u}{\partial x}+\tau_{yy}\frac{\partial y}{\partial y}+\tau_{zz}\frac{\partial w}{\partial z}+ \\ &+\tau_{xy}\left(\frac{\partial u}{\partial y}+\frac{\partial v}{\partial x}\right)+\tau_{yz}\left(\frac{\partial v}{\partial z}+\frac{\partial w}{\partial y}\right)+\tau_{zx}\left(\frac{\partial w}{\partial x}+\frac{\partial u}{\partial z}\right), \end{aligned} \tag{3.87}$$

where the stresses have been expressed above.

For an introduction into the modelling of the turbulent or apparent stresses and heat transfers—the closure of the fluctuation terms—and for turbulence models for practical applications see, e.g., [9, 20]. Section 9.6 gives some general information.

3.6 Problems

Problem 3.1. From Fig. 3.4 we find at T = 1,500 K for the specific heat $0.2c_p/R = 0.842$ and for the Prandtl number $Pr = 0.771$. How large is the ratio of specific heats γ? Verify eq. (3.9). Compute γ and Pr also for T = 400 K and 1,000 K and verify further the equation.

Problem 3.2. Consider a slender, blunt-nosed body in a supersonic wind tunnel. The reservoir conditions of the tunnel are such that at the nozzle exit a Mach number $M_\infty = 6$ is reached at T_∞ = 200 K. The vibrational degrees of freedom are exited, but frozen during the nozzle expansion process. Assume that neither de-excitation nor additional excitation of vibrational energy happens and that γ = const. = 1.33. How large is the total temperature at the body nose for $\gamma = 1.33$ and $\gamma = 1.4$?

Problem 3.3. Derive the conservative forms of the y- and the z-component of the Navier-Stokes equations.

Problem 3.4. Slip flow can be present in boundary layers of hypersonic flight vehicles or of vacuum machinery [4]. Show with the help of eq. (A.6) that the heat flux in the gas at the wall normal to a surface contains besides the ordinary heat-flux term also a slip-flow term. Assume two-dimensional flow in Cartesian coordinates with the y-axis being the coordinate normal to the wall.

Problem 3.5. What is the use of equations in dimensionless form?

Problem 3.6. A turbulent flow is said to be two-dimensional when the average flow is two-dimensional. The objective of this problem is to show that the hypothesis of two-dimensional flow must be applied after the averaging process has been applied.

We consider an incompressible turbulent flow. The instantaneous velocity $\tilde{u}$ is decomposed as :

$$\tilde{u} = u + u'.$$

In the same way we have for $\tilde{v}$ and $\tilde{w}$:

$$\tilde{v} = v + v',$$
$$\tilde{w} = w + w'.$$

1. Write the continuity equation for the instantaneous flow.
2. We assume that the instantaneous flow is two-dimensional. Write the continuity equation for the instantaneous flow, for the average flow and for the fluctuating flow.
3. In reality, the instantaneous flow is always three-dimensional. Write the continuity equation for the average flow and for the fluctuating flow. Simplify the equations when the average flow is two-dimensional.

Problem 3.7. The objective of the problem is to show that if the turbulence fluctuations were irrotational there would be no effect on the average flow. We do that for incompressible flow.

1. We assume that the flow is irrotational. Show the following identity

$$\frac{\partial}{\partial x_i}\langle u_i' u_j' \rangle = \frac{\partial k}{\partial x_j}.$$

   This identity has been derived by S. Corrsin and A.L. Kistler [21]. In this equation k is the turbulent kinetic energy

$$k = \frac{1}{2}\langle u_i' u_i' \rangle.$$

2. Show that the turbulent terms in the average momentum equation can be absorbed in the pressure term.
3. Deduce that the average flow is not influenced by turbulence.

References

1. Bird, R.B., Stewart, W.E., Lightfoot, E.N.: Transport Phenomena, 2nd edn. John Wiley & Sons, New York (2002)
2. N.N.: U. S. Standard Atmosphere. Government Printing Office, Washington, D.C. (1976)
3. Vincenti, W.G., Kruger, C.H.: Introduction to Physical Gas Dynamics. John Wiley, New York (1965); reprint edn. Krieger Publishing Comp., Melbourne (1975)
4. Hirschel, E.H.: Basics of Aerothermodynamics, AIAA, Reston, VA. Progress in Astronautics and Aeronautics, vol. 204. Springer, Heidelberg (2004)

5. Hirschel, E.H., Weiland, C.: Design of Hypersonic Flight Vehicles: Some Lessons from the Past and Future Challenges. CEAS Space J. 1, 3–22 (2011)
6. Hansen, C.F.: Approximations for the Thermodynamic and Transport Properties of High-Temperature Air. NACA TR R-50 (1959)
7. Jischa, M.: Konvektiver Impuls-, Wärme- und Stoffaustausch. Vieweg & Sohn, Braunschweig Wiesbaden (1982)
8. Shapiro, A.H.: Basic Equations of Fluid Flow. In: Streeter, V.L. (ed.) Handbook of Fluid Dynamics, pp. 2-1–2-19. McGraw-Hill, New York (1961)
9. Cebeci, T., Cousteix, J.: Modeling and Computation of Boundary-Layer Flows, 2nd edn. Horizons Publ., Springer, Long Beach, Heidelberg (2005)
10. Hirschel, E.H., Groh, A.: Wall-Compatibility Condition for the Solution of the Navier-Stokes Equations. J. Computational Physics 53(2), 346–350 (1984)
11. Schlichting, H., Gersten, K.: Boundary Layer Theory, 8th edn. Springer, Heidelberg (1999)
12. Eckert, E.R.G., Drake, R.M.: Heat and Mass Transfer, 2nd edn. MacGraw-Hill, New York (1950)
13. Holt, M.: Dimensional Analysis. In: Streeter, V.L. (ed.) Handbook of Fluid Dynamics, pp. 15-1–15-25. McGraw-Hill, New York (1961)
14. Oskam, B.: Navier-Stokes Similitude. NLR Memorandum AT-91, Amsterdam, The Netherlands (1991)
15. Pai, S.I.: Laminar Flow. In: Streeter, V.L. (ed.) Handbook of Fluid Dynamics, pp. 5-1–5-34. McGraw-Hill, New York (1961)
16. Pope, S.B.: Turbulent Flows. Cambridge University Press, Cambridge (2000)
17. Moin, P., Mahesh, K.: Direct Numerical Simulation: A Tool in Turbulence Research. Annual Review of Fluid Mechanics 30, 539–578 (1998)
18. Lesieur, M., Metais, O., Comte, P.: Large-Eddy Simulations of Turbulence. Cambridge University Press, Cambridge (2005)
19. Favre, A., Kovasznay, L.S.G., Dumas, R., Gaviglio, J., Coantic, M.: La turbulence en mécanique des fluides. Gauthier Villars, Paris (1976)
20. Wilcox, D.C.: Turbulence Modelling for CFD, 3rd edn. DCW Industries, La Cañada (2000)
21. Corrsin, S., Kistler, A.L.: The Free-Stream Boundaries of Turbulent Flows. NACA Techn. Note 3133 (1954)

4

Boundary-Layer Equations for Three-Dimensional Flow

In this chapter the boundary-layer equations, both for laminar and turbulent weakly interacting three-dimensional flow, are derived and discussed. Assumed as before is Newtonian fluid, calorically and thermally perfect gas, and steady flow. The basic considerations are made in Cartesian coordinates. With the boundary-layer equations the characteristic properties and the compatibility conditions for attached viscous flow are treated. This is easier to accomplish than with the Navier-Stokes equations. The results apply for the latter, too. The boundary-layer equations in general notation for surface-oriented non-orthogonal curvilinear coordinates, the small cross-flow equations, and the equations in contravariant formulation are given in Appendix A.2. The latter permit a convenient treatment of cases with geometrical complexity and a compact formulation of higher-order equations.

4.1 Preliminary Notes

4.1.1 Coordinate Convention

The derivation of the equations is made in Cartesian coordinates. Such coordinates were also used in the preceding chapter, but we change now the convention. In two-dimensional boundary-layer theory the rectilinear coordinate normal to the surface is denoted with y, and accordingly the normal velocity component with v. That notation is now used in order to ease the reading of the derivation. Hence in this chapter—in contrast to Chapter 3—the coordinates x and z are the coordinates tangential to the body surface, and y is the coordinate normal to it. Accordingly u and w are the tangential velocity components, and v is the component normal to the body surface. This coordinate convention will also be employed in the following chapters. Illustrating figures, however, may have other conventions. This will be noted in each case.

4.1.2 The Boundary-Layer Criteria

The derivation of the boundary-layer equations takes into account the basic properties of boundary layers—observed first by Prandtl [1]—in the form of

E.H. Hirschel, J. Cousteix, and W. Kordulla, *Three-Dimensional Attached Viscous Flow*,
DOI: 10.1007/978-3-642-41378-0_4,

the boundary-layer criteria. The criteria, that a given attached viscous flow is a—weakly interacting—boundary layer, can be formulated in the following way:

1. The extent of the boundary layer in direction normal to the body surface (boundary-layer thickness), compared to a characteristic body length, is very small.
2. The velocity component in the boundary layer normal to the body surface, compared to a characteristic velocity, is very small.
3. The inviscid external pressure field is "impressed" on the boundary layer. Two cases must be distinguished:
   a) If the body surface is flat, the wall-normal pressure gradient in the boundary layer is very small, for very high Reynolds numbers it tends to zero. This is the case usually considered in first-order boundary-layer theory.
   b) If the body surface is curved, the wall-normal pressure gradient in the boundary layer is not small due to the centrifugal forces which the surface curvature induces. This is the general case—considered in second-order or higher-order boundary-layer theory. The flat-plate case is the limiting case for very large Reynolds numbers.

4.2 First-Order Boundary-Layer Equations for Steady Laminar Flow

We derive in the following the first-order boundary-layer equations for steady, compressible, three-dimensional flow past a flat surface. We assume laminar flow, but note that the resulting equations also hold for turbulent flow, if we treat the equations as Reynolds-averaged, Section 4.3.

The boundary-layer equations are derived from the Navier-Stokes equations, and are complemented by the continuity and the energy equation, Section 3.2. They cannot be derived from first principles, even if the method of matched asymptotic expansion seems to suggest this.[1]

Prandtl's observation, [1], is basically that the different boundary-layer thicknesses and v in y-direction are inversely proportional to the square root of the Reynolds number, if the flow is laminar.

We use this observation by introducing the so-called boundary-layer stretching, or magnification [2], which brings y and v—non-dimensionalized with reference quantities L_{ref} and v_{ref}, respectively—to $O(1)$:

[1] The method of matched asymptotic expansions derives the boundary-layer equations from the Navier-Stokes equations in a formal manner [2], see also Sub-Section 6.2.2. The Navier-Stokes equations themselves are derived formally from the Boltzmann equation, using the Chapman-Enskog expansion, see, e.g., [3].

$$\tilde{y} = \frac{y\sqrt{Re_{ref}}}{L_{ref}}, \tag{4.1}$$

$$\tilde{v} = \frac{v\sqrt{Re_{ref}}}{v_{ref}}, \tag{4.2}$$

with the reference Reynolds number defined by

$$Re_{ref} = \frac{\rho_{ref} v_{ref} L_{ref}}{\mu_{ref}}. \tag{4.3}$$

The tilde above denotes variables, which were non-dimensionalized and stretched. However, we use it in the following also for variables, which are only non-dimensionalized.

All other variables are simply made dimensionless with appropriate reference quantities, and then assumed to be $O(1)$: velocity components u and w with v_{ref}, lengths x and z with L_{ref}, temperature T with T_{ref}, density ρ with ρ_{ref}, pressure p with $\rho_{ref} v_{ref}^2$ instead of p_{ref},[2] the transport coefficients μ and k with μ_{ref} and k_{ref}, respectively, and finally the specific heat at constant pressure c_p with $c_{p_{ref}}$. Each resulting dimensionless variable is marked by a tilde, for instance:

$$\tilde{u} = \frac{u}{v_{ref}}. \tag{4.4}$$

We introduce boundary-layer stretching and non-dimensionalization first into the continuity equation, eq. (3.13). We do this for illustration in full detail. We replace u with $\tilde{u} v_{ref}$, eq. (4.4), v with $\tilde{v} v_{ref}/\sqrt{Re_{ref}}$, eq. (4.2), and so on, and find:

$$\frac{\partial \tilde{\rho} \rho_{ref} \tilde{u} v_{ref}}{\partial \tilde{x} L_{ref}} + \frac{\partial \tilde{\rho} \rho_{ref} \tilde{v} v_{ref}/\sqrt{Re_{ref}}}{\partial \tilde{y} L_{ref}/\sqrt{Re_{ref}}} + \frac{\partial \tilde{\rho} \rho_{ref} \tilde{w} v_{ref}}{\partial \tilde{z} L_{ref}} = 0. \tag{4.5}$$

Since all reference quantities, and also Re_{ref} are constants, we find immediately the stretched and dimensionless continuity equation which has the same form—this does not hold for the other equations—as the original equation:

$$\frac{\partial \tilde{\rho} \tilde{u}}{\partial \tilde{x}} + \frac{\partial \tilde{\rho} \tilde{v}}{\partial \tilde{y}} + \frac{\partial \tilde{\rho} \tilde{w}}{\partial \tilde{z}} = 0. \tag{4.6}$$

Consider now the Navier-Stokes equations, Sub-Section 3.2.2. We introduce non-dimensional and stretched variables, as we did with the continuity

[2] This has the advantage that the equations describe in this form both compressible and incompressible flows.

equation. We write explicitly all terms of $O(1)$ and bundle together all terms, which are of order of $O(Re_{ref}^{-1})$, $O(Re_{ref}^{-2})$:

$$\tilde{\rho}\tilde{u}\frac{\partial \tilde{u}}{\partial \tilde{x}} + \tilde{\rho}\tilde{v}\frac{\partial \tilde{u}}{\partial \tilde{y}} + \tilde{\rho}\tilde{w}\frac{\partial \tilde{u}}{\partial \tilde{z}} = -\frac{\partial \tilde{p}}{\partial \tilde{x}} + \frac{\partial}{\partial \tilde{y}}\left(\tilde{\mu}\frac{\partial \tilde{u}}{\partial \tilde{y}}\right) + O\left(\frac{1}{Re_{ref}}\right), \tag{4.7}$$

$$\frac{1}{Re_{ref}}\left(\tilde{\rho}\tilde{u}\frac{\partial \tilde{v}}{\partial \tilde{x}} + \tilde{\rho}\tilde{v}\frac{\partial \tilde{v}}{\partial \tilde{y}} + \tilde{\rho}\tilde{w}\frac{\partial \tilde{v}}{\partial \tilde{z}}\right) = -\frac{\partial \tilde{p}}{\partial \tilde{y}} + O\left(\frac{1}{Re_{ref}}, \frac{1}{Re_{ref}^2}\right), \tag{4.8}$$

$$\tilde{\rho}\tilde{u}\frac{\partial \tilde{w}}{\partial \tilde{x}} + \tilde{\rho}\tilde{v}\frac{\partial \tilde{w}}{\partial \tilde{y}} + \tilde{\rho}\tilde{w}\frac{\partial \tilde{w}}{\partial \tilde{z}} = -\frac{\partial \tilde{p}}{\partial \tilde{z}} + \frac{\partial}{\partial \tilde{y}}\left(\tilde{\mu}\frac{\partial \tilde{w}}{\partial \tilde{y}}\right) + O\left(\frac{1}{Re_{ref}}\right). \tag{4.9}$$

Finally we treat the energy equation, eq. (3.24) for perfect gas. Again we introduce non-dimensional and stretched variables, as we did above. We also write explicitly all terms of $O(1)$, and bundle together all terms, which are of smaller order of magnitude:

$$\begin{aligned} \tilde{c}_p\left(\tilde{\rho}\tilde{u}\frac{\partial \tilde{T}}{\partial \tilde{x}} + \tilde{\rho}\tilde{v}\frac{\partial \tilde{T}}{\partial \tilde{y}} + \tilde{\rho}\tilde{w}\frac{\partial \tilde{T}}{\partial \tilde{z}}\right) &= \frac{1}{Pr_{ref}}\left\{\frac{\partial}{\partial \tilde{y}}\left(\tilde{k}\frac{\partial \tilde{T}}{\partial \tilde{y}}\right)\right\} + \\ + E_{ref}\left\{\left[\tilde{u}\frac{\partial \tilde{p}}{\partial \tilde{x}} + \tilde{v}\frac{\partial \tilde{p}}{\partial \tilde{y}} + \tilde{w}\frac{\partial \tilde{p}}{\partial \tilde{z}}\right] + \tilde{\mu}\left[\left(\frac{\partial \tilde{u}}{\partial \tilde{y}}\right)^2 + \left(\frac{\partial \tilde{w}}{\partial \tilde{y}}\right)^2\right]\right\} + \\ + O\left(\frac{1}{Re_{ref}}\right) + O\left(\frac{E_{ref}}{Re_{ref}}\right) + O\left(\frac{E_{ref}}{Re_{ref}^2}\right). \end{aligned} \tag{4.10}$$

In this equation Pr_{ref} is the reference Prandtl number:

$$Pr_{ref} = \frac{\mu_{ref} c_{p_{ref}}}{k_{ref}}, \tag{4.11}$$

and E_{ref} the reference Eckert number:

$$E_{ref} = (\gamma_{ref} - 1) M_{ref}^2. \tag{4.12}$$

We arrive at the classical first-order boundary-layer equations in three dimensions by neglecting all terms of $O(1/Re_{ref})$ and $O(1/Re_{ref}^2)$ in eqs. (4.7) to (4.10). We note especially the implication of eq. (4.8): $\partial \tilde{p}/\partial \tilde{y} \to 0$ for $R_{ref} \to \infty$.

We write now the variables without tilde, understanding that the equations can be read in either way, non-dimensional, stretched or non-stretched, and dimensional and non-stretched, then without the similarity parameters Pr_{ref} and E_{ref}. We introduce in addition the "boundary-layer shear stresses"

τ_x and τ_z and the "boundary-layer heat flux" q_y in order to be compatible with the formulation for turbulent flow in the next section:

$$\frac{\partial \rho u}{\partial x} + \frac{\partial \rho v}{\partial y} + \frac{\partial \rho w}{\partial z} = 0, \tag{4.13}$$

$$\rho u \frac{\partial u}{\partial x} + \rho v \frac{\partial u}{\partial y} + \rho w \frac{\partial u}{\partial z} = -\frac{\partial p}{\partial x} + \frac{\partial \tau_x}{\partial y}, \tag{4.14}$$

$$0 = -\frac{\partial p}{\partial y}, \tag{4.15}$$

$$\rho u \frac{\partial w}{\partial x} + \rho v \frac{\partial w}{\partial y} + \rho w \frac{\partial w}{\partial z} = -\frac{\partial p}{\partial z} + \frac{\partial \tau_z}{\partial y}, \tag{4.16}$$

$$\begin{aligned} c_p \left(\rho u \frac{\partial T}{\partial x} + \rho v \frac{\partial T}{\partial y} + \rho w \frac{\partial T}{\partial z} \right) &= -\frac{1}{Pr_{ref}} \frac{\partial q_y}{\partial y} + \\ + E_{ref} \left[u \frac{\partial p}{\partial x} + w \frac{\partial p}{\partial z} + \left(\tau_x \frac{\partial u}{\partial y} \right) + \left(\tau_z \frac{\partial w}{\partial y} \right) \right], \end{aligned} \tag{4.17}$$

with the boundary-layer shear-stress components

$$\tau_x = \mu \frac{\partial u}{\partial y}, \tag{4.18}$$

$$\tau_z = \mu \frac{\partial w}{\partial y}, \tag{4.19}$$

and the boundary-layer heat flux:

$$q_y = -k \frac{\partial T}{\partial y}. \tag{4.20}$$

With the above equations we can determine the unknowns u, v, w, and T. The unknowns density ρ, viscosity μ, thermal conductivity k, and specific heat at constant pressure c_p are to be found with the equation of state $p = \rho RT$, and the respective relations given in Section 3.1. If the boundary-layer flow is turbulent, the apparent turbulent shear stresses are introduced, Section 4.3.

Since $\partial p/\partial y$ is zero, eq. (4.15), the pressure field of the external inviscid flow field, represented by $\partial p/\partial x$ and $\partial p/\partial z$, is impressed on the boundary layer. This assumes that the interaction of the boundary layer with the external inviscid flow field is weak and can be neglected. The result is that in the boundary layer $\partial p/\partial x$ and $\partial p/\partial z$ are constant in y-direction. This holds for first-order boundary layers on flat surfaces.

If the body surface is curved, the pressure gradient in the boundary layer in direction normal to the surface is not zero. The inviscid pressure field is still imposed on the boundary layer. This and other so-called higher-order effects are treated in Chapter 6.

The above equations are first-order boundary-layer equations, based on Cartesian coordinates. In surface-oriented non-orthogonal curvilinear coordinates, Appendix A.2, factors and additional terms have to be considered, involving the metric properties of the coordinate system. It should be noted that the equations for the general coordinates are formulated such that also the velocity components are transformed. This is in contrast to modern Euler and Navier-Stokes/RANS methods formulated for general coordinates. There only the geometry is transformed and not the velocity components, Appendix A.1.

4.3 Boundary-Layer Equations for Steady Turbulent Flow

In turbulent flow, it is more difficult to derive the boundary layer equations from the Navier-Stokes equations because the various scales of the motion endow the boundary layer with a double layer structure. In this section, we give a stringent derivation. In order to ease its reading, we consider steady, two-dimensional and incompressible flow. The result holds for compressible three-dimensional boundary layers, too.

4.3.1 Averaged Navier-Stokes Equations

We use again an orthonormal axis system, Sub-Section 2.2.1. The x-axis is along the wall and the y-axis normal to it. All quantities are dimensionless. Regarding the velocity components and the pressure we initially change the nomenclature from u, v, p to U, V, P in order to obtain a general point of origin. The coordinates x and y are non-dimensionalized by the reference length L_{ref}, the velocity components by a reference velocity V_{ref}, the pressure P by $\rho_{ref} V_{ref}^2$, the turbulent stresses by $\rho_{ref} V_{ref}^2$. In fact, the mean flow scales are chosen to define the reference quantities V and L. The non-dimensional velocity components and pressure are marked with a bar: $\bar{U}, \bar{V}, \bar{P}$.

In steady, two-dimensional, incompressible flow, the Reynolds-Averaged Navier-Stokes (RANS equations or Reynolds equations) are:

$$\frac{\partial \overline{U}}{\partial x} + \frac{\partial \overline{V}}{\partial y} = 0, \tag{4.21a}$$

$$\overline{U}\frac{\partial \overline{U}}{\partial x} + \overline{V}\frac{\partial \overline{U}}{\partial y} = -\frac{\partial \overline{P}}{\partial x} + \frac{\partial}{\partial x}\left(\frac{1}{Re}\frac{\partial \overline{U}}{\partial x} + \overline{T}_{xx}\right) + \frac{\partial}{\partial y}\left(\frac{1}{Re}\frac{\partial \overline{U}}{\partial y} + \overline{T}_{xy}\right), \tag{4.21b}$$

$$\overline{U}\frac{\partial \overline{V}}{\partial x} + \overline{V}\frac{\partial \overline{V}}{\partial y} = -\frac{\partial \overline{P}}{\partial y} + \frac{\partial}{\partial x}\left(\frac{1}{Re}\frac{\partial \overline{V}}{\partial x} + \overline{T}_{xy}\right) + \frac{\partial}{\partial y}\left(\frac{1}{Re}\frac{\partial \overline{V}}{\partial y} + \overline{T}_{yy}\right), \tag{4.21c}$$

where Re denotes the reference Reynolds number

$$Re = \frac{\rho_{ref} V_{ref} L_{ref}}{\mu_{ref}}$$

and the $\overline{T}_{ij}$ the reduced turbulent shear-stress components

$$\overline{T}_{ij} = - < \overline{U}'_i \overline{U}'_j > .$$

4.3.2 Scales

The results presented in this section are based, to a considerable extent, on a large amount of experimental data. From this a consistent theoretical description has been developed in order to reproduce the observations and in which the notion of *turbulence scales* plays an essential role. Therefore, the issue is addressed without resting on a well-posed mathematical frame as it is the case of laminar flow.

In a standard manner, with the method of matched asymptotic expansions, the flow is decomposed in two regions: the external inviscid region and the boundary layer. The former is treated separately and provides us with the necessary data to calculate the boundary layer. The boundary layer is described by a two-layer structure, [4]–[6], consisting of: i) an outer layer characterized by the boundary layer thickness δ and ii) an inner layer whose thickness is of order $\dfrac{\nu}{u_\tau}$ with $\nu = \mu/\rho$, and u_τ denoting the *friction velocity*

$$u_\tau = \sqrt{\frac{\tau_w}{\rho}}, \tag{4.22}$$

and τ_w the wall shear stress.

The *turbulence velocity scale*—denoted by $\boldsymbol{u}$—is identical in the outer region and in the inner region of the boundary layer and is of the order of the friction velocity u_τ. In the outer region, the turbulence length scale, of the order of δ, is denoted by $\boldsymbol{\ell}$ whereas in the inner region, the length scale is $\nu/\boldsymbol{u}$.

In the outer region, we assume that the time scale of the transport due to turbulence $(\boldsymbol{\ell}/\boldsymbol{u})$ is of the same order as the time scale of the mean flow convection. We can view this hypothesis as the counterpart—for turbulent flows—of the hypothesis used for a laminar boundary layer according to which the viscosity time scale is of the same order as the convection time scale. If the reference quantities V and L are chosen as velocity and length scales of the mean flow, we deduce

$$\frac{\boldsymbol{\ell}}{L} = \frac{\boldsymbol{u}}{V}. \tag{4.23}$$

The asymptotic analysis introduces the small parameters ε and $\hat{\varepsilon}$ which define, with dimensionless variables, the order of the thicknesses of the outer and inner layers:

$$\varepsilon = \frac{\boldsymbol{\ell}}{L}, \tag{4.24}$$

$$\hat{\varepsilon} = \frac{\nu}{\boldsymbol{u}L}. \tag{4.25}$$

Using eq. (4.23), we have

$$\varepsilon\hat{\varepsilon}Re = 1. \tag{4.26}$$

With the skin-friction law, eq. (4.36), the following relation between the gauge ε and the Reynolds number holds:

$$\varepsilon = \mathrm{O}\left(\frac{1}{\ln Re}\right). \tag{4.27}$$

In particular, we deduce that, for any positive n,

$$\varepsilon^n \succ \hat{\varepsilon} \succ \frac{1}{Re}.$$

The variables appropriate to the study of each region are

$$\text{outer region} : \eta = \frac{y}{\varepsilon}, \tag{4.28a}$$

$$\text{inner region} : \hat{y} = \frac{y}{\hat{\varepsilon}}. \tag{4.28b}$$

4.3.3 Structure of the Flow

The whole flow is described by a three-layer structure: the external region which is inviscid to first order, and the outer and inner region of the boundary layer.

The results are stated here assuming that surface curvature effects are negligible.

External Inviscid Region. In this region, the expansions are

$$\begin{aligned}
\overline{U} &= \bar{u}_0(x, y) + \varepsilon\bar{u}_1(x, y) + \cdots, \\
\overline{V} &= \bar{v}_0(x, y) + \varepsilon\bar{v}_1(x, y) + \cdots, \\
\overline{P} &= \bar{p}_0(x, y) + \varepsilon\bar{p}_1(x, y) + \cdots, \\
\overline{T}_{ij} &= 0.
\end{aligned}$$

It follows that $\bar{u}_0$, $\bar{v}_0$, $\bar{p}_0$ satisfy the Euler equations and $\bar{u}_1$, $\bar{v}_1$, $\bar{p}_1$ satisfy the linearized Euler equations.

Matching velocity v to order ε with the outer region of the boundary layer yields:

$$\bar{v}_{0w} = 0,$$
$$\bar{v}_{1w} = \lim_{\eta\to\infty}\left[v_0 - \eta\left(\frac{\partial \bar{v}_0}{\partial y}\right)_w\right],$$

where the index "w" denotes the wall.

The first condition enables us to calculate the flow defined by $\bar{u}_0$, $\bar{v}_0$, $\bar{p}_0$. Taking into account eq. (4.30b) and the continuity equation, the second condition gives $\bar{v}_{1w} = 0$. Then, with the condition that $\bar{u}_1$, $\bar{v}_1$ and $\bar{p}_1$ vanish at infinity, we have everywhere in the external region:

$$\bar{u}_1 = 0, \quad \bar{v}_1 = 0, \quad \bar{p}_1 = 0.$$

Outer Region of the Boundary Layer. In the outer region of the boundary layer, the expansions are

$$\overline{U} = u_0(x,\eta) + \varepsilon u_1(x,\eta) + \cdots, \tag{4.29a}$$
$$\overline{V} = \varepsilon\left[v_0(x,\eta) + \varepsilon v_1(x,\eta) + \cdots\right], \tag{4.29b}$$
$$\overline{P} = p_0(x,\eta) + \varepsilon p_1(x,\eta) + \cdots, \tag{4.29c}$$
$$\overline{T}_{ij} = \varepsilon^2 \tau_{ij,1}(x,\eta) + \cdots. \tag{4.29d}$$

The expansion of $\overline{V}$ is chosen in such a way that the continuity equation keeps its standard form to any order. The expansion of the turbulent stresses imply that their dominant order of magnitude is ε^2, i.e. the friction velocity is actually a *turbulence velocity scale.*

The equations for u_0, v_0 and p_0 are

$$\frac{\partial u_0}{\partial x} + \frac{\partial v_0}{\partial \eta} = 0,$$
$$u_0\frac{\partial u_0}{\partial x} + v_0\frac{\partial u_0}{\partial \eta} = -\frac{\partial p_0}{\partial x},$$
$$0 = \frac{\partial p_0}{\partial \eta}.$$

A solution which matches with the inviscid flow is

$$u_0 = u_e, \tag{4.30a}$$
$$v_0 = -\eta\frac{\mathrm{d}u_e}{\mathrm{d}x}, \tag{4.30b}$$

where u_e is the inviscid flow velocity at the wall:

$$u_e = \bar{u}_{0w}.$$

Moreover, the pressure p_0 is constant over the thickness of the outer region and is equal to the inviscid flow pressure at the wall:

$$p_0 = \bar{p}_{0w}.$$

Therefore, we have

$$\frac{\mathrm{d}p_0}{\mathrm{d}x} = -u_e \frac{\mathrm{d}u_e}{\mathrm{d}x}.$$

Neglecting wall curvature effects, the equations for u_1, v_1 and p_1 are

$$\frac{\partial u_1}{\partial x} + \frac{\partial v_1}{\partial \eta} = 0, \tag{4.31a}$$

$$u_1 \frac{\mathrm{d}u_e}{\mathrm{d}x} + u_e \frac{\partial u_1}{\partial x} - \eta \frac{\mathrm{d}u_e}{\mathrm{d}x} \frac{\partial u_1}{\partial \eta} = -\frac{\partial p_1}{\partial x} + \frac{\partial \tau_{xy,1}}{\partial \eta}, \tag{4.31b}$$

$$0 = \frac{\partial p_1}{\partial \eta}. \tag{4.31c}$$

With the hypothesis that surface curvature effects are negligible, it can be shown that $p_1 = 0$.

It is noted that the expansion given above amounts to consider the velocity defect $(u_e - u)/u_\tau$ as the pertinent velocity function. When the Reynolds number goes to infinity, the friction velocity u_τ goes to 0, but the ratio $(u_e - u)/u_\tau$ remains finite and non zero. In similarity solutions, it is assumed that this velocity defect is only a function of $\dfrac{y}{\delta}$.

Inner Region of the Boundary Layer. It is necessary to introduce an inner region, otherwise the no-slip condition at the wall is not satisfied. In this region, the expansions are

$$\overline{U} = \varepsilon \hat{u}_1(x, \hat{y}) + \cdots, \tag{4.32a}$$

$$\overline{V} = \hat{\varepsilon}(\varepsilon \hat{v}_1 + \cdots), \tag{4.32b}$$

$$\overline{P} = \hat{p}_0 + \varepsilon \hat{p}_1 + \cdots, \tag{4.32c}$$

$$\overline{T}_{ij} = \varepsilon^2 \hat{\tau}_{ij,1} + \cdots. \tag{4.32d}$$

The expansion chosen for $\overline{U}$ shows that the order of the stream-wise velocity is ε. With dimensionalized variables, this means that the velocity scale is the friction velocity. This essential hypothesis, consistent with experimental results, implies the logarithmic matching between the outer and inner region of the boundary layer.

The pressure $\hat{p}_0$ is constant along a normal to the wall and is equal to the pressure p_0 in the outer region:

$$\hat{p}_0 = p_0 = \bar{p}_{0w}.$$

The equations for $\hat{u}_1$, $\hat{v}_1$ and $\hat{p}_1$ are

$$\frac{\partial \hat{u}_1}{\partial x} + \frac{\partial \hat{v}_1}{\partial \hat{y}} = 0, \tag{4.33a}$$

$$0 = \frac{\partial}{\partial \hat{y}}\left(\hat{\tau}_{xy,1} + \frac{\partial \hat{u}_1}{\partial \hat{y}}\right), \tag{4.33b}$$

$$0 = \frac{\partial \hat{p}_1}{\partial \hat{y}}. \tag{4.33c}$$

The matching of the pressure to order ε between the outer and inner regions of the boundary layer gives $\hat{p}_1 = 0$.

From eq. (4.33b), the total stress—sum of the viscous stress and of the turbulent stress—is constant along a normal to the wall.

The matching between the outer region and the inner region on the velocity $\overline{U}$ (expansions given by eqs. (4.29a) and (4.32a)) raises a difficulty due to the absence of a term of order O(1) in the inner expansion. The solution rests upon a *logarithmic evolution* of the velocity in the overlap region:

$$u_1 = A \ln \eta + C_1 \quad \text{as} \quad \eta \to 0, \tag{4.34a}$$

$$\hat{u}_1 = A \ln \hat{y} + C_2 \quad \text{as} \quad \hat{y} \to \infty. \tag{4.34b}$$

The law for $\hat{u}_1$ corresponds to the universal law of the wall, where A and C_2 do not depend on the conditions under which the boundary layer develops (Reynolds number, pressure gradient). Constant A corresponds to the inverse of von Kármán's constant. It is noted that, in terms of more standard notations, the analysis of the inner region of the boundary layer leads to the law of the wall:

$$\frac{u}{u_\tau} = f\left(\frac{y u_\tau}{\nu}\right) \tag{4.35}$$

which is considered as a universal function, i.e. a function which does not depend on the conditions in which the boundary layer develops.

In the overlap region, the equality of velocity in the outer and the inner region gives

$$u_e + \varepsilon(A \ln \eta + C_1) = \varepsilon(A \ln \hat{y} + C_2),$$

or

$$\frac{u_e}{\varepsilon} = A \ln \frac{\varepsilon}{\hat{\varepsilon}} + C_2 - C_1. \tag{4.36}$$

This equation represents the *skin-friction law*. Expressed in terms of dimensionalized variables, this law takes the standard form

$$\frac{u_e}{u_\tau} = \frac{1}{\chi} \ln \frac{u_\tau \delta}{\nu} + B, \tag{4.37}$$

where $\chi \simeq 0.4$ is von Kármán's constant and B depends on the pressure gradient.

This relation and the logarithmic variation of velocity in the overlap region are the keys of the asymptotic structure of the turbulent boundary layer.

4.3.4 Boundary-Layer Equations

Generally, for practical applications, we consider a set of equations which encompasses the equations developed in Sub-Section 4.3.3 for the outer region and for the inner region of the boundary layer. The boundary layer equations then read:

$$\frac{\partial u}{\partial x} + \frac{\partial v}{\partial y} = 0, \tag{4.38}$$

$$\rho u \frac{\partial u}{\partial x} + \rho v \frac{\partial u}{\partial y} = -\frac{\mathrm{d}p}{\mathrm{d}x} + \frac{\partial \tau}{\partial y}, \tag{4.39}$$

$$0 = \frac{\partial p}{\partial y}, \tag{4.40}$$

with τ denoting the total stress, i.e. the sum of the viscous stress and of the apparent turbulent stress:

$$\tau = \mu \frac{\partial u}{\partial y} - \rho < u'v' > . \tag{4.41}$$

Formally, the equations for a turbulent boundary layer are the same as those for a laminar boundary layer. The difference is the expression of the stress which takes into account the turbulence effect.

Since eqs. (4.38) to (4.40) include equations for the outer and the inner region of the boundary layer, an asymptotic expansion of them gives again the equations for these two regions.

In compressible flow, a similar extension is used to write the turbulent boundary layer equations. For the three-dimensional boundary layers equations, eqs. (4.14) to (4.17), we obtain for the boundary-layer shear-stress components and the boundary-layer heat flux:

$$\tau_x = \mu \frac{\partial u}{\partial y} - < \rho u'v' >, \tag{4.42}$$

$$\tau_z = \mu \frac{\partial w}{\partial y} - < \rho w'v' >, \tag{4.43}$$

$$q_y = -k \frac{\partial T}{\partial y} + < \rho h'v' > . \tag{4.44}$$

Regarding the closure of the turbulent fluctuation terms, the possible turbulence models, see the remarks at the end of Section 3.5, page 72.

4.4 Characteristic Properties of Attached Viscous Flow

We identify the characteristic properties of steady three-dimensional attached viscous flow. In order to ease the derivation, we look at the characteristic properties of the boundary-layer equations rather than those of the Navier-Stokes equations. For convenience we assume incompressible flow. Following [7], we introduce characteristic manifolds $\varphi(x, y, z)$, for instance like

$$\frac{\partial}{\partial x} = \frac{\partial \varphi}{\partial x}\frac{d}{d\varphi} = \varphi_x \frac{d}{d\varphi} \tag{4.45}$$

into eqs. (4.13) to (4.16).

After introduction of the kinematic viscosity $\nu = \mu/\rho$ and some manipulation the characteristic form is found:

$$C = \begin{vmatrix} \varphi_x & \varphi_y & \varphi_z \\ (\Delta - \nu\varphi_y^2) & 0 & 0 \\ 0 & 0 & (\Delta - \nu\varphi_y^2) \end{vmatrix} = \varphi_y(\Delta - \nu\varphi_y^2)^2 = 0, \tag{4.46}$$

with the abbreviation

$$\Delta = u\varphi_x + v\varphi_y + w\varphi_z. \tag{4.47}$$

The pressure gradients $\partial p/\partial x$ and $\partial p/\partial z$ do not enter the characteristic form, because the pressure field is imposed on the boundary layer, i. e. $\partial p/\partial x$ and $\partial p/\partial z$ are forcing functions.

Eq. (4.47) corresponds to the projection of the gradient of the manifold onto the streamline, and represents the boundary layer streamlines as characteristic manifolds. To prove this, we write the total differential of φ for a given characteristic manifold

$$d\varphi = \varphi_x dx + \varphi_y dy + \varphi_z dz = 0, \tag{4.48}$$

and combine it with the definition of streamlines in three dimensions

$$\frac{dx}{u} = \frac{dy}{v} = \frac{dz}{w}, \tag{4.49}$$

in order to find

$$d\varphi = u\varphi_x + v\varphi_y + w\varphi_z = \Delta = 0. \tag{4.50}$$

Thus it is shown that streamlines are characteristics, too [7]. They are considered as sub-characteristics. The consequence of this is that the flow properties along the normal of a point P(x, z) on the body surface depend on upstream flow properties. The—analytical—domain of dependence is given by the streamlines including the skin-friction line. On the other hand, the point P properties influence the downstream flow properties, i.e. the influence of an event at P is spread over a downstream domain.

In a three-dimensional flow hence both the domain of dependence and that of influence are defined by the strength of the skewing of the stream surface, Fig. 4.1 a) (note the different coordinate notation in that picture!). The skin-friction line alone is not representative. Of course also here lateral molecular or turbulent transport happens.

If a space-marching boundary-layer method is used for the determination of the flow field, it must take into account the analytical domain of dependence. If the flow is to be determined at the point $P(x, y)$, the numerical domain of dependence, Fig. 4.1 b) must include the analytical one.

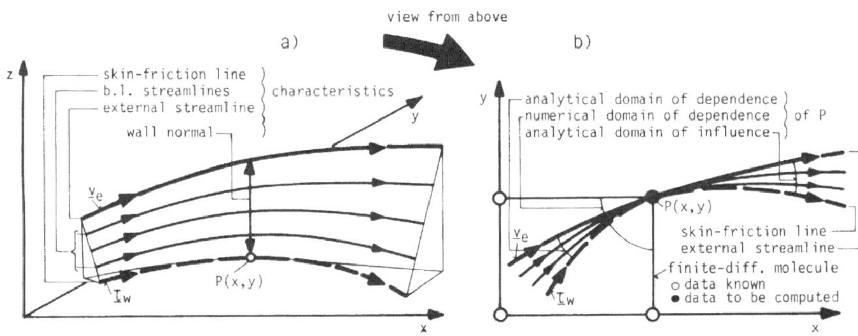

Fig. 4.1. Three-dimensional boundary layer with skewed stream surface (schematically): **a**) the streamlines as characteristics, **b**) domains of dependence and of influence of flow properties in $P(x, y)$. Note that in this figure, [8], x and y are the surface coordinates, and z is the coordinate normal to the surface!

This is the so-called Courant-Friedrichs-Lewy (CFL) condition, see, e.g., [9].[3] The CFL condition was published in 1928 [10]. For a three-dimensional boundary-layer solution it makes necessary a dedicated orientation of the surface coordinates, as well as special discretization approaches (finite-difference molecules), Appendix A.3.2. In 1957 G.S. Raetz discussed this problem in the frame of his studies [11]. In the literature it is known as the "Raetz principle".

We have shown above that $\triangle = 0$, i.e., it represents the streamlines as sub-characteristics. Hence in eq. (4.46) remains a five-fold set of characteristics $\varphi_y = 0$ in y-direction. These are typical for boundary-layer equations. These characteristics are complemented by two-fold characteristics in y-direction coming from the energy equation, eq. (4.17), in its incompressible form which

[3] In time-marching Navier-Stokes/RANS solutions, the CFL condition takes into account the temporal domain of dependence [9].

we do not demonstrate here. These results are valid for compressible flow, too, and also for second-order boundary-layer equations.

The above discussion shows that boundary-layer equations of first, also of second order, in two or three dimensions, are parabolic, and hence pose a mixed initial value/boundary condition problem. Where the boundary-layer flow enters the domain under consideration, initial conditions must be prescribed. At the surface of the body, $y = 0$, and at the outer edge of the boundary layer, $y = \delta$, boundary conditions are to be described for u, w, and T, hence the six-fold characteristics in y-direction. For the surface-normal velocity component v only a boundary condition at the body surface, $y = 0$, must be described, which reflects the seventh characteristic.[4]

We have introduced the boundary layer as phenomenological model of attached viscous flow. This model is valid everywhere on the surface of a flight vehicle, where strong interaction phenomena are not present like separation, shock/boundary-layer interaction, hypersonic viscous interaction, et cetera.

If three-dimensional attached viscous flow—laminar or turbulent, incompressible or compressible—is boundary-layer like, we can now, based on the above analysis, give a summary of its global characteristic properties:

1. Attached viscous flow is governed primarily by the external inviscid flow field via its pressure field, and by the surface conditions.
2. In space it has parabolic character, i. e. the boundary conditions in general dominate its properties (seven-fold characteristics in direction normal to the surface), the influence of the initial conditions usually is weak and decreases with increasing downstream distance.
3. Events in attached viscous flow are felt only downstream, as long as they do not invalidate the boundary-layer criteria.[5] This means, for instance, that a surface disturbance or surface suction or normal injection can have a magnitude at most of $O(1/\sqrt{Re_{ref}})$. Otherwise the attached viscous flow loses its boundary-layer properties (strong interaction).

   The downstream effect is—as mentioned above—spread over the domain of influence. We illustrate this in Fig. 4.2 with the result of a numerical study [12] (note the different coordinate notation in that picture!). Within a three-dimensional boundary-layer domain—the basic flow case being the same as considered in Section 2.3—a small area is heated with the maximum temperature twice as high as the wall temperature in the remaining domain. Outside of the heated area the heat flux is prescribed with $q_{gw} = 0$. The flow is incompressible, the energy equation is reduced to the convection and the conduction transport terms. Downstream of the heated area the heat is transported in direction of the flow. The figure

[4] The boundary conditions are basically those discussed in Section 3.3. The outer boundary conditions are now at $y = \delta$ instead of $y \to \infty$.

[5] However, disturbances can also to a degree propagate upstream, Sub-Section 6.2.3.

shows well, here for one chosen initial condition, how the effect is spread over the domain of influence, bounded by the external inviscid streamline and the skin-friction line.

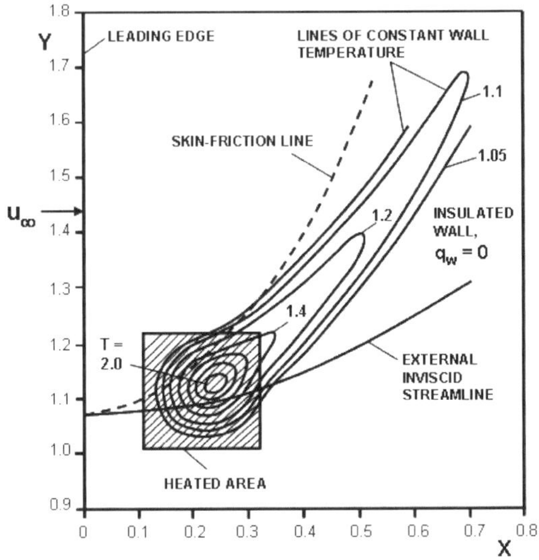

Fig. 4.2. Partially heated wall [12]: the effect of the heated area is spread downstream over the domain of influence. Note that in this figure x and y are the surface coordinates, z is the coordinate normal to the surface.

4. In attached viscous flow an event is felt upstream only if it influences the pressure field via, e.g., a disturbance of $O(1)$ or if strong temperature gradients in main-flow direction are present ($\partial(k\partial T/\partial x)/\partial x$ and $\partial(k\partial T/\partial z)/\partial z = O(Re_{ref})$. The displacement properties of an attached boundary layer are of $O(1/\sqrt{Re_{ref}})$, and hence influence the pressure field only weakly (weak interaction).
5. Separation causes locally strong interaction and may change the onset boundary-layer flow, however only via a global change of the pressure field. Strong interaction phenomena have only small upstream influence, i. e. their influence is felt predominantly downstream (locality principle) and via the global change of the pressure field, see also Section 1.2.
6. In two-dimensional attached viscous flow the domain of influence of an event is defined by the convective transport along the streamlines in downstream direction. Due to lateral molecular or turbulent transport it assumes a wedge-like pattern with small spreading angle.

4.5 Wall Compatibility Conditions

In the continuum flow regime the no-slip condition at the surface of a body is the cause for the development of the boundary layer. The following consideration is valid for both the Navier-Stokes/RANS and the boundary-layer equations. A flat body surface is assumed, the coordinate system is, as before, the Cartesian system.

At the surface we have $u_{wall} = w_{wall} = 0$, and also that the normal velocity component at the body surface is zero: $v_{wall} = 0$, although $|v_{wall}/v_{ref}| \lessapprox O(1/Re_{ref})$ would be permitted. We formulate:

$$u|_{y=0} = 0, \; v|_{y=0} = 0, \; w|_{y=0} = 0, \tag{4.51}$$

which also means that for $y = 0$ all derivatives of u, v, w in x- and z-direction are zero.

The y-derivatives of u and w at the wall are not zero, because we consider attached viscous flow.[6] The classical wall compatibility conditions for three-dimensional attached viscous flow follow from eqs. (4.14) and (4.16). They connect the second-order derivatives of the tangential flow components u and w at the surface with the respective pressure gradients:

$$\frac{\partial}{\partial y}(\mu \frac{\partial u}{\partial y})|_{y=0} = \frac{\partial p}{\partial x}, \tag{4.52}$$

$$\frac{\partial}{\partial y}(\mu \frac{\partial w}{\partial y})|_{y=0} = \frac{\partial p}{\partial z}. \tag{4.53}$$

For the first derivatives of u and w we obtain in external streamline coordinates, Figs. 2.3 and 2.4 (note that in these figures the coordinate notations are different from those used here) for the main-flow direction:

$$\frac{\partial u}{\partial y}|_{y=0} > 0, \tag{4.54}$$

and for the cross-flow direction:

$$\frac{\partial w}{\partial y}|_{y=0} \lessgtr 0. \tag{4.55}$$

The first derivative at the surface of the normal velocity component v in y-direction is found from the continuity equation, eq. (3.13):

$$\frac{\partial v}{\partial y}|_{y=0} = 0. \tag{4.56}$$

---

[6] Of course, due to a given flow field and a given coordinate orientation, the y-derivative of the cross-flow component locally can be zero. In such a case the coordinate orientation must be changed.

We generalize now eqs. (4.52) and (4.53), permitting suction or blowing ($v_{wall} \lessgtr 0$) by adding the respective terms from eqs. (4.14) and (4.16), as well as a temperature gradient in direction normal to the wall. We find

$$\frac{\partial^2 u}{\partial y^2}|_{y=0} = \left\{ \frac{1}{\mu} \left[\rho v \frac{\partial u}{\partial y} + \frac{\partial p}{\partial x} - \frac{\partial \mu}{\partial T} \frac{\partial T}{\partial y} \frac{\partial u}{\partial y} \right] \right\}_{y=0}, \tag{4.57}$$

$$\frac{\partial^2 w}{\partial y^2}|_{y=0} = \left\{ \frac{1}{\mu} \left[\rho v \frac{\partial w}{\partial y} + \frac{\partial p}{\partial z} - \frac{\partial \mu}{\partial T} \frac{\partial T}{\partial y} \frac{\partial w}{\partial y} \right] \right\}_{y=0}. \tag{4.58}$$

The functions of the tangential velocity components $u(y)$ and $w(y)$ and their derivatives at the outer edge of the boundary layer $y = \delta$ are found with the assumption—in the frame of first-order boundary-layer theory—that the boundary-layer equations approach asymptotically at the edge of the boundary layer the (two-dimensional) Euler equations. From eqs. (4.14) and (4.16) we get:

$$u|_{y=\delta} = u_e, \ \frac{\partial u}{\partial y}|_{y=\delta} = 0, \ \frac{\partial^2 u}{\partial y^2}|_{y=\delta} = 0, \tag{4.59}$$

$$w|_{y=\delta} = w_e, \ \frac{\partial w}{\partial y}|_{y=\delta} = 0, \ \frac{\partial^2 w}{\partial y^2}|_{y=\delta} = 0. \tag{4.60}$$

The normal velocity component $v(y)$ is not defined at the outer edge of the boundary layer, nor its second derivative. From the continuity equation, eq. (4.13), we find only the compatibility condition:

$$\frac{\partial \rho v}{\partial y}|_{y=\delta} = -\left(\frac{\partial \rho u}{\partial x} + \frac{\partial \rho w}{\partial z} \right)_{y=\delta} = -\left(\frac{\partial \rho_e u_e}{\partial x} + \frac{\partial \rho_e w_e}{\partial z} \right). \tag{4.61}$$

The compatibility conditions permit to make assertions about the shape of boundary-layer velocity profiles. We demonstrate this with the profile of the tangential velocity component of a two-dimensional flat-surface boundary layer, which is assumed to be representative also for the stream-wise profiles of weakly three-dimensional boundary layers, Fig. 2.3. We assume that the velocity profile has no overshoot at the boundary-layer edge.[7]

We consider three possible values of $\partial^2 u/\partial y^2|_{wall}$: < 0 (case 1), $= 0$ (case 2), > 0 (case 3), Fig. 4.3. We see that the second derivative (curvature) is negative above the broken line for all profiles given in Fig. 4.3 a). Hence the second derivative will approach in any case $\partial^2 u/\partial y^2|_{y=\delta} = 0$ with a negative value, Fig. 4.3 a). It can be shown by further differentiation of the x-momentum equation, eq. (4.14), that for incompressible flow also

$$\frac{\partial^3 u}{\partial y^3}|_{y=0} = 0. \tag{4.62}$$

[7] Such an (small) overshoot can be expected due to the displacement property of the boundary layer [13].

With these elements the function $\partial^2 u(y)/\partial y^2$ can be sketched qualitatively, Fig. 4.3 a). Because we consider attached viscous flow, $\partial u/\partial y|_{wall} > 0$ holds in all three cases, Fig. 4.3 b). We obtain finally the result that boundary-layer flow in the cases 1 and 2 has profiles $u(y)$ without a point of inflection, and in the case 3 has a profile $u(y)$ with a point of inflection.

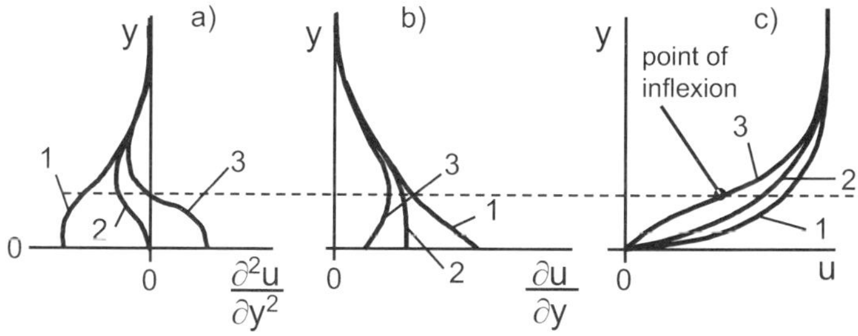

Fig. 4.3. Shape (qualitatively) of **a**) second derivative, **b**) first derivative, and **c**) function of the tangential velocity component $u(y)$ of a two-dimensional boundary layer, or the stream-wise profile of a weakly three-dimensional boundary layer [14]. Case 1: $\partial^2 u/\partial y^2|_{wall} < 0$, case 2: $\partial^2 u/\partial y^2|_{wall} = 0$, case 3: $\partial^2 u/\partial y^2|_{wall} > 0$.

The interpretation of this result, Table 4.1, is found through a term by term examination of eq. (4.57). It was assumed that $\partial u/\partial y|_w$ is always positive, because we consider attached flow only. Also viscosity μ and density ρ are positive. Since we deal with a gas, the derivative of the viscosity with respect to the temperature is always positive: $\partial\mu/\partial T > 0$ (in liquids, especially in water, it is negative).

We see in Table 4.1 the individual terms in eq. (4.57) that may or may not cause a point of inflection of the tangential velocity profile $u(y)$. In any case an adverse pressure gradient may cause a point of inflection in the boundary-layer profile[8], and also heating of the boundary layer, i. e., transfer of heat from the body surface into the flow, or surface-normal blowing.

In a real flow situation several of the flow features considered in Table 4.1 may be present. Accordingly the sum of the terms in the bracket of eq. (4.57) is the determining factor. The individual terms may weaken or cancel their combined influence, or may enhance it. The factor $1/\mu$ in front of the square bracket is a modifier, which reduces $\partial^2 u/\partial y^2|_w$, if the surface is hot, and enlarges it, if the surface is cold.

We add a note on radiation-cooled outer surfaces of hypersonic flight vehicles. The radiation-adiabatic or radiation-equilibrium wall generally is a good approximation of the thermal state of the surface in reality [14]. It

[8] It is the classical interpretation that an adverse pressure gradient leads to a profile $u(y)$ with point of inflection, but zero and favorable pressure gradient not. With our generalization we see that also other factors can lead to a point of inflection of the profile $u(y)$.

Table 4.1. Influence of terms in eq. (4.57) on $\partial^2 u/\partial y^2|_{y=0}$.

Term	Flow feature	$\partial^2 u/\partial y^2\|_{y=0}$	Point of inflection
$v_{wall} > 0$	blowing	> 0	yes
$v_{wall} = 0$	non-permeable surface	0	no
$v_{wall} < 0$	suction	< 0	no
$\partial p/\partial x > 0$	decelerated flow	> 0	yes
$\partial p/\partial x = 0$	flat-plate flow	0	no
$\partial p/\partial x < 0$	accelerated flow	< 0	no
$\partial T/\partial y\|_{wall} > 0$	cooling of boundary layer	< 0	no
$\partial T/\partial y\|_{wall} = 0$	adiabatic wall	0	no
$\partial T/\partial y\|_{wall} < 0$	heating of boundary layer	> 0	yes

assumes that no heat enters the wall, like in the classical case of the adiabatic wall. However, due to the radiation cooling exists a—usually strong—gradient $\partial T/\partial y|_{wall} > 0$ in the gas at the wall. Hence in this case, zero or small heat flux into the wall does not mean $\partial T/\partial y|_{wall} = 0$.

The point of inflection of the velocity profile $u(y)$ is an important indicator of (inviscid) boundary-layer instability, Sub-Section 9.2.3. The extended compatibility conditions moreover permit to gain qualitative insight into how an attached viscous flow reacts on the different flow features listed in Table 4.1. This ranges from laminar-turbulent transition to the separation susceptibility of the flow. See also the discussion in Section 2.3.

In Chapter 7 we will show that only in a few—singular—points streamlines actually impinge on or leave the body surface. This implies that in attached viscous flow close to a non-permeable wall the flow is parallel to the body surface.

We use now the wall compatibility conditions to determine the flow angle θ in the limit $y \to 0$. We do this once more only for the profile of the tangential velocity component of two-dimensional boundary layers, which also holds for the main-flow profiles of three-dimensional boundary layers, Fig. 2.3.

With the no-slip wall boundary condition eq. (4.51), the assumption of attached viscous flow with $\partial u/\partial y|_{wall} > 0$, and condition eq. (4.56), we find by means of a Taylor expansion around a point on the surface for small distances y from the surface:

$$u \sim y, \; v \sim y^2, \tag{4.63}$$

and hence

$$\tan\theta = \frac{v}{u} \sim y. \tag{4.64}$$

The result is that when the surface is approached in attached viscous flow, the flow in the limit becomes parallel to it:

$$y \to 0 : \theta \to 0. \tag{4.65}$$

If we consider y and v as with $\sqrt{Re_{ref}}$ stretched entities, the result tells us further that the whole boundary-layer flow for $Re_{ref} \to \infty$ becomes parallel to the surface.[9] This observation is decisive in the derivation of the Orr-Sommerfeld equation, Sub-Section 9.2.2: the flow is assumed to be parallel to the surface. "Non-parallel effects" as well as "surface-curvature effects", however, are a topic in stability theory.

4.6 Problems

Problem 4.1. What was Prandtl's assumption leading to the boundary-layer equations?

Problem 4.2. What is the friction velocity and how does it relate to the inner layer thickness of a turbulent boundary layer? How does it relate to the law of the wall?

Problem 4.3. The topic of this problem is the law of the wall. The turbulent boundary layer is made of two distinct layers. One layer, which is close to the wall, is very thin compared to the other one. In the wall layer, the flow is directly influenced by viscosity. A turbulence velocity scale is the friction velocity u_τ, Sub-Section 4.3.2:

$$u_\tau = \sqrt{\frac{\tau_w}{\rho}}.$$

The pertinent variables in the wall region are

$$u^+ = \frac{u}{u_\tau}, \quad y^+ = \frac{y u_\tau}{\nu}.$$

We note that y^+ is a Reynolds number based on the distance to the wall and the friction velocity.

From dimensional analysis it can be seen that

$$u^+ = f(y^+),$$

which is called the law of the wall, Sub-Section 4.3.3.

1. Show that the law of the wall becomes

$$u^+ = y^+$$

very close to the wall.

[9] A purely inviscid flow anyway is parallel to the surface.

2. In the turbulent part of the wall layer, a length scale is the distance to the wall. The inverse of a time scale for the average velocity is $\partial u/\partial y$. It is assumed that the time scale of turbulence and the time scale of the average flow have the same order.
   Deduce that the law of the wall takes a logarithmic form.

Problem 4.4. The law of the wall for a rough wall. Let k be the characteristic height of a surface roughness In the wall layer of the boundary layer we have from dimensional analysis

$$u^+ = u^+(y^+, k^+).$$

The wall variables are formed with the friction velocity u_τ as the velocity scale:

$$u_\tau = \sqrt{\frac{\tau_w}{\rho}}.$$

The wall variables are:

$$u^+ = \frac{u}{u_\tau}, \quad y^+ = \frac{yu_\tau}{\nu}, \quad k^+ = \frac{ku_\tau}{\nu}.$$

In the fully turbulent part of the wall layer, the turbulence length scale is the distance from the wall. As for a smooth surface (Problem 4.2), we assume that the time scale of the average flow is the same as the time scale of turbulence.

1. Show that the velocity in the turbulent part of the wall layer takes the form
$$u^+ = \frac{1}{\chi}\ln y^+ + B_1(k^+).$$
2. Show that the logarithmic law of the wall can be written as
$$u^+ = \frac{1}{\chi}\ln \frac{y}{k} + B_2.$$
   Give the expression of B_2.
3. In the case of a smooth wall, the logarithmic law of the wall is
$$u^+ = \frac{1}{\chi}\ln y^+ + c.$$
   Show that the logarithmic law of the wall in the case of a rough wall is shifted by a quantity Δu^+. Give the expression of Δu^+.

Problem 4.5. How many characteristics do the three-dimensional boundary layer equations have?

Problem 4.6. If a slip-flow boundary layer exists (see Problem 3.4) how does this affect the wall compatibility conditions?

References

1. Prandtl, L.: Über Flüssigkeitsbewegung bei sehr kleiner Reibung. In: Proceedings 3rd Intern., Math. Congr., Heidelberg, pp. 484–491 (1904)
2. Van Dyke, M.: Perturbation Methods in Fluid Mechanics. Academic Press, New York (1964)
3. Vincenti, W.G., Kruger, C.H.: Introduction to Physical Gas Dynamics. John Wiley, New York (1965), Reprint edn. Krieger Publishing Comp., Melbourne (1975)
4. Mellor, G.L.: The Large Reynolds Number Asymptotic Theory of Turbulent Boundary Layers. Int. J. Eng. Sci. 10, 851–873 (1972)
5. Yajnik, K.S.: Asymptotic Theory of Turbulent Shear Flows. J. Fluid Mech. 42(Pt. 2), 411–427 (1970)
6. Panton, R.L.: Review of Wall Turbulence as Described by Composite Expansions. Applied Mechanics Reviews 58, 1–36 (2005)
7. Courant, R., Hilbert, D.: Methods of Mathematical Physics, vol. II. John Wiley-Interscience, New York (1962)
8. Hirschel, E.H.: Evaluation of Results of Boundary-Layer Calculations with Regard to Design Aerodynamics. AGARD R-741, 5-1–5-29 (1986)
9. Hirsch, C.: Numerical Computation of Internal and External Flows. Fundamentals of Numerical Discretization, vol. 1. John Wiley, New York (1997)
10. Courant, R., Friedrichs, K.O., Lewy, H.: Über die partiellen Differenzengleichungen der mathematischen. Physik. Math. Ann. 100, 32–74 (1928); On the Partial Difference Equations of Mathematical Physics. IBM Journal, 215–234 (1967)
11. Raetz, G.S.: A Method of Calculating Three-Dimensional Laminar Boundary Layers of Steady Compressible Flows. Northrop Aircraft, Inc., Rep. No. NAI-58-73, BLC-144 (1957)
12. Krause, E., Hirschel, E.H.: Exact Numerical Solutions for Three-Dimensional Boundary Layers. In: Hold, M. (ed.) Proc. 2nd Int. Conf. on Num. Methods in Fluid Dynamics, Berkeley, USA, September 15-19. Leture Notes in Physics, vol. 8, pp. 132–137. Springer (1970)
13. Lighthill, M.J.: On Displacement Thickness. J. Fluid Mechanics 4, 383–392 (1958)
14. Hirschel, E.H.: Basics of Aerothermodynamics, AIAA, Reston, VA. Progress in Astronautics and Aeronautics, vol. 204. Springer, Heidelberg (2004)

5

Boundary-Layer Integral Parameters

This chapter is devoted to the formulation of boundary-layer integral parameters for three-dimensional attached viscous flow. These parameters are basically the same as in two-dimensional flow, but there are some notable differences, of most interest those regarding the boundary-layer displacement thickness. We give this thickness, as well as the momentum flow and the energy flow displacement thickness in Cartesian coordinates. For the formulation of the displacement thickness in general curvilinear coordinates see Appendix A.2.4.

5.1 General Considerations

Boundary-layer integral parameters are of practical importance not only in the frame of boundary-layer theory and methods, where they may appear in integral solution techniques as dependent variables. See, in this respect the classical Kármán-Pohlhausen method for two-dimensional boundary layers, [1, 2], and for three-dimensional ones, e. g., [3]. They are often used to facilitate the correlation of experimental results, for instance regarding laminar-turbulent instability and transition, flow separation, surface roughness effects, etc. [4].

The integral parameters reflect the influence of the boundary layer upon the hypothetical inviscid flow past a body, see also Section 1.2. In this way they characterize globally the development of the attached viscous flow past that body.

The concept of the familiar (mass-flow) displacement thickness in two dimensions is not valid for three-dimensional flow. It was extended to the latter by F.K. Moore [5] and by M.J. Lighthill [6]. The concepts of momentum- and energy-loss thicknesses used in two-dimensional flow, however, cannot be extended to three-dimensional flow. We follow here the concept of J. Kux, who introduces instead momentum- and energy-flow displacement thicknesses [7], see also [8]. These quantities appear naturally as dependent variables in the three-dimensional integral boundary-layer equations. The three concepts have in common that partial differential equations have to be integrated, in contrast to the situation in two-dimensional boundary layers, for which only quadratures have to be performed in order to find the integral properties.

E.H. Hirschel, J. Cousteix, and W. Kordulla, *Three-Dimensional Attached Viscous Flow*,
DOI: 10.1007/978-3-642-41378-0_5, © Springer-Verlag Berlin Heidelberg 2014

In the following three sections we present the mass-flow, the momentum-flow, and the energy-flow displacement thickness in Cartesian coordinates. For the mass-flow displacement thickness we use the notation δ_1 instead of δ^* which is often found in the literature, see, e.g., [9]. The momentum-flow displacement thickness is called δ_2 and the energy-flow displacement thickness δ_3. Note again that the latter two are conceptually different from the momentum thickness θ, the energy, and the energy dissipation thickness. We define the boundary-layer thickness δ as usual as 99 per cent thickness both for laminar and turbulent flow, in the latter case in the sense of the Reynolds-averaged consideration. If the inviscid flow is not known, for instance in Navier-Stokes solutions or in experiments, the boundary-layer edge can be defined by vanishing boundary-layer vorticity $|\underline{\omega}| \leqq \epsilon$ [8], provided the external inviscid flow is irrotational.

We give the equations for first-order boundary layers in Cartesian coordinates. Like in the preceding chapters, the surface-parallel coordinates are x and z, the surface-normal coordinate is y, u and w are the tangential velocity components, and v is the component normal to the body surface.

5.2 Mass-Flow Displacement Thickness and Equivalent Inviscid Source Distribution

The equation for the displacement thickness δ_1 is found by a control-volume consideration, [5], which we do not give here, but see Problem 5.4. That consideration results in the first-order differential equation:

$$\frac{\partial}{\partial x}[\rho_e u_e(\delta_1 - \delta_{1x})] + \frac{\partial}{\partial z}[\rho_e w_e(\delta_1 - \delta_{1z})] = \rho_0 v_0. \tag{5.1}$$

The term on the right-hand side of the equation, $\rho_0 v_0$, is a wall source term, taking into account possible wall-normal suction or blowing. It vanishes for an impermeable wall. Whereas the displacement thickness in general is always positive, eq. (5.1) can yield also negative values. We can observe such a result in Section 10.1.

The quantities δ_{1x} and δ_{1z} are the familiar two-dimensional definitions of displacement thickness with respect to the mass-flow components ρu and ρw in x- and z-direction:

$$\delta_{1x} = \int_0^\delta (1 - \frac{\rho u}{\rho_e u_e})\, dy, \tag{5.2}$$

$$\delta_{1z} = \int_0^\delta (1 - \frac{\rho w}{\rho_e w_e})\, dy. \tag{5.3}$$

Eq. (5.1) can only be solved after the solution of the boundary-layer equations has been obtained or measurements have been made, because the evaluation of eqs. (5.2) and (5.3) requires the knowledge of the velocity and density profiles, as well as the properties of the external inviscid flow field.

For two-dimensional flow in x-direction, $\partial/\partial z = 0$, and eq. (5.1) reduces to the two-dimensional definition of the displacement thickness, eq. (5.2), if $v_0 = v(y = 0) = 0$.

The (mass-flow) displacement thickness is sometimes used in inviscid-viscous interaction calculations to simulate the effective "viscous" shape of a body. It represents the displacement of the external inviscid flow due to the attached viscous flow. In the case of weak interaction, three steps are made: (1) original calculation of the inviscid flow, (2) boundary-layer calculation with the original pressure distribution, (3) calculation of the inviscid flow by taking into account the displacement effect. For the new—virtual—body shape with changed metric properties a final boundary-layer calculation can be made. Repeating the steps may not lead to convergence.

A better way to take into account the displacement effect of the boundary layer is to use the equivalent inviscid source distribution. It is a transpiration mass flux $(\rho_w v_w)|_{inv}$ on the original body surface, which serves as wall boundary condition in order to reflect the influence of the viscous flow. This was proposed by M.J. Lighthill [6] and by W.J. Piers et al. [10]. No equation like eq. (5.1) has to be solved, but the boundary-layer profiles are still needed [6, 10]:

$$(\rho_w v_w)|_{inv} = \frac{\partial}{\partial x}[\rho_e u_e \delta_{1x}] + \frac{\partial}{\partial z}[\rho_e w_e \delta_{1z}] + \rho_0 v_0. \tag{5.4}$$

The surface-normal mass-flow component $(\rho_w v_w)|_{inv}$ is to be imposed at the body surface. Because it is initially unknown, the use of eq. (5.4) involves the same steps as needed with the displacement thickness. But because the shape of the body remains unchanged in the course of the inviscid-viscous coupling procedure, the metric properties of the body, see Appendix C, remain unchanged, too. They have to be determined as usual only once at the beginning of the simulation process.

5.3 Momentum-Flow Displacement Thickness

The momentum flux is a vector entity. Therefore two momentum-flow displacement thicknesses δ_{2x} and δ_{2z} are defined in three-dimensional flow. Following Kux, [7], we give, however without derivation, see [8], the respective first-order partial differential equations—in Cartesian coordinates and assuming a non-permeable wall—for their determination:

$$\frac{\partial}{\partial x}[\rho_e (u_e)^2 (\delta_{2xx} - \delta_{2x})] + \frac{\partial}{\partial z}[\rho_e u_e w_e (\delta_{2xz} - \delta_{2x})] = 0, \tag{5.5}$$

$$\frac{\partial}{\partial x}[\rho_e w_e u_e (\delta_{2zx} - \delta_{2z})] + \frac{\partial}{\partial z}[\rho_e (w_e)^2 (\delta_{2zz} - \delta_{2z})] = 0, \tag{5.6}$$

with

$$\delta_{2xx} = \int_0^\delta (1 - \frac{\rho(u)^2}{\rho_e(u_e)^2})\, dy, \tag{5.7}$$

$$\delta_{2xz} = \int_0^\delta (1 - \frac{\rho u w}{\rho_e u_e w_e})\, dy = \delta_{2zx}, \tag{5.8}$$

$$\delta_{2zz} = \int_0^\delta (1 - \frac{\rho(w)^2}{\rho_e(w_e)^2})\, dy. \tag{5.9}$$

These equations are to be handled like those for the mass-flow displacement thickness.

5.4 Energy-Flow Displacement Thickness

The energy flux is a scalar. Therefore only one equation results for the energy-flow displacement thickness δ_3, like for the displacement thickness. The energy E can be the kinetic energy per unit mass $0.5\,q^2$, with $q^2 = u^2 + v^2 + w^2$ being the kinetic energy per unit volume $0.5\,\rho\,q^2$, or the total enthalpy $h_t = h + 0.5\,q^2$.

The equation for a non-permeable wall reads, following [7]:

$$\frac{\partial}{\partial x}[E_e u_e(\delta_3 - \delta_{3x})] + \frac{\partial}{\partial y}[E_e w_e(\delta_3 - \delta_{3z})] = 0, \tag{5.10}$$

with

$$\delta_{3x} = \int_0^\delta (1 - \frac{Eu}{E_e u_e})\, dy, \tag{5.11}$$

$$\delta_{3z} = \int_0^\delta (1 - \frac{Ew}{E_e w_e})\, dy. \tag{5.12}$$

Also this equation is to be handled like that for the mass-flow displacement thickness.

5.5 Problems

Problem 5.1. Given is a compressible boundary layer. The Prandtl number is $Pr = 0.75$. When integrating eqs. (5.2) and (5.3), what boundary-layer thickness has to be chosen for the upper bound of the integrals? How is that thickness to be found?

Problem 5.2. Recover from eq. (5.1) the displacement thickness for two-dimensional flat-plate flow along the x-axis. Assume zero wall-source strength. What is to be regarded if the flow past a blunt body is considered.

Problem 5.3. We consider a two-dimensional, incompressible, steady flow at large Reynolds number near a non-permeable wall. We look for the definition of the fictitious inviscid flow equivalent to the real flow, Fig. 5.1.

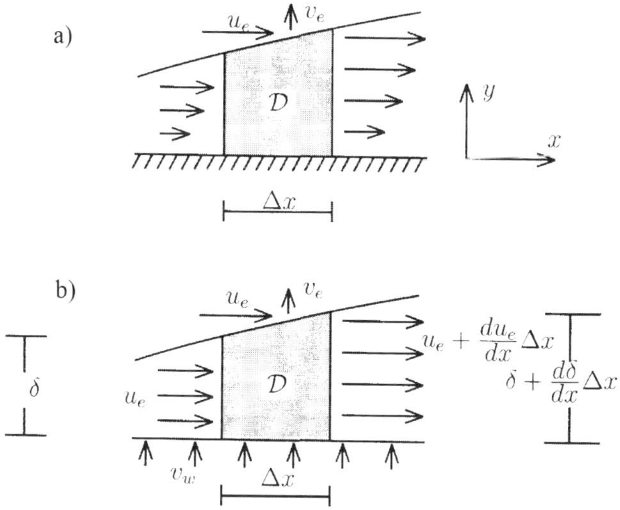

Fig. 5.1. Definition of an inviscid flow equivalent to the real flow: a) real flow, b) fictitious inviscid flow with $v_w = f(x)$ being the equivalent inviscid source velocity.

Along a normal to the wall, between the wall ($y = 0$) and the line $y = \delta(x)$, the stream-wise velocity of the fictitious inviscid flow is constant and equal to its value $u_e(x)$ in the real flow at the edge of the boundary layer $y = \delta$. At $y = \delta$, the component of velocity normal to the wall v_e of the fictitious inviscid flow is equal to its value in the real flow at $y = \delta$. At the wall, the component of velocity normal to the wall in the inviscid flow is v_w, the equivalent inviscid source velocity. In the real flow, this velocity is $v_w = 0$.

1. Write the mass conservation in the volume $\mathcal{D}$ of Fig. 5.1 for the real flow on one hand and for the fictitious inviscid flow on the other hand.
2. By using the equality of the velocities v_e in the real flow and in the equivalent inviscid flow, show that the velocity v_w is given by

$$v_w = \frac{d(u_e\delta_1)}{dx},$$

where δ_1 is the displacement thickness.

Then, the velocity v_w represents the velocity which must be prescribed at the wall to obtain a fictitious inviscid flow equivalent to the real flow. The equivalence is achieved from the point of view of mass flow.

3. By considering a control volume $\mathcal{V}$, as shown in Fig. 5.2, show that the equation of streamlines in the equivalent inviscid flow is

$$y = \delta_1 + \frac{C}{u_e},$$

where C is a constant depending on the considered streamline. Deduce that the line $y = \delta_1$ is a streamline of the equivalent inviscid flow.

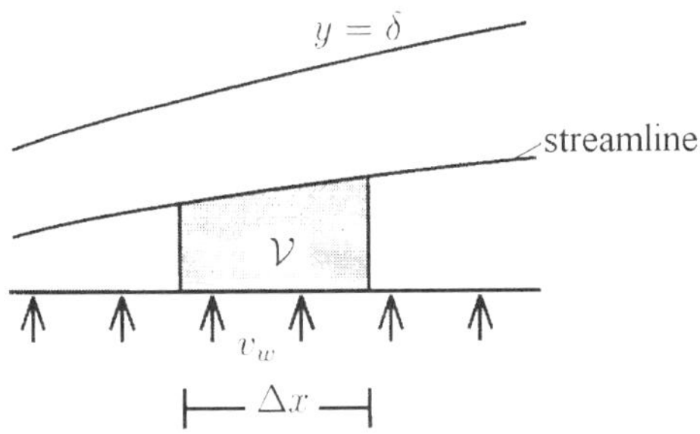

Fig. 5.2. Control volume in the equivalent inviscid flow limited by the wall and a given streamline.

4. Draw qualitatively the streamlines of the inviscid flow and indicate the lines $y = \delta_1$ and $y = \delta$.

Problem 5.4. Derive the equation for the determination of the displacement thickness δ_1 of three-dimensional boundary layers. Use the control-volume approach, assume for convenience incompressible flow and Cartesian coordinates. Remember further the definition of δ_1 in two dimensions:

$$\delta_1 = \int_0^\delta (1 - \frac{u}{u_e})dy,$$

which can be written as

$$\int_0^{\delta_1} u_e dy = \int_0^\delta u_e dy - \int_0^\delta u dy,$$

respectively as

$$\int_0^\delta u dy = \int_0^\delta u_e dy - \int_0^{\delta_1} u_e dy.$$

References

1. Pohlhausen, K.: Zur näherungsweisen Integration der Differentialgleichung der laminaren Reibungsschicht. ZAMM 1, 252–268 (1921)

2. Holstein, H., Bohlen, T.: Ein einfaches Verfahren zur Berechnung laminarer Reibungsschichten, die dem Näherungsansatz von K. Pohlhausen genügen. Lilienthal-Bericht S 10, 5–16 (1940)
3. Cousteix, J.: Analyse théorique et moyens de prévision de la couche limite turbulente tridimensionelle. Doctoral thesis, University of Paris VI, Paris, France (1974); Theoretical Analysis and Prediction Methods for a Three-Dimensional Turbulent Boundary-Layer. ESA TT-238 (1976)
4. Schlichting, H., Gersten, K.: Boundary Layer Theory, 8th edn. Springer, Heidelberg (1999)
5. Moore, F.K.: Displacement Effect of a Three-Dimensional Boundary Layer. NACA Rep. 1124 (1953)
6. Lighthill, M.J.: On Displacement Thickness. J. Fluid Mechanics 4, 383–392 (1958)
7. Kux, J.: Über dreidimensionale Grenzschichten an gekrümmten Wänden (About Three-Dimensional Boundary Layers at Curved Surfaces). Doctoral thesis, Universität Hamburg, Germany, Institut für Schiffbau der Universität Hamburg, Bericht Nr. 273 (1971)
8. Hirschel, E.H., Kordulla, W.: Shear Flow in Surface-Oriented Coordinates. NNFM, vol. 4. Vieweg, Braunschweig Wiesbaden (1981)
9. Cebeci, T., Cousteix, J.: Modeling and Computation of Boundary-Layer Flows, 2nd edn. Horizons Publ., Springer, Long Beach, Heidelberg (2005)
10. Piers, W.J., Schipholt, G.J., Van den Berg, B.: Calculation of the Flow Around a Swept Wing Taking Into Account the Effect of the Three-Dimensional Boundary Layer. NLR TR 75076 (1975)

6

Viscous Flow and Inviscid Flow—Connections and Interactions

In the present book mainly attached viscous flow of boundary-layer type is treated. That is flow which interacts weakly with the external inviscid flow. This chapter is devoted to a general discussion of the connections and interactions between attached viscous flow and inviscid flow. We wish to give the reader insight into different possible phenomena, and an overview of theoretical approaches to what—from the point of view of boundary-layer theory—are called higher-order effects. These encompass weak and strong interaction phenomena.

After the introductory remarks the so-called interaction theory is sketched. Its beginnings are traced and then weak and strong interaction are treated. It is not intended to give complete derivations, but instead to show the major concepts and results. An overview of viscous-inviscid interaction methods and the presentation of some examples of numerical simulations of weak and strong interaction phenomena close the chapter.

6.1 Introductory Remarks—The Displacement Effect

Viscous flow and inviscid flow past flight vehicle configurations rightfully can be treated as separate flow entities, however, not in any respect. In Section 1.2 we have discussed aspects of connections between them, and in Sections 2.3, 4.4, and 4.5 modes of interactions.

In this chapter we address two topics regarding connection and interaction phenomena: (a) understanding, both qualitatively and by means of rigorous theoretical analysis, and (b) mathematical models for their numerical simulation. In addition, important for design work is: (c) awareness of the implications of the respective phenomena.

Topic (b) does not concern single-domain methods, i.e. Navier-Stokes or RANS methods, Section 1.3, page 11 ff., where the connection between the two flow entities is inherent. It concerns two- and three-domain methods, that is, couplings of Euler or potential-flow methods with boundary-layer methods, or their additional coupling in zonal methods with NS/RANS methods.

In this book topic (c) generally is not treated explicitly. However, because of its importance, we give here a short qualitative discussion of the implications of one particular effect of attached viscous flow, the displacement effect.

E.H. Hirschel, J. Cousteix, and W. Kordulla, *Three-Dimensional Attached Viscous Flow*,
DOI: 10.1007/978-3-642-41378-0_6, © Springer-Verlag Berlin Heidelberg 2014

When considering boundary-layer flow, this effect usually is seen only in view of the displacement of the external inviscid flow past a vehicle configuration or a configuration element. In reality there is more connected to it.

Many aerodynamic properties of a flight vehicle can be described in the frame of inviscid theory, and, for sub-critical flight—regarding the flight Mach number—even in the frame of potential theory. The relevant property of the flow field past the vehicle is the surface-pressure distribution.

Considering the drag, the picture changes. Up to more than half of the total drag of a sub-sonic flight vehicle is viscous drag, see, e.g., [1].[1] The viscous drag consists of the skin-friction drag and of the viscosity-effects induced pressure drag, the latter usually simply called pressure drag or also form drag. The sum of the skin-friction drag and the pressure drag of an airfoil often is called profile drag.

The pressure drag basically is a consequence of the displacement effect of the attached viscous flow. It is connected to the flow-off separation at the trailing edges of the wing, at stabilization and control surfaces, and at the aft end of the fuselage.

Regarding the displacement effect we summarize:

1. The displacement of the external inviscid flow due to the attached viscous flow, the boundary layer, is the effect usually coming into one's mind when considering displacement effects. It leads to a change (weak interaction) of the external flow field, especially of the external pressure field. That change may or may not be negligible regarding a vehicle's aerodynamic properties.
2. The pressure drag is a consequence of the displacement effect of attached viscous flow, too. It can simply be understood by considering the flow past an airfoil. With inviscid flow the upper-side and the lower-side streamlines are closing at the trailing edge and the stagnation pressure is recovered. With viscous flow, the boundary layers on the upper and the lower side, inhibit, due to their finite thickness, the closing of the inviscid streamlines, and therefore the stagnation pressure cannot be recovered. Hence the pressure drag appears.

   The pressure drag, like the skin-friction drag is present at any structural element of an airplane. It appears where the flow leaves the surface, either by flow-off or by squeeze-off separation. Flow-off separation at a trailing edge is a strong-interaction phenomenon. It can be described by exact

[1] The other large part of the total drag is the induced drag of lifting configuration elements, in particular of the wing. At flight above the critical Mach number the wave drag appears. In design work other drag forms, with usually smaller contributions to the total drag are considered, for instance the excrescence drag. In flight comes to these kinds of drag the trim drag of the air vehicle.

theory for an infinitely thin trailing edge, Sub-Section 6.2.3.[2] This also holds for some cases with local (squeeze-off) separation.

3. At a lifting airfoil or wing the boundary layer at the suction (upper) side toward the trailing edge is exposed to a stronger adverse pressure gradient than that at the pressure (lower) side. Hence the boundary-layer thickness as well as the displacement thickness on the suction side is larger than on the pressure side. The consequence of the different displacement thicknesses is a slight upward deflection of the external inviscid flow behind the trailing edge—out of the bi-sector direction—which is called boundary-layer decambering of the airfoil or wing. The decambering leads to a lift loss compared to the lift with inviscid flow.[3]
4. At supercritical, but still subsonic flight conditions a particular effect is present. The shock wave/boundary-layer interaction at the end of the embedded supersonic pocket—a strong-interaction phenomenon—causes an extra thickening of the down-stream boundary layer [4]. The enlarged displacement thickness enlarges the pressure drag. This and the wave-drag increment due to the shock wave, the boundary-layer decambering and the associated shock-wave decambering combined, cause the detrimental transonic drag divergence (strong increase of drag) and the lift divergence (decrease of lift). The swept wing and the supercritical airfoil are means to shift in particular the drag divergence to as high as possible (sub-sonic) flight Mach numbers.
5. Another displacement effect is observed at very slender, sharp-nosed configurations at high supersonic and at hypersonic Mach numbers and low Reynolds numbers. Even if the upper side of the vehicle would be a free-stream surface (a surface not inclined against the free stream [5]) with sharp leading edge, the displacement effect of the viscous flow, which initially is not of boundary-layer type, induces a weak oblique shock wave (hypersonic viscous interaction, see, e.g., [6]). The result is, at least, an increment of the wave drag on the vehicle.
6. Finally a displacement effect is mentioned which appears at supersonic and hypersonic flight of blunt-nosed vehicles. The curved bow shock at the vehicle nose leads to an entropy layer, see, e.g., [6]. This usually is considered only with regard to its vorticity and to the entropy-layer swallowing of the boundary layer and its influence on laminar-turbulent transition. Actually the velocity defect of the wall-near inviscid flow leads

[2] In reality a trailing edge has finite, however small thickness, see, e.g., Fig. 10.24. The theory for sharp trailing edges nevertheless provides valid insight.

[3] A similar effect, the shock-wave decambering, can be present at supercritical flight speed [2]. If the shock wave at the end of the supersonic pocket on the suction side is stronger than that of the pressure side (there may be no shock wave present at all) the larger total pressure loss at the suction side also leads to an upward deflection of the flow behind the trailing edge out of the bi-sector. This is of importance for computation models of transonic airfoils and wings [3].

to an additional displacement effect. This can be described with the help of perturbation theory [7]. The entropy-layer displacement effect is similar to the displacement effect of the boundary layer.

In a large flight speed and altitude domain vehicle flow fields with mutually interacting effects can be described in the frame of weak interaction approaches, for a hypersonic flight case see the second example in Sub-Section 6.4.1. One should mention, however, that today single-domain solution methods, viz. Navier-Stokes or RANS methods are widely employed, which in principle automatically cover these effects.

Only a few of the relevant interaction phenomena are treatable by rigorous approaches. What we call interaction theory basically is perturbation theory, more correctly the method of matched asymptotic expansions, see, e.g., [7]. We discuss in the following sections weak and strong interaction, the latter in the mode of viscous-inviscid interaction.

6.2 Interaction Theory

6.2.1 About the Beginnings

Interaction theory in a sense is a bottom-up concept. Originally Prandtl's boundary-layer concept regarded simply viscous flow past a semi-infinite flat plate. In discussing the solution for this case in 1935 [8], he recognized the possibility of improving the boundary-layer solution which he had proposed in 1904. He suggested the implementation of an iterative process between the inviscid flow and the boundary layer in which the body is thickened by the displacement thickness.

Later, when a sound mathematical basis was introduced with the method of matched asymptotic expansions, Prandtl's boundary layer was seen as a first step of successive approximations. Higher-order boundary-layer solutions have been developed with emphasis on second-order approximations [9, 10].

In these developments the interaction is hierarchical. The calculation of a first-order inviscid flow enables one to determine the first-order boundary layer. Then, the boundary layer reacts on the inviscid flow through a second-order inviscid approximation. Finally, the second-order boundary layer can be determined. It is said that *weak interaction* takes place.

When the interaction becomes stronger, it is necessary to break the hierarchy, and the inviscid flow and the boundary-layer flow must be calculated simultaneously: *strong interaction* is present.

In particular, this occurs when separation is imminent.[4] Soon after the introduction of the boundary-layer concept, numerical solutions of the boundary-layer equation ran into difficulties when, in the presence of adverse

[4] At the time of Prandtl the flow-off separation at a sharp trailing edge of an airfoil or wing was not yet under consideration.

pressure gradients, the skin-friction decreases and vanishes. This problem has been analyzed by S. Goldstein [11] and by L. Landau and E. Lifschitz [12].

In a general manner, the question first has been raised by Goldstein how to determine the solution of the boundary-layer equation downstream of a station where the velocity profile is given [13]. Among different results, he has shown that generally the solution of the boundary-layer equation is singular, if the velocity profile has a zero derivative at the wall, i.e. the wall shear-stress is zero. In such a case it is not possible to continue a boundary-layer calculation downstream of the point of zero wall shear-stress.

Goldstein suggested that *inverse methods* could solve the separation singularity problem. In such methods, instead of the external velocity, the distribution of a quantity associated with the boundary layer is prescribed, for example the distribution of the displacement thickness. Then, the velocity at the edge of the boundary layer becomes an unknown which is determined from the solution of the boundary-layer equation. As long as the distribution of the displacement thickness is a regular function, the solution of the boundary-layer equation is regular, even in the presence of separated flow.

This result has been verified numerically by D. Catherall and W. Mangler [14]. Goldstein's idea was a step towards viscous-inviscid interaction methods in which strong interaction can be described.

6.2.2 Weak Interaction

We consider steady and for convenience incompressible, two-dimensional flow. An asymptotic expansion of the solution of the Navier-Stokes equations is obtained by introducing a small parameter ϵ

$$\epsilon = Re^{-1/2}, \tag{6.1}$$

related to the Reynolds number

$$Re = \frac{U_0 L}{\nu}, \tag{6.2}$$

where U_0 denotes a velocity scale, ν the kinematic viscosity, and L a length scale.

In asymptotic methods, it is assumed that the Reynolds number is very large compared to unity and even goes to infinity. The small parameter ε is formed with the inverse of the square root of the Reynolds number for convenience. Any other negative power could have been chosen, but the presentation of results is simpler with the power $-1/2$, and is appropriate for laminar flow.

First, the Navier-Stokes equations for two-dimensional flow are written in terms of non-dimensional variables. The velocity components are reduced—non-dimensionalized, Section 4.2—by U_0 and the coordinates in both directions are reduced by L. This means that the length scale is the same in the two space directions.

When $Re \to \infty$, it is shown that the Navier-Stokes equations reduce to the Euler equations. This is the first-order approximation. Contrary to a regular perturbation problem, the first approximation is not valid everywhere. It is valid in a large part of the flow, in the inviscid flow region. Near a wall, it is not valid because the no-slip condition can not be applied. It is said that a *singular perturbation* problem occurs.

To restore the no-slip condition at the wall, it is required to introduce another approximation which is valid only near the wall, in the boundary layer. In this region, the length scale along the normal to the wall is no longer L, but εL. This length is much smaller than L and constitutes a scale for the thickness of the boundary layer.

In the boundary layer, the appropriate variable in the direction normal to the wall is $Y = y/\varepsilon$ where y is the physical coordinate reduced by L.[5] In essence, the theory consists of looking for two complementary sets of expansions, one is called the outer expansion—in the region outside the boundary layer—and the other one is called the inner expansion, inside the boundary layer. These expansions are sought as series in terms of the small parameter ε. To get a closed problem, a matching between these two expansions is needed.

In terms of reduced variables, the outer expansions are sought as

$$u(x, y; \varepsilon) = u_1(x, y) + \varepsilon u_2(x, y) + \cdots, \tag{6.3}$$

$$v(x, y; \varepsilon) = v_1(x, y) + \varepsilon v_2(x, y) + \cdots, \tag{6.4}$$

$$p(x, y; \varepsilon) = p_1(x, y) + \varepsilon p_2(x, y) + \cdots. \tag{6.5}$$

These expansions are introduced in the Navier-Stokes equations and the terms are sorted according to the powers of ε. It can be shown that the first-order solution satisfies the Euler equations and the second-order solution satisfies the linearized Euler equations. Viscous terms are present in the next approximation.

The first-order equations are solved with the conditions that the flow tends to a uniform flow at infinity and with a slip condition along the wall.

The inner expansions are expressed in the orthogonal coordinate system shown in Fig. 6.1: yL is the physical distance along a normal to the body surface, xL is the physical distance along the body surface. Along the normal to the wall, the metric coefficient is unity and along the x-axis the metric coefficient is $h = 1 + yK$, where K denotes the surface curvature reduced by L^{-1}.

The inner expansions are sought as

$$u(x, y; \varepsilon) = U_1(x, Y) + \varepsilon U_2(x, Y) + \cdots, \tag{6.6}$$

$$v(x, y; \varepsilon) = \varepsilon V_1(x, Y) + \varepsilon^2 V_2(x, Y) + \cdots, \tag{6.7}$$

$$p(x, y; \varepsilon) = P_1(x, Y) + \varepsilon P_2(x, Y) + \cdots, \tag{6.8}$$

[5] In Section 4.2 we have termed this as non-dimensionalization and stretching.

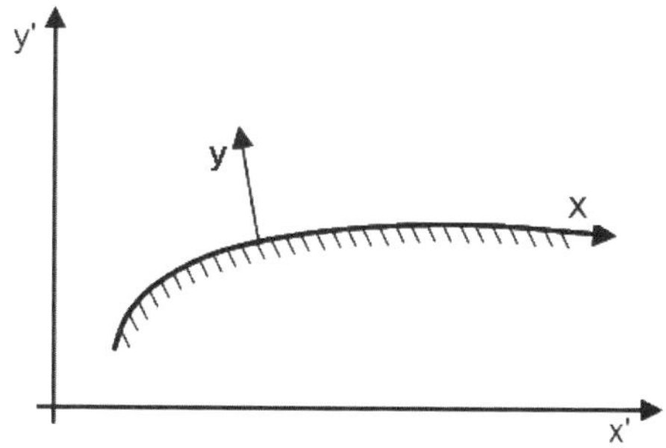

Fig. 6.1. Orthogonal boundary-layer coordinate system for two-dimensional flow past a curved body surface.

where Y is the mentioned magnified (stretched) coordinate in the direction normal to the wall $Y = y/\varepsilon$. The expansion for v is chosen to get a non-trivial continuity equation.

These expansions are substituted in the Navier-Stokes equations and the continuity equation written in the coordinate system of Fig. 6.1 and the different terms are sorted according to the powers of ε. The first-order approximation obeys the standard Prandtl boundary-layer formulation where the density and the viscosity coefficient—being constant and reduced—are unity:

$$\frac{\partial U_1}{\partial x} + \frac{\partial V_1}{\partial Y} = 0, \tag{6.9}$$

$$U_1 \frac{\partial U_1}{\partial x} + V_1 \frac{\partial U_1}{\partial Y} = -\frac{\partial P_1}{\partial x} + \frac{\partial^2 U_1}{\partial Y^2}, \tag{6.10}$$

$$0 = \frac{\partial P_1}{\partial Y}. \tag{6.11}$$

The first-order equations are solved with the following boundary conditions. At the wall, the no-slip condition is applied. At the outer edge of the boundary layer, the matching with the external flow shows that

$$\lim_{Y \to \infty} U_1 = u_1(x, 0).$$

At the boundary-layer edge, the velocity U_1 becomes equal to the wall velocity calculated from the first-order inviscid solution. A matching condition on pressure shows that $P_1 = p_1(x, 0)$.

The next step is to formulate the second-order inviscid equations. The boundary conditions are: i) the velocity vanishes at infinity because the first-order inviscid solution satisfies the condition of uniform flow at infinity, and ii) at the wall, the matching between the inner and outer expansions shows that

$$v_2(x, 0) = \frac{1}{\varepsilon} \frac{\mathrm{d}}{\mathrm{d}x} \left(u_1(x, 0)\, \delta_1(x) \right), \tag{6.12}$$

where δ_1 denotes the displacement thickness calculated from the first-order boundary-layer solution

$$\delta_1 = \varepsilon \int_0^\infty \left(1 - \frac{U_1(x,Y)}{u_1(x,0)}\right) \mathrm{d}Y. \tag{6.13}$$

For the second-order inviscid solution, the velocity $v_2(x,0)$ normal to the wall simulates the boundary-layer displacement effect. In other words, at second-order, the inviscid flow is calculated by taking into account the effect of the boundary layer. The velocity $v_2(x,0)$ is called the *blowing velocity* of the equivalent inviscid source distribution, Section 5.2.

Once the second-order inviscid solution has been obtained, the second-order boundary-layer equations are considered. In the coordinate system of Fig. 6.1, the second-order boundary-layer equations are:

$$\frac{\partial U_2}{\partial x} + \frac{\partial}{\partial Y}(V_2 + V_1 Y K) = 0, \tag{6.14}$$

$$U_1 \frac{\partial U_2}{\partial x} + U_2 \frac{\partial U_1}{\partial x} + V_1 \frac{\partial U_2}{\partial Y} + V_2 \frac{\partial U_1}{\partial Y} + K V_1 \frac{\partial}{\partial Y}(Y U_1) = -\frac{\partial P_2}{\partial x} + \frac{\partial^2 U_2}{\partial Y^2} + K\left(Y \frac{\partial^2 U_1}{\partial Y^2} + \frac{\partial U_1}{\partial Y}\right), \tag{6.15}$$

$$K U_1^2 = \frac{\partial P_2}{\partial Y}. \tag{6.16}$$

In the absence of any other second-order effect than the displacement effect, in particular for the flat plate, i.e. zero surface curvature ($K = 0$), the second-order boundary layer can be calculated with the no-slip condition at the wall. At the outer edge of the boundary layer a matching condition gives:

$$\lim_{Y \to \infty} U_2(x,Y) = u_2(x,0).$$

The term $u_2(x,0)$ represents the change of velocity in the external flow due to the displacement effect. This term is called the displacement speed. Without curvature effects, the pressure P_2 is obtained from the matching between the inner and outer expansions: $P_2 = p_2(x,0)$.

Other second-order effects, of relative order $\varepsilon = Re^{-1/2}$ with respect to the first-order boundary layer, are taken into account through the solution of the second-order boundary-layer equations. For example, the surface curvature is represented in the second-order boundary-layer equations by the presence of various terms in the continuity equation (6.14) and in the streamwise momentum equation (6.15). From the transverse momentum equation (6.16) it is seen that the pressure P_2 is no longer constant across the boundary layer. The pressure P_2 is obtained by integrating eq. (6.16) and by using a matching condition:

$$P_2 = YKu_1^2(x,0) + K\int_Y^\infty \left(u_1^2(x,0) - U_1^2(x,Y)\right) \mathrm{d}Y + p_2(x,0). \qquad (6.17)$$

The boundary conditions are also modified by the effect of surface curvature. At the wall, the no-slip condition applies and at the outer edge of the boundary layer, a matching condition gives:

$$\lim_{Y\to\infty} \left[U_2(x,Y) + Ku_1(x,0)Y\right] = u_2(x,0).$$

The effect of possible vorticity in the external flow is also a second-order effect. If the vorticity in the external flow is of $O(1)$, the ratio with the vorticity in the boundary layer is of $O(Re^{-1/2})$. In the stream-wise momentum equation (6.15), the vorticity of the external flow is present through the pressure gradient term $\partial P_2/\partial x$. The integration of eq. (6.16) with a matching condition for the pressure shows that $\partial P_2/\partial x$ contains a term expressing the interaction between the vorticity of the external flow at the wall and the displacement effect. At the wall, the no-slip boundary condition applies and, at the outer edge of the boundary layer, a matching condition gives:

$$\lim_{Y\to\infty} \left[U_2(x,Y) - Y\left(\frac{\partial u_1}{\partial y}\right)_{y=0}\right] = u_2(x,0).$$

It is noted that external vorticity affects the second-order boundary layer only by its value at the wall, in the equations and in the boundary conditions.

When both effects of surface curvature and external vorticity are present, the condition at the edge of the boundary layer is:

$$\lim_{Y\to\infty} \left[U_2(x,Y) - Y\left(\frac{\partial u_1}{\partial y}\right)_{y=0} + Ku_1(x,0)Y\right] = u_2(x,0).$$

The above considerations can be extended to three-dimensional compressible boundary-layer flow in surface-oriented non-orthogonal curvilinear coordinates with uniform or non-uniform external inviscid flow. Then we distinguish generally two main classes of second-order effects: (1) effects influencing the formulation of the boundary-layer equations, (2) effects due to the interaction with the external inviscid flow:[6]

- To class (1) belong the effects due to the curvature of the body surface. These can be divided into longitudinal and transversal curvature effects. This distinction is problematic, because in three-dimensional flow past complicated body geometries it makes no sense to distinguish between longitudinal and transversal effects. It would be more relevant to look locally

[6] In the discussion of these classes we follow the considerations of F. Monnoyer [15].

at the two principal curvatures of the surface, see, e.g., [16], [17]. These, however, are not necessarily related to the main-flow and the cross-flow directions. The consequence is that their effects on the boundary-layer flow cannot be treated separately. Therefore it can be accepted to consider them simply as curvature effects without further distinction.

If the boundary-layer thickness δ is small compared to the smallest radius of curvature R_{min} of the body surface under consideration, we can treat the flow as first-order problem, Appendix A.2.1. If this is not the case, the most typical class (1) effect appears: centrifugal forces produce a pressure gradient across the boundary layer, as was shown above.[7] The surface curvature contributes to several terms in the second-order equations. It also has to be taken into account at the matching boundary.

– Class (2) effects are the displacement effect, the effect of a vorticity or total pressure/entropy surface-normal gradient in the external inviscid flow and also the effect of a possible total enthalpy gradient in the external flow. The displacement effect is the most significant of these effects, see the discussion in Section 6.1. The possible means to take it into account are discussed in Section 5.2. Total pressure/entropy gradients appear in supersonic and hypersonic flows behind a curved bow-shock surface. In such cases also low-density effects like slip flow and temperature jump, see, e.g., [6], show up as second-order effects.

In closing this discussion we note that references to the formulation of second- and higher-order boundary-layer equations are given in Appendix A.2.5.

6.2.3 Strong Interaction

In the early 1950s M.J. Lighthill made a major contribution to the field of strong interaction [18]. The problem he analyzed is how a disturbance in a flat plate boundary layer, for example a small deflection of the wall, affects the boundary layer when the external inviscid flow is supersonic.

Lighthill proposed a small perturbation theory in which the perturbation flow is structured into three layers. In the region farthest from the wall, the perturbations obey the linearized equations of inviscid supersonic flow. In the region corresponding to the usual boundary layer, the perturbations obey the small perturbation equations of a parallel, inviscid, compressible flow.

Close to the wall, a viscous layer is introduced in order to satisfy the no-slip conditions at the wall. In this latter region, the Orr-Sommerfeld equation is the governing equation which also provides the evolution of the stability of an incompressible boundary layer, Sub-Section 9.2.2. The solution shows that the perturbations can propagate upstream. The order of magnitude of

[7] In Section 6.4 we show this explicitly and implicitly with examples of computed flow fields.

the length of interaction is $LRe^{-3/8}$, where L is the distance of the prescribed disturbance from the flat plate leading edge and Re is the Reynolds number based on L. (In practice this means that the interaction still is local.)

In this problem, we say that there is a strong interaction between the inviscid flow and the boundary layer because the usual hierarchy has disappeared. The outer flow and the boundary-layer flow must be treated simultaneously.

Lighthill's problem addresses the general question of viscous-inviscid interaction, i.e. the interaction between the boundary layer and the inviscid flow region. Lighthill's analysis shed new light on the understanding of this phenomenon. His analysis has been complemented with the triple-deck theory. A discussion of the problems associated with separation and with the structure of separated flows, in relation with the triple-deck theory in particular, can be found in [19].

The triple-deck theory is attributed to K. Stewartson and P.G. Williams, [20]–[22], and to V.Ya. Neyland [23, 24]. A.F. Messiter has also contributed to this theory by analyzing the flow near the trailing edge of a flat plate [25]. Stewartson and Williams, [21], considered that their theory is a non-linear extension of the theory proposed by Lighthill.

In fact, without reducing the value of Lighthill's contribution, the triple-deck theory has been the most important advance in the study of boundary layers after Prandtl's theory. Triple-deck theory led to significant progress in the understanding of many types of flows [26, 27]. In this theory, the mathematical tool is the method of matched asymptotic expansions which permits a systematic study of various flow phenomena involving a strong interaction.

Such theoretical results lead to a better understanding of the interaction between the boundary layer and the inviscid flow. They also help us to interpret the solutions of the Navier-Stokes equations.

6.3 Viscous-Inviscid Interaction Methods

Following these theoretical breakthroughs, practical methods have been devised to solve the interaction between the boundary layer and the inviscid flow, in particular with the goal of calculating separated flows [28]–[34]. A partial justification of the interaction methods is provided by the triple-deck theory [35, 36].

Rational mathematical arguments to support the concept of interactive boundary-layer methods have been provided by applying the successive complementary expansion method to high Reynolds number flows [37]. The basis of this technique is to seek a uniformly valid approximation in the whole flow field by using generalized expansions instead of regular expansions as in the method of matched asymptotic expansions. The result of this theory is that the inviscid flow equations and the boundary-layer equations interact, which implies that there is no hierarchy between the two sets of equations. Another result of this theory is that the second-order interactive boundary-layer model contains the triple-deck model.

The viscous-inviscid interaction methods have been developed for two-dimensional laminar flow first. For three-dimensional flow, equivalent results have been obtained, at least when the crosswise length scale is not much shorter than the stream-wise length scale. Interactive boundary-layer methods hence are also available for three-dimensional flows [38]–[40].

To be more specific, we describe different types of viscous-inviscid interaction methods. We consider a typical problem of aerodynamics which is to calculate the flow past an airfoil for angles of attack up to maximum lift. The flow is supposed to be incompressible.

According to the standard boundary-layer theory, the calculation of the interaction is performed *sequentially*. At first, the inviscid flow is calculated around the real airfoil by applying a slip condition at the wall. Afterwards, the boundary layer is calculated with, as input, the stream-wise wall velocity determined by the inviscid flow. Finally, the inviscid flow is corrected by taking into account the displacement effect. The procedure is called *direct-direct*: direct for the inviscid flow and direct for the boundary layer, Fig. 6.2.

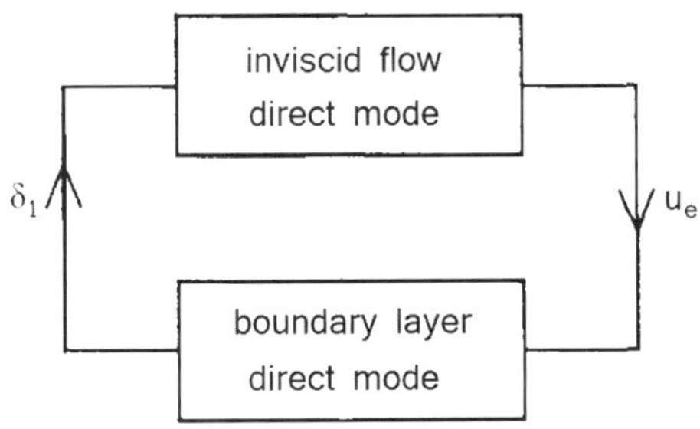

Fig. 6.2. Schematic of viscous-inviscid interaction: direct-direct mode.

In the presence of separation, this procedure is no longer valid because the solution of the boundary-layer equations is singular and it is not possible to calculate the boundary layer downstream of the separation point. To solve this problem, inverse methods can be used, Fig. 6.3. These methods can be associated with inverse methods to calculate inviscid flow: the input is the pressure calculated from the boundary-layer displacement effect and the result is the shape of the body corresponding to the pressure distribution—in fact the real shape modified by the displacement effect. In practice, this type of method, called *inverse-inverse*, is not easy to implement and therefore other procedures have been developed [41].

Semi-inverse methods are very efficient computation tools [28, 29, 31, 32, 39]. These methods consist of solving the boundary-layer equations in the inverse mode and the inviscid flow equations in the direct mode, Fig. 6.4. For a given distribution of the displacement thickness, the boundary-layer equations yield a distribution of velocity $u_{e\,\mathrm{BL}}(x)$. For the same displacement

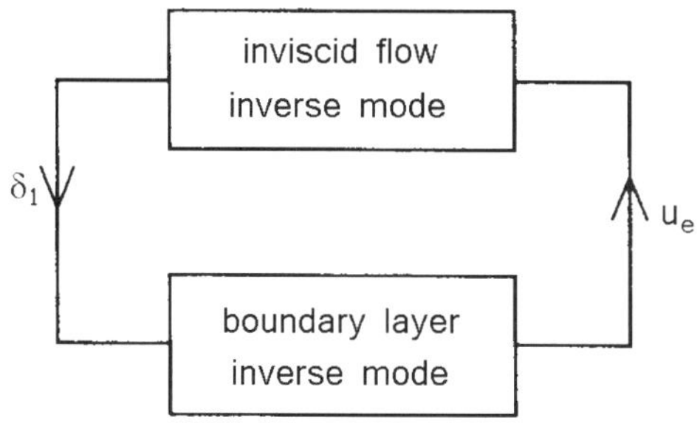

Fig. 6.3. Schematic of viscous-inviscid interaction: inverse-inverse mode.

thickness distribution, the inviscid flow equations yield a distribution of the wall velocity $u_{e\,\mathrm{IN}}(x)$. Generally, for any distribution of the displacement thickness, the two velocity distributions are not identical. Iterative procedures have been devised to obtain $u_{e\,\mathrm{BL}}(x) = u_{e\,\mathrm{IN}}(x)$. For example, J.E. Carter, [28, 29], proposed to determine the new estimate of the displacement thickness at iteration $(n+1)$ by

$$\delta_1^{n+1}(x) = \delta_1^n(x)\left[1 + \omega\left(\frac{u_{e\,\mathrm{BL}}^n(x)}{u_{e\,\mathrm{IN}}^n(x)} - 1\right)\right], \tag{6.18}$$

where ω is a relaxation factor.

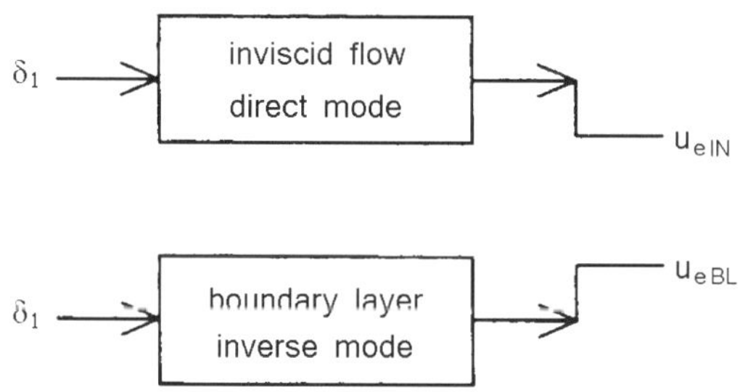

Fig. 6.4. Schematic of viscous-inviscid interaction: semi-inverse mode.

Another approach has been developed by A.E.P. Veldman [34]. In agreement with the triple-deck theory, the inviscid flow and the boundary layer are strongly coupled and there is no hierarchy between the systems of equations. In a simultaneous method, the external velocity $u_e(x)$ and the displacement thickness $\delta_1(x)$ are calculated simultaneously from the set of viscous and inviscid equations.

For example, let us consider a flow on a flat plate perturbed by a small local deformation of the wall. The external velocity u_e is given by

$$u_e(x) = u^0 + \delta u_e(x),$$

where u^0 is the velocity induced by the shape of the real wall calculated from the (linearized) Euler equations and $\delta u_e(x)$ is the perturbation due to the boundary layer. This perturbation is expressed by a Hilbert integral:

$$\delta u_e = \frac{1}{\pi} \oint_{x_a}^{x_b} \frac{v_b}{x - \xi} \, d\xi, \tag{6.19}$$

where the integral denotes Cauchy's principal value integral.

In eq. (6.19) $v_b = v_b(\xi)$ is the equivalent inviscid source distribution given by

$$v_b(\xi) = \frac{d}{d\xi} \left[u_e(\xi) \delta_1(\xi) \right]$$

which simulates the boundary-layer effect in the domain (x_a, x_b). The Hilbert integral and the boundary-layer equations are solved simultaneously with an iterative method [34].

This method has been extended to the calculation of the flow past wings with compressibility effects [30].

6.4 Examples

6.4.1 Second-Order Boundary-Layer Effects

We discuss two examples of applications of a three-dimensional second-order boundary-layer method. The results are compared with Navier-Stokes results and in the second example also with experimental data.

The first example demonstrates the effect of surface curvature and boundary-layer displacement at a 1:6 ellipsoid [42]. The flow conditions are $M_\infty = 0.6$, $Re_L = 10^6$, axisymmetric flow, laminar-turbulent transition is enforced at $x/L = 0.10$. The angle of attack is $\alpha = 0°$. Fig. 6.5 shows the results of three computation cases.

Presented are the pressure profiles in direction normal to the surface at three stations ahead and two behind the location $x/L = 0.70$, where the zonal boundary is located. The single-domain solution of the Navier-Stokes equations in the upper figure (a) shows in downstream direction a smooth transition to pressure profiles $p(x^3)$ more strongly curved directly at the body surface.[8] The three-domain solution with a first-order boundary-layer method (b) shows ahead of the zonal boundary the surface-normal method-inherent constant pressure in the boundary layer. The three-domain solution with a second-order boundary-layer method (SOBOL) (c) finally shows a nearly perfect agreement with the single-domain Navier-Stokes solution.

[8] Note the strong growth of the boundary-layer thickness in downstream direction which is due to the rising pressure and to the reverse Mangler effect, Section 8.7. Separation happens at $x/L \approx 0.92$.

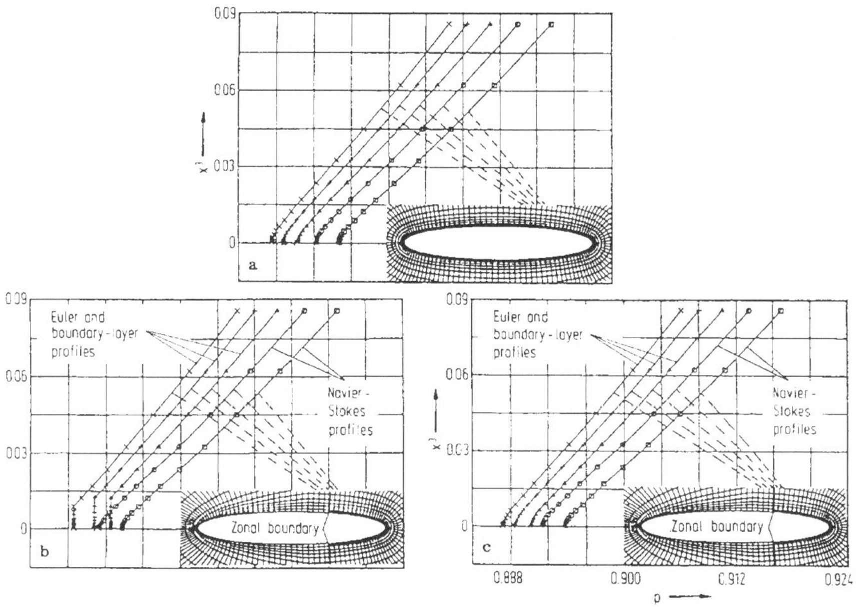

Fig. 6.5. Influence of surface curvature and displacement effect on the surface-normal pressure distribution in different computation models for an ellipsoid flow [42]. Upper figure: a) single-domain Navier-Stokes solution. Lower figures: three-domain solution with Euler/first-order, left b), and Euler/second-order boundary-layer method, right c), coupled each with a Navier-Stokes method. x^3 is the surface-normal coordinate with $x^3 = 0$ at the body surface. In all cases the flow comes from the left.

The second example concerns second-order effects in the flow past a blunt re-entry vehicle configuration at hypersonic speed. The effects are due to the entropy layer, the surface curvature and the boundary-layer displacement.

M. van Dyke discusses in [10] the influence of the entropy layer in hypersonic flow past such bodies. The cases treated in his book have a certain relationship to the there also treated case of a flat-plate boundary layer in shear flow. The external inviscid flow in all cases has a non-zero vorticity profile in direction normal to the surface.

The entropy layer at a blunt-nosed body in hypersonic flow is due to the curved bow-shock surface, see, e.g., [6]. It is equivalent to a total-pressure change in direction normal to the body surface. By means of Crocco's theorem that is related to a vorticity profile which resembles the velocity profile of a slip-flow boundary layer. In the literature usually only the symmetric case is considered, Fig. 6.6 a). There indeed this profile is present. The stagnation-point streamline passes the normal-shock location of the bow-shock surface.

The situation is different in the asymmetric case, which is shown in Fig. 6.6 b). At an axisymmetric body at angle of attack, for instance a re-entry capsule, or at an asymmetric body, for instance a winged re-entry vehicle at

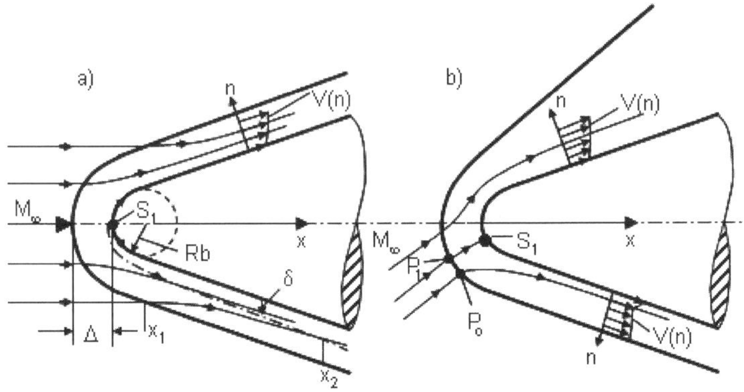

Fig. 6.6. Schematic of velocity gradients across streamlines due to the entropy layer in the inviscid flow field behind a curved bow-shock surface [6]: a) symmetric, b) asymmetric situation with wake-like entropy layer on the windward side.

angle of attack, the streamline passing the normal-shock location P_0 of the bow-shock surface is different from the stagnation-point streamline which passes through P_1. However, the streamline passing through P_0 suffers from the highest total-pressure loss. Hence the entropy layer in this case has a wake-like structure (lower side of Fig. 6.6 b)). This structure seems to appear always on the fuller side of the body turned towards the free-stream.[9] In [5] a more detailed discussion of this case can be found.

The following discussion concerns results of simulations of the flow field past the HERMES configuration, Fig. 6.7.[10] The computational simulations had to deal with such an entropy layer structure, and in addition with a thick boundary layer on partially strongly curved surface portions with the Mangler effect, Section 8.7, being present.

In Fig. 6.7 the heat-flux comparison of a three-dimensional coupled Euler/second-order boundary layer (SOBOL) result—the heat flux in the gas at the wall—with a Navier-Stokes result (Dornier) and wind-tunnel data (ONERA S4) shows rather good agreements [43].

The heat-flux iso-lines in the left part of the figure show the high heating in the nose region and along the attachment lines, see also Section 10.4. SOBOL was applied only to the windward side of the vehicle. In the right part of the figure the differences at $x \gtrapprox 0.30$ m are due to the deflected body flap which was not modelled in the SOBOL solution. The not shown surface-pressure and wall shear-stress distributions compare very well for the SOBOL and the Navier-Stokes solution, too.

We note that second-order boundary-layer methods are well suited to calculate highly complex flow fields where higher-order effects are present. This holds for fuselage and wing flow fields. Problems may be encountered

[9] The respective figure in [6] shows it the other way around.

[10] For details of this former European re-entry vehicle project see, e.g., [5].

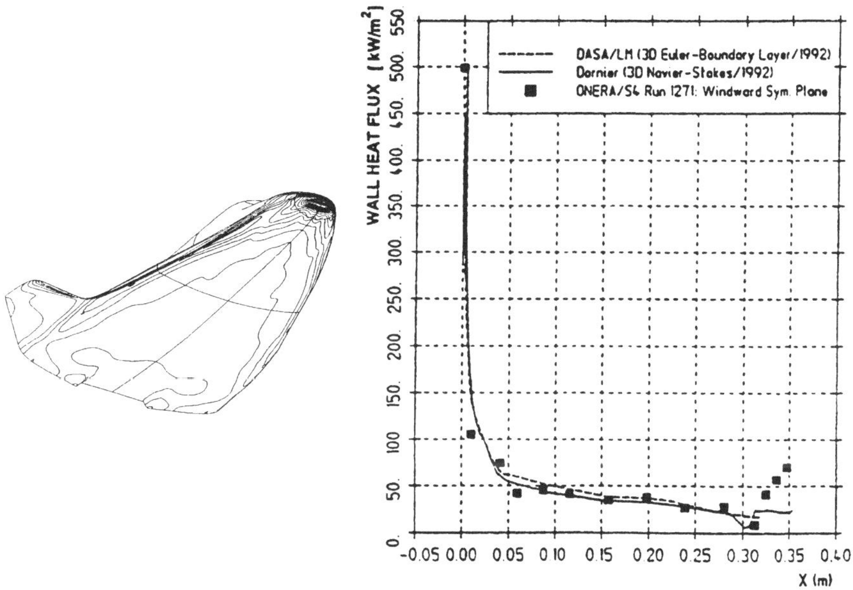

Fig. 6.7. Heat flux in the gas at the wall of a 1:40 scale wind-tunnel model of the HERMES configuration [43]. $M = 10$, $\alpha = 30°$, $\eta_{bf} = 10°$, $T_w = 300$ K, laminar flow, perfect gas, comparison of Euler/second-order boundary layer (SOBOL) solution, Navier-Stokes solution, and wind-tunnel data. Left side: the windward side of the vehicle with computed wall heat-flux iso-lines. Right side: computed and measured heat flux along the lower symmetry line.

with the determination of the initial data for the boundary-layer solution. At fuselages this concerns the forward stagnation point, and at wings the attachment line. With single-domain Navier-Stokes/RANS methods these problems do not exist.

6.4.2 Viscous-Inviscid Interaction Effects

Results obtained with a viscous-inviscid interaction scheme, [30, 44], demonstrate that interactive boundary-layer theory can be applied well to complex two-dimensional and three-dimensional flow problems.

Fig. 6.8 shows the comparison of computed and measured lift and drag coefficients for the NACA 0012 airfoil with a chord Reynolds number $Re = 3 \cdot 10^6$. With wake modelling included, it is possible to get a perfect agreement between measured and computed data up to nearly maximum lift for both the lift and the drag coefficient. Separation in this case occurs already for $\alpha > 10°$.

A comparison of computed and measured lift and drag coefficients in a cross-section of a swept and tapered RAE wing, [45], with retracted slat but deflected flap is given in Fig. 6.9 [44]. First of all we note that inviscid theory

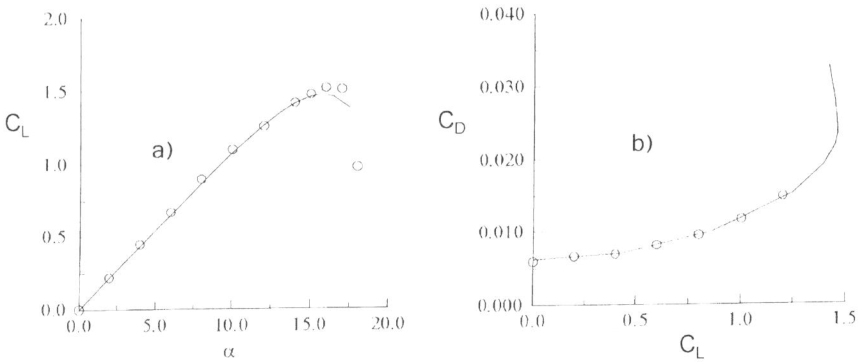

Fig. 6.8. NACA 0012 airfoil, $Re = 3{\cdot}10^6$ [44]. Comparison of computed (solid lines) and measured (circles) data: (a) lift coefficient C_L as function of the angle of attack α, (b) drag coefficient C_D as function of C_L.

yields a higher lift curve with a somewhat larger $dC_L/d\alpha$ than is found in the experiment. Including viscous effects in the computation method yields an almost good agreement nearly up to maximum lift, with the computed lift curve now lying only marginally higher. However, maximum lift is not captured, in contrast to the two-dimensional example, Fig. 6.8.

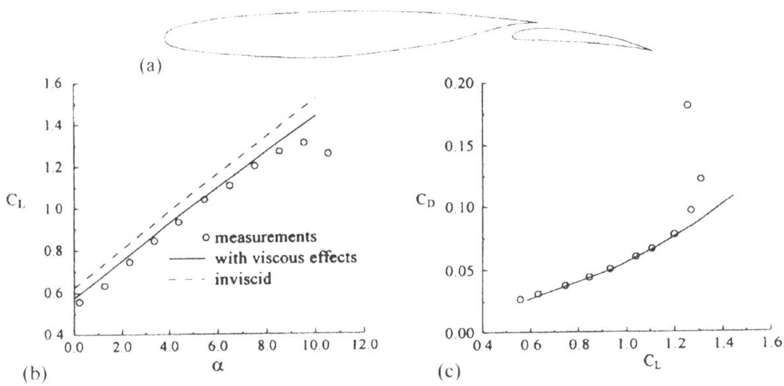

Fig. 6.9. RAE wing, slat retracted, flap deflected ($\eta = 10°$), $Re = 1.35{\cdot}10^6$ [44]. (a) wing cross section. Comparison of computed and measured data: (b) lift coefficient C_L as function of angle of attack α, and (c) drag coefficient C_D as function of C_L.

In [44] this is attributed to the close location of the flap to the wing. This leads to the merging of the wing wake with the boundary layer which develops on the flap's upper surface. This merging was not modelled in the computation method. Another phenomenon of influence might be the presence of a separation bubble on the wing.

These two examples show that viscous-inviscid interaction schemes are versatile and powerful computation tools, even if today single-domain Navier-Stokes/RANS methods more and more are taking over in aerodynamic design. The possibility to change the modelling level, however makes viscous-inviscid interaction schemes also to interesting problem-diagnosis tools.

6.5 Problems

Problem 6.1. Why is it advantageous for an airplane to have a wing with large aspect ratio, small chord depth, slender fuselage and generally small wetted surface?

Problem 6.2. Why has a modern transonic airplane a swept wing and a supercritical airfoil?

Problem 6.3. Why is flight at high subsonic Mach number desirable? Discuss it with Breguet's range equation. The range equation in its simplest form connects the range R with the parameters flight speed v_∞, aerodynamic performance (lift-to-drag ratio) C_L/C_D, specific impulse I_{sp}, with the structural parameters mass empty m_E, payload mass m_P and fuel mass m_F:

$$R = v_\infty \frac{C_L}{C_D} I_{sp} \ln\left(1 + \frac{m_F}{m_E + m_P}\right) .$$

Problem 6.4. We consider the inverse approach for solving the boundary-layer equations. For calculating the boundary layer, an integral method is used. The method is based on two equations, the integral form of the continuity equation and of the momentum equation [44]:

$$\frac{1}{u_e}\frac{\mathrm{d}}{\mathrm{d}x}\left[u_e(\delta - \delta_1\right] = c_E ,$$

$$\frac{\mathrm{d}\delta_2}{\mathrm{d}x} + (H+2)\frac{\delta_2}{u_e}\frac{\mathrm{d}u_e}{\mathrm{d}x} = \frac{c_f}{2} ,$$

where δ is the boundary layer thickness, δ_1 is the displacement thickness, δ_2 is the momentum thickness, H is the shape factor ($H = \delta_1/\delta_2$), c_f is the skin-friction coefficient ($c_f = 2\tau_w/(\rho u_e^2)$) and u_e is the boundary layer edge velocity. The entrainment coefficient c_E

$$c_E = \frac{\mathrm{d}\delta}{\mathrm{d}x} - \frac{v_e}{u_e}$$

is proportional to the mass flow which enters the boundary layer through its edge.

To study the properties of an integral method based on the above equations (for two-dimensional incompressible steady turbulent flows) assume that the entrainment coefficient c_E and local skin-friction coefficient c_f are known functions of H and Re_{δ_2}. The Reynolds number Re_{δ_2} is defined as

$$Re_{\delta_2} = \frac{u_e \delta_2}{\nu}.$$

Assume also that $H^* = (\delta - \delta_1)/\delta_2$ is a function of H given by

$$H^* = \frac{\alpha H^2 + H}{H - 1}, \; \alpha = 0.631.$$

1. Express the first two equations as a system S of two equations where the derivatives are $\frac{\mathrm{d}\delta_1}{\mathrm{d}x}$, $\frac{\mathrm{d}\delta_2}{\mathrm{d}x}$ and $\frac{\mathrm{d}u_e}{\mathrm{d}x}$.

   Hint. Differentiate

   $$\delta - \delta_1 = \delta_2 H^*$$

   and write it in the form

   $$\frac{d}{dx}(\delta - \delta_1) = H^* \frac{d\delta_2}{dx} + \delta_2 H^{*'} \frac{d}{dx}\left(\frac{\delta_1}{\delta_2}\right), \; H^{*'} = \frac{dH^*}{dH}$$

   and

   $$\frac{d}{dx}(\delta - \delta_1) = (H^* - H H^{*'})\frac{d\delta_2}{dx} + H^{*'}\frac{d\delta_1}{dx}.$$

2. Assume that $u_e(x)$ is known. System S is a system for $\frac{\mathrm{d}\delta_1}{\mathrm{d}x}$ and $\frac{\mathrm{d}\delta_2}{\mathrm{d}x}$. Analyze the determinant Δ of this system as a function of H.
3. Write an equation for $\frac{\mathrm{d}H^*}{\mathrm{d}x}$ from the first two equations. Show that if there exists a point $x = x_s$ for which $\Delta = 0$. The derivative $\frac{\mathrm{d}H^*}{\mathrm{d}x}$ at this point is not zero in general. Deduce that it is not possible to integrate the first two equations beyond $x = x_s$, where x_s represents the x-location corresponding to the boundary layer separation.
4. Now, it is assumed that δ_1 is known. System S is a system for $\frac{\mathrm{d}\delta_2}{\mathrm{d}x}$ and $\frac{\mathrm{d}u_e}{\mathrm{d}x}$. Analyze the determinant Δ' of this system as a function of H and show that Δ' is never null. This way of solving the boundary layer equations is called the inverse approach.

Problem 6.5. We consider a case of viscous-inviscid coupling. An integral method (see Problem 6.4) based on the solution of the first two equations given there can be used to calculate the flow in a plane diffuser, assuming that the core of the flow is an inviscid one-dimensional flow.

1. Assuming that the entrainment coefficient c_E and local skin-friction coefficient c_f are known functions of H and Re_{δ_2}, and $H^* \equiv (\delta - \delta_1)/\delta_2$ is related to H by the equation given in Problem 6.4, show that the first two equations given there can be written as

$$(H^* - HH^{*\prime})\frac{d\delta_2}{dx} + H^{*\prime}\frac{d\delta^*}{dx} + \frac{\delta - \delta^*}{u_e}\frac{du_e}{dx} = c_E,$$
$$\frac{d\delta_2}{dx} + \delta_2\frac{H+2}{u_e}\frac{du_e}{dx} = \frac{c_f}{2},$$

where

$$H^{*\prime} = \frac{\mathrm{d}H^*}{\mathrm{d}H} = \frac{\alpha H^2 - 2\alpha H - 1}{(H-1)^2}.$$

2. With $2h$ denoting the total height of the diffuser and the function $h(x)$ known, assume that the velocity in the inviscid core u_e is constant in a cross-section of the diffuser. Show that the conservation of mass flow in the diffuser

$$u_e(h - \delta_1) = \text{ const.}$$

can be written as

$$-u_e\frac{\mathrm{d}\delta_1}{\mathrm{d}x} + (h - \delta_1)\frac{\mathrm{d}u_e}{\mathrm{d}x} + u_e\frac{\mathrm{d}h}{\mathrm{d}x} = 0.$$

3. The calculation of flow in a diffuser with the integral method consists of solving the following system of differential equations:

$$(H^* - HH^{*\prime})\frac{d\delta_2}{dx} + H^{*\prime}\frac{d\delta^*}{dx} + \frac{\delta - \delta^*}{u_e}\frac{du_e}{dx} = c_E,$$
$$\frac{d\delta_2}{dx} + \delta_2\frac{H+2}{u_e}\frac{du_e}{dx} = \frac{c_f}{2},$$
$$-u_e\frac{\mathrm{d}\delta_1}{\mathrm{d}x} + (h - \delta_1)\frac{\mathrm{d}u_e}{\mathrm{d}x} = -u_e\frac{\mathrm{d}h}{\mathrm{d}x}.$$

In the above system, the unknowns are $\frac{\mathrm{d}\delta_1}{\mathrm{d}x}$, $\frac{\mathrm{d}\delta_2}{\mathrm{d}x}$ and $\frac{\mathrm{d}u_e}{\mathrm{d}x}$. Show that the determinant of this system is zero when

$$\frac{h}{\delta_2} = f(H)$$

with

$$f(H) = \frac{(H^* - HH^{*\prime})(H+1)}{H^{*\prime}}.$$

4. Study the function $f(H)$ and deduce that the system is non-singular if $h/\delta_2 < 28.9$. In Problem 6.4 the boundary layer separation occurs at a location where $H^{*\prime} = 0$. In the present problem this location is not

singular (if $h/\delta_2 < 28.9$). The reason is that the coupling between the boundary layer and the inviscid flow is taken into account. Note that the restriction $h/\delta_2 < 28.9$ is partly due to the assumed shape of the function $H^*(H)$ and also to the hypothesis that the inviscid core velocity is constant in a cross-section. When the height of the diffuser is too large compared to the boundary layer, the coupling between the inviscid flow and the boundary layer is no longer effective to avoid the singular behavior of the boundary layer equations at separation.

References

1. Nicolai, L.M.: Fundamentals of Aircraft Design. METS, Inc., San Jose (1975)
2. Hirschel, E.H.: Vortex Flows: Some General Properties, and Modelling, Configurational and Manipulation Aspects. AIAA-Paper 96-2514 (1996)
3. Hirschel, E.H., Lucchi, C.W.: On the Kutta-Condition for Transonic Airfoils. MBB-UFE122-Aero-MT-651, Ottobrunn, Germany (1983)
4. McCormick, B.W.: Aerodynamics, Aeronautics and Flight Mechanics, 2nd edn. Wiley & Sons, Hoboken (1995)
5. Hirschel, E.H., Weiland, C.: Selected Aerothermodynamic Design Problems of Hypersonic Flight Vehicles, AIAA, Reston, Va. Progress in Astronautics and Aeronautics, vol. 229. Springer, Heidelberg (2009)
6. Hirschel, E.H.: Basics of Aerothermodynamics, AIAA, Reston, VA. Progress in Astronautics and Aeronautics, vol. 204. Springer, Heidelberg (2004)
7. Van Dyke, M.: Perturbation Methods in Fluid Mechanics. Academic Press, New York (1964)
8. Prandtl, L.: The Mechanics of Viscous Fluids. In: Durand, W.F. (ed.) Aerodynamic Theory, vol. III, pp. 34–208. Springer, Heidelberg (1935)
9. Van Dyke, M.: Higher Approximations in Boundary-Layer Theory. Part 1. General Analysis. J. of Fluid Mech. 14, 161–177 (1962)
10. Van Dyke, M.: Higher Approximations in Boundary-Layer Theory. Part 2. Application to Leading Edges. J. of Fluid Mech. 14, 481–495 (1962)
11. Goldstein, S.: On Laminar Boundary-Layer Flow Near a Position of Separation. Quarterly J. Mech. and Appl. Math. 1, 43–69 (1948)
12. Landau, L., Lifschitz, E.: Mécanique des fluides. Les Editions Ellipses, Paris, France (1994)
13. Goldstein, S.: Concerning Some Solutions of the Boundary Layer Equations in Hydrodynamics. Proc. Camb. Phil. Soc. XXVI(Pt. I), 1–30 (1930)
14. Catherall, D., Mangler, W.: The Integration of a Two-Dimensional Laminar Boundary Layer Past the Point of Vanishing Skin Friction. J. Fluid. Mech. 26(1), 163–182 (1966)
15. Monnoyer, F.: Calculation of Three-Dimensional Viscous Flow on General Configurations Using Second-Order Boundary-Layer Theory. ZFW 14, 95–108 (1990)
16. Aris, R.: Vectors, Tensors and the Basic Equations of Fluid Mechanics. Dover, New York (1989)
17. Hirschel, E.H., Kordulla, W.: Shear Flow in Surface-Oriented Coordinates. NNFM, vol. 4. Vieweg, Braunschweig Wiesbaden (1981)

18. Lighthill, M.J.: On Boundary-Layer and Upstream Influence: II. Supersonic Flows Without Separation. Proc. R. Soc., Ser. A 217, 478–507 (1953)
19. Schlichting, H., Gersten, K.: Boundary Layer Theory, 8th edn. Springer, New York (2000)
20. Stewartson, K.: On the Flow Near the Trailing Edge of a Flat Plate, II. In: Mathematika, vol. 16, pp. 106–121 (1969)
21. Stewartson, K., Williams, P.G.: Self-Induced Separation. Proc. Roy. Soc. London A 312, 181–206 (1969)
22. Stewartson, K.: Multistructured Boundary Layers of Flat Plates and Related Bodies. Adv. Appl. Mech. 14, 145–239 (1974)
23. Neyland, V.Y.: Theory of Laminar Boundary-Layer Separation in Supersonic Flow. Izv. Akad. Nauk SSSR Mekh. Zhidk. Gaza. 4, 53–57 (1969); Fluid Dyn. 4, 33–35 (1969)
24. Sychev, V.V.: Concerning Laminar Separation. Izv. Akad. Nauk. SSSR Mekh. Zhidk. Gaza 3, 47–59 (1972); Fluid. Dyn. 7, 407–417 (1972)
25. Messiter, A.F.: Boundary-Layer Flow Near the Trailing Edge of a Flat Plate. SIAM J. Appl. Math. 18, 241–257 (1970)
26. Guiraud, J.P.: Going on With Asymptotics. In: Gatignol, R., Bois, P.A., Dériat, E., Rigolot, A. (eds.) Asymptotic Modelling in Fluid Mechanics. Lecture Notes in Physics, vol. 442, pp. 257–307. Springer, Heidelberg (1995)
27. Zeytounian, R.K.: Asymptotic Modelling of Fluid Flow Phenomena. Kluwer Academic Publishers, Dordrecht (2002)
28. Carter, J.E., Wornom, S.F.: Solutions for Incompressible Separated Boundary-Layers Including Viscous-Inviscid Interaction. In: Aerodynamic Analysis Requiring Advanced Computers. NASA SP-347, pp. 125–150 (1975)
29. Carter, J.E.: A New Boundary Layer Inviscid Iteration Technique for Separated Flow. AIAA-Paper 79-1450 (1979)
30. Cebeci, T.: An Engineering Approach to the Calculation of Aerodynamic Flows. Horizons Publishing Inc., Springer, Long Beach, Heidelberg (1999)
31. Le Balleur, J.C.: Couplage visqueux-non visqueux: analyse du probléme incluant décollements et ondes de choc. La Recherche Aérospatiale 6, 349–358 (1977)
32. Le Balleur, J.C.: Couplage visqueux-non visqueux: méthode numérique et applications aux écoulements bidimensionnels transsoniques et supersoniques. La Recherche Aérospatiale 2, 65–76 (1978)
33. Lock, R.C.: A Review of Methods for Predicting Viscous Effects on Aerofoils and Wings at Transonic Speed. AGARD CP No. 291 (1981)
34. Veldman, A.E.P.: New, Quasi-Simultaneous Method to Calculate Interacting Boundary Layers. AIAA J. 19(1), 79–85 (1981)
35. Rothmayer, A.P., Smith, F.T.: Numerical Solution of Two-Dimensional, Steady Triple-Deck Problems. In: Johnson, R.W. (ed.) The Handbook of Fluid Dynamics, ch. 25. CRC Press, Springer, Boca Raton, Heidelberg (1998)
36. Veldman, A.E.P.: Matched Asymptotic Expansions and the Numerical Treatment of Viscous-Inviscid Interaction. J. Engineering Mathematics 39, 189–206 (2001)
37. Cousteix, J., Mauss, J.: Asymptotic Analysis and Boundary Layers. Scientific Computation, vol. XVIII. Springer, Heidelberg (2007)
38. Lazareff, M., Le Balleur, J.C.: Méthode de couplage fluide parfait fluide visqueux en tridimensionnel avec calcul de la couche limite par méthode multi-zones. In: Proc. 28éme réunion ss grpe AAAS, groupe sectoriel Franco-Soviétique Aéronautique, Moscou, also T.P. ONERA 1986-134 (1986)

39. Le Balleur, J.C., Girodroux-Lavigne, P.: Calculation of Fully Three-Dimensional Separated Flows with an Unsteady Viscous-Inviscid Interaction Method. In: Proc. 5th Int. Symp. on Numerical and Physical Aspects of Aerodynamical Flows. California State University, Long Beach CA, USA, also T.P. ONERA no 1992-1 (1992)
40. Le Balleur, J.C., Lazareff, M.: A Multi-Zonal Marching Integral Method for 3-D Boundary Layer with Viscous-Inviscid Interaction. In: Proc. 9'eme Congrés International des Méthodes Numériques en Mécanique des Fluides, Saclay, France, also T.P. ONERA 1984-67 (1984)
41. Sychev, V.V., Ruban, A.I., Sychev, V.V., Korolev, G.L.: Asymptotic Theory of Separated Flows. Cambridge University Press, Cambridge (1998)
42. Wanie, K.M., Schmatz, M.A., Monnoyer, F.: A Close Coupling Procedure for Zonal Solutions of Navier-Stokes, Euler and Boundary-Layer Equations. ZFW 11, 347–359 (1987)
43. Mundt, C., Monnoyer, F., Höld, R.: Computational Simulation of the Aerothermodynamic Characteristics for the Reentry of HERMES. AIAA-Paper 93-5069 (1993)
44. Cebeci, T., Cousteix, J.: Modeling and Computation of Boundary-Layer Flows, 2nd edn. Horizons Publ., Springer, Long Beach, Heidelberg (2005)
45. Lovell, D.A.: A Wind-Tunnel Investigation of the Effects of Flap Span and Deflection Angle, Wing Planform and a Body on the High-Lift Performance of a 28° Swept Wing. RAE CP 1372 (1977)

7

Topology of Skin-Friction Fields

Topological considerations of velocity and skin-friction fields are a powerful tool for flow-field diagnostics and interpretation. We studied in Section 4.4 characteristic properties of three-dimensional attached viscous flow and in Chapter 6 aspects of the interaction between attached viscous flow and its external inviscid flow. In the present chapter we investigate the topology of skin-friction fields of attached viscous flow. Mainly attachment points and lines, but also to a certain extent separation points and lines are considered.

First some general topics are treated, like the concept of limiting streamlines and issues of three-dimensional attachment and separation. An introduction to singular points of skin-friction fields follows. The classical approach to that topic is considered in detail. Then singular lines, i.e. attachment and separation lines are studied. Classical flow topology looks only at the pattern—the phase portrait (see Sub-Section 7.2.3)—of the velocity, respectively the skin-friction field at and around a singular point. Regarding singular lines, we take a broader view and include the consideration of flow-field properties.

The theory of singular points is well developed. That, however, is not true to the same extent for the theory of singular lines. Nevertheless, there is enough analytical, numerical and experimental evidence available to arrive at a coherent picture.

Steady flow is assumed throughout, the body surface may be flat or curved, the flow may be compressible or incompressible, laminar or—time-averaged—turbulent. Regarding the coordinate convention, we keep that of Chapter 4. The surface-parallel coordinates are x and z, with the velocity components accordingly u and w, the surface-normal coordinate is y and the surface-normal velocity component v.

7.1 Introduction

7.1.1 General Remarks

Classical flow topology looks mainly at the patterns of skin-friction lines in the vicinity of singular points. We basically follow that approach, but look also at singular lines, i.e. attachment and separation lines. Consequently, we

E.H. Hirschel, J. Cousteix, and W. Kordulla, *Three-Dimensional Attached Viscous Flow*,
DOI: 10.1007/978-3-642-41378-0_7, © Springer-Verlag Berlin Heidelberg 2014

have to consider, in addition, aspects of the topology of pressure fields and other entities.

We consider in this book in particular attached viscous flow past finite bodies. Hence in this chapter the topology of especially attached-flow skin-friction fields is considered. Of major interest are the situations at attachment points and lines, of minor interest those at separation points and lines, and also those at detachment points and lines.

The reader may wonder what *detachment* points and lines are. We introduce this term in order to distinguish between the viscous and the inviscid picture of flow leaving a finite body, Sub-Section 7.1.5. This appears to be desirable in view of the fact that separation of three-dimensional flow is not defined as simply and unambiguously as that of two-dimensional flow.

Different from the situation in two-dimensional flow the attachment and the separation of three-dimensional viscous flow is not characterized by vanishing wall shear stress, except in singular points on the body surface (attachment points, separation points).[1] However, usually a distinctive pattern of skin-friction lines can be observed at attachment lines, namely that the skin-friction lines diverge to both sides away from the attachment line. At separation lines, the skin-friction lines converge from both sides towards the separation line.

Both the attachment line and the separation line themselves are skin-friction lines. This cannot be proven in the frame of local topology considerations. For separation lines D.J. Peake and M. Tobak note that the convergence of the skin-friction lines (towards the separation line) is a necessary, but not a sufficient condition [1]. An example is the detachment line at the leeward side of a fuselage-like body at angle of attack.

We look now first at the concept of limiting streamlines and then at some general issues of three-dimensional attachment and separation.

7.1.2 The Concept of Limiting Streamlines

The concept of limiting streamlines was introduced by W.R. Sears [2]. We mention it here, because it is often used in the literature. Sears compares in the laminar boundary-layer flow over a yawed cylinder the "limiting streamline" at $y = 0$ (actually he calls the normal coordinate z) with the streamline of the external inviscid flow, in his case the potential flow. For us the surface-normal coordinate is y, and the surface-parallel coordinates are x and z, with the velocity components accordingly u and w. In our notation Sears's definition of the limiting streamline reads:

$$\frac{dz}{dx} = \frac{w}{u}\Big|_{lim\ y \to 0}. \tag{7.1}$$

[1] In singular points in general the wall shear stress vanishes to zero.

Using the rule of l'Hospital and taking into account the isotropy of the (molecular) viscosity at the wall, we obtain from this

$$\frac{w}{u}|_{y \to 0} = \frac{\tau_{wz}}{\tau_{wx}} \tag{7.2}$$

which is the definition of the skin-friction line.[2] In three-dimensional attached viscous flow the skin-friction line has not the same shape and direction as the external inviscid streamline, as was discussed in Section 4.4 and as is illustrated in Figs. 4.1 and 4.2. This holds for all streamlines in the boundary layer.

If the term limiting streamline is taken in the sense of the word, [1], it has its merits in the analysis of, for instance, oil-streak flow-visualization techniques [3]. The oil sheet has a finite thickness—this may hold also for other visualization indicators—hence the streak as visual indicator may represent a limiting streamline. If the sheet is thin enough, the skin-friction line is a projection of the limiting streamline on the body surface.

7.1.3 General Issues of Three-Dimensional Attachment

Under attachment we generally understand the impingement of the inviscid free stream on a body surface. The body surface itself is covered by a sheet of viscous flow, the attached viscous flow, which in general is of boundary-layer type. However, in a separation region also separated viscous flow can (re-) attach on the body surface. Prominent examples are the separation bubbles which can appear in two-dimensional and also in three-dimensional form.

For an easier understanding we consider first the attachment process assuming inviscid flow throughout. The free stream impinges in the form of a streamline at, for instance, the forward attachment point, i.e. the nose point, of a fuselage. At this point, the stagnation point, the surface velocity is zero. The flow at the body surface then evolves—with non-zero velocity along the streamlines—exclusively from this attachment point. No other streamlines than the stagnation-point streamline impinge on the body surface.

The attachment line is a location along the body surface where the arriving flow diverges to the left and the right side of it without impinging on the body surface. The attachment line itself is a streamline which originates at the forward stagnation point. The attachment line is part of a dividing surface which separates the to the sides diverging flow parts.

Prominent examples are the attachment line at the leading edge of a swept wing and at the lower symmetry line of a (round or nearly round) fuselage at angle of attack. The attachment point and the attachment line in inviscid flow can well be prescribed by solutions of the Euler equations, the potential equation or the linearized potential equation (panel method). Examples of solutions of the latter two are given with Fig. 7.2 on page 137, and with

[2] Wall streamline is also a term frequently used. We advocate that it should be avoided, like the term limiting streamline.

Fig. 7.10 on page 151. We call such attachment lines loosely primary attachment lines.

If the flow on the body surface is viscous—attached viscous flow—the inviscid flow, impinging or arriving at the body surface, impresses its pattern almost fully on the viscous flow. The role of the surface streamlines of the inviscid flow is now taken over by the skin-friction lines.

A (primary) attachment line can begin at a singular point. Examples are given in Fig. 7.2 (inviscid flow), in Fig. 10.16 on page 264 (viscous flow, with the singular point being an embedded attachment point), and in Fig. 10.26 a) on page 275 (viscous flow). The latter example is typical for the flow past a flat or almost flat surface with angle of attack against the free stream, being the characteristic of a delta wing. At a round fuselage only one primary attachment line appears. In this case, however, the attachment line is split into two primary attachment lines, see also the examples discussed at the end of Section 7.3.

However, a primary attachment line must not necessarily begin at a singular point. An example is given with Fig. 10.18 on page 266 (viscous flow). In analogy to open type separation, see Sub-Section 7.2.1, we call this open type attachment.

Attachment lines can appear also at other locations of a body surface, usually in connection with separation phenomena. Then we call them embedded attachment lines. If they appear in a regular pattern, we may call them secondary, tertiary and so on attachment lines. Fig. 10.24 on page 272 shows an embedded attachment line of open type. Two secondary and on tertiary attachment lines are shown in Fig. 10.27 on page 276. All these lines are of open type, too.

7.1.4 General Issues of Three-Dimensional Separation

The separation line in three-dimensional flow is the location, where the boundary layer, actually two converging boundary layers, separates from the body surface. The separation line is part of a dividing surface, which separates the converging boundary layers on each side of the separation line. A very detailed discussion of the distinction between attached and separated flows is given in [1], too.

At finite bodies, separation naturally occurs always—the attached viscous flow past the body finally leaves the body, Fig. 7.1. The situation in three-dimensional flow generally is very complicated compared to that in two-dimensional flow, which is a special case.

The classical definition of separation in two-dimensional flow, which is based on the observation that the wall-shear stress vanishes at the separation point, $\tau_{wall} = 0$, is not sufficient[3]. We do not give here an overview over the criteria found in the literature, see, e.g., [1]. We simply state that a possible definition reads like [5, 6]:

[3] This is one of three criteria discussed, for instance, by E.A. Eichelbrenner [4].

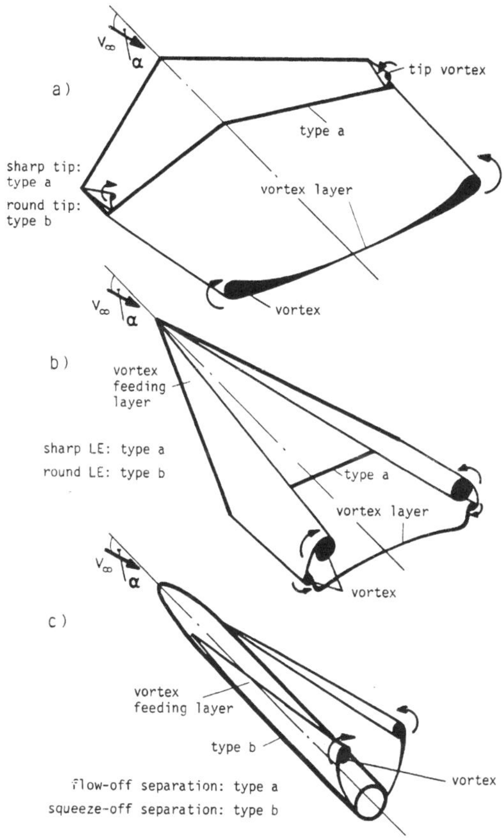

Fig. 7.1. Schematic of kinds of separation and the resulting vortex layers and vortices, [5], at a) large-aspect-ratio wing with small leading-edge sweep, b) slender wing with large leading-edge sweep, c) fuselage. Small or moderate angles of attack, the situation at the afterbody of the fuselage is not considered, LE: leading edge.

Separation is present, if vorticity is transported away from the body surface by convection, and subsequently vortex sheets and vortices are formed. Locally the boundary-layer criteria are violated.

We take now a maybe somewhat naive but pragmatic view and distinguish globally between two kinds of separation: *flow-off separation* and *squeeze-off separation*, which appear in three-dimensional as well as in two-dimensional viscous flow. In these two kinds, always two boundary layers are involved in the separation process.[4] The reader should be aware that this is a very simple view. Extended discussions of flow separation patterns can be found, e.g., in [1, 7, 8], and very detailed, for instance, in [9, 10].

[4] This situation is often found, but there are possible singular points at the body surface where the situation is different, see next section.

Consider the flow cases shown in Fig. 7.1. They are based on the assumption of small to moderate angles of attack.

Note for the delta wing that the appearance of the lee-side vortex pair depends not only on the angle of attack, but also on the leading-edge sweep and the free-stream Mach number, see, e.g, [11]. Secondary vortex systems, as they can appear at the delta wing and the fuselage are not indicated. The situation at the afterbody of the fuselage is also not indicated. It depends strongly on the configuration details and can be very complex.

Flow-off separation happens at acute edges, like wing trailing edges, Fig. 7.1 a) and b), or sharp highly-swept wing leading edges, Fig. 7.1 b). If a wing tip, Fig. 7.1 a), would be sharp-edged, also there flow-off separation would be present. Two boundary layers coming from the upper and the lower side of the wing flow off the edge and merge, forming a wake. The wake may contain only kinematically inactive vorticity, as is the case with an airfoil in steady flow. The wake behind a lifting wing, as shown in Figs. 7.1 a) and b), is called a vortex layer and contains both kinematically inactive and active vorticity [12], see also Section 10.2. This vortex layer further downstream of the wing curls up and forms the two discrete trailing vortices [13], Section 10.2. The tip vortices merge into these vortices.

Squeeze-off separation is the separation form which usually is considered. It appears at round body flanks, like round wing tips, Fig. 7.1 a), highly swept round leading edges, Fig. 7.1 b), or fuselages at moderate angle of attack, Fig. 7.1 c).

The first squeeze-off separation line when coming from the front of a configuration usually is called the primary separation line. The separation line can begin at a singular point. However, that is not necessarily so, see Sub-Section 7.2.1. The simplest form of squeeze-off separation is the so-called open type separation, [14], which does not begin at a singular point.

Examples of primary open type separation lines are given in Fig. 7.21 on page 167 and in Fig. 10.27 on page 276.

Separation lines can appear as embedded separation lines at other locations of a body surface. If they appear—like attachment lines—in a regular pattern, we may call them secondary, tertiary and so on separation lines. Figs. 10.23, page 271, and 10.24, page 272 show such embedded separation lines. Two secondary separation lines are shown in Fig. 10.27 on page 276. All these separation lines are of open type.

7.1.5 Detachment Points and Lines

In closing this section we come back to the terms *detachment* points and lines. In inviscid flow—other than in viscous flow—the flow leaves the surface of a finite body without transporting kinematically active or inactive vorticity away from the body surface. The streamlines, however, form similar patterns as we find them in viscous flow.

The term *detachment line* was introduced in [15]. Consider the inviscid flow past a wing-like flat ellipsoid at angle of attack, Fig. 7.2. The flow field and the streamlines were computed, [16], with exact potential-flow theory [17].

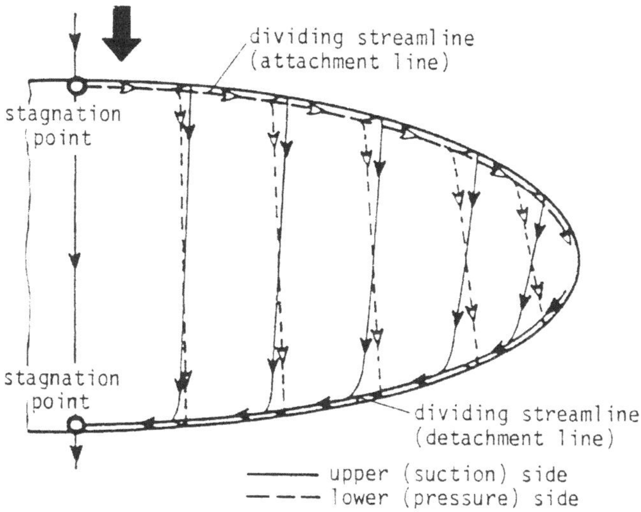

Fig. 7.2. Streamlines of the inviscid velocity field past a wing-like 3:1:0.125 ellipsoid at angle of attack $\alpha = 15°$ [15]. View towards the upper side.

The ellipsoid is seen from above. The forward stagnation point and the forward dividing streamline—the attachment line—lie, because of the positive angle of attack, at the lower side of the ellipsoid. The streamline pattern is symmetrical around the lateral axis (which is the major axis). Hence we find both the rear stagnation point—which we call detachment point—and the rear dividing streamline—which we call detachment line—on the upper side of the ellipsoid.

This is an extreme example. In applied aerodynamics less pronounced detachment lines typically appear at the lee side of fuselages at small to moderate angles of attack. There a thickening of the viscous layer happens without separation of the flow. This might be, for instance, of interest for design considerations.

A simple detachment point situation is shown in Fig. 7.4 on page 144. At the rear stagnation point, N_2, the inviscid flow leaves the body surface, by definition without transporting kinematically active or inactive vorticity away from the body.

7.2 Singular Points

7.2.1 Introduction

The pattern of skin-friction lines—also that of inviscid streamlines—on a body surface can be considered as continuous vector field, orthogonal to it lying surface vortex lines [18].[5] Through each point on the surface passes one and only one skin-friction line.

However, there are locations, where this is not true: singular points and singular lines (attachment and separation lines).[6] Singular points are isolated locations in the skin-friction field where the skin friction as well as the surface vorticity become zero. The number of skin-friction lines associated with such a point is different from one, Sub-Section 7.2.3.

The skin friction is non-zero along attachment and separation lines. Along an attachment line an infinite number of skin-friction lines diverges from it. In contrast to this along separation lines an infinite number of skin-friction lines converges towards it.

M.J. Lighthill defined as necessary condition for separation that the particular skin-friction line, to which the other skin-friction lines converge, must originate from a singular point, which is a saddle [18].[7] This was refuted by K.C. Wang in the 1970s by his finding of the phenomenon of 'open type separation' [14]. In this case the separation line does not begin at a singular point, but at some location in the skin-friction field. Peake and Tobak call that 'local separation' [1].

The theory of critical points, as singular points are also called, was realized by R. Legrende as being *the* rational tool for the analysis of three-dimensional separated flow [8]. It goes back to H. Poincaré's work on singular points of systems of differential equations [20]. Legrende made pioneering work in the clarification of separation phenomena, together with H. Werlé, who developed and applied with great success visualization techniques in a water tunnel.

We give here the basic derivation of singular points and in Section 7.4 that of singular lines. Steady flow is assumed, the body surface may be flat or curved, the flow may be compressible or incompressible, laminar or—time-averaged— turbulent, heat flux may be present, the body surface is non-permeable and the no-slip condition holds. We proceed—following the approach of K. Oswatitsch [21]—from the continuity equation and the Navier-Stokes equations (not from the boundary-layer equations!) as they are given in Chapter 3 and from the wall-compatibility conditions given in Section 4.5.

---

[5] H.J. Lugt differentiates between vorticity lines and vortex lines [19]. What is called here vortex lines is in his nomenclature vorticity lines.

[6] We do not always mention detachment points and lines. What holds also for them in the following discussion is self-evident.

[7] A more detailed discussion, see, e.g., [8], uses the concept of the 'separator' (or 'separatrix') which separates skin-friction lines into families with different origin. This holds for both attachment and separation lines.

We make the derivation in Cartesian coordinates on a flat surface. This does not degrade the general validity of the results.

7.2.2 Flow-Field Continuation around a Surface Point

We consider the compressible viscous flow in the neighborhood of a point $P_0(x_0, y_0, z_0)$ on the body surface. We employ again the notation used in Chapter 4: the surface-normal coordinate is y and the surface-normal velocity component v, the surface-parallel coordinates are x and z, with the velocity components accordingly u and w. We develop local solutions of the Navier-Stokes equations with Taylor expansions around the point P_0 on the body surface.

The expansion of the velocity vector $\underline{V} = (u, v, w)^T$ reads for $P \neq P_0$

$$\begin{aligned} \underline{V}(P) = \underline{V}\,(P_0) + x\frac{\partial \underline{V}}{\partial x}\,(P_0) + ... + z\frac{\partial \underline{V}}{\partial z}\,(P_0) + \frac{x^2}{2}\frac{\partial^2 \underline{V}}{\partial x^2}\,(P_0) + \\ + ... + \frac{z^2}{2}\frac{\partial^2 \underline{V}}{\partial z^2}\,(P_0) + O(x^3, ...). \end{aligned} \quad (7.3)$$

At the body surface all components of the velocity vector are zero:

$$\underline{V}|_{y=0} = \underline{V}(u, v, w)|_{y=0} = 0, \quad (7.4)$$

as well as the derivatives tangential to the surface (α, $\beta \geqq 0$):

$$\frac{\partial^{\alpha+\beta}\underline{V}}{\partial x^\alpha \partial z^\beta}|_{y=0} = 0. \quad (7.5)$$

From the continuity equation, eq. (3.13), we obtain

$$\frac{\partial v}{\partial y}|_{y=0} = \frac{\partial^2 v}{\partial x\, \partial y}|_{y=0} = \frac{\partial^2 v}{\partial z\, \partial y}|_{y=0} = 0 \quad (7.6)$$

and

$$\frac{\partial^2 v}{\partial y^2}|_{y=0} = -\left[\frac{\partial^2 u}{\partial x\, \partial y} + \frac{\partial^2 w}{\partial y\, \partial z} + \frac{1}{\rho}\left(\frac{\partial u}{\partial y}\frac{\partial \rho}{\partial x} + \frac{\partial w}{\partial y}\frac{\partial \rho}{\partial z}\right)\right]_{y=0}. \quad (7.7)$$

We write the wall shear-stress components in x- and in z-direction:

$$\tau_x = \mu\frac{\partial u}{\partial y}|_{y=0}, \;\; \tau_z = \mu\frac{\partial w}{\partial y}|_{y=0} \quad (7.8)$$

and find their derivatives at $y = 0$ in the two directions x and z to

$$\begin{aligned}\frac{\partial \tau_x}{\partial x} &= \frac{1}{\mu}\frac{\partial \mu}{\partial x}\tau_x + \mu\frac{\partial^2 u}{\partial x\,\partial y}, \quad \frac{\partial \tau_x}{\partial z} = \frac{1}{\mu}\frac{\partial \mu}{\partial z}\tau_x + \mu\frac{\partial^2 u}{\partial y\,\partial z}, \\ \frac{\partial \tau_z}{\partial x} &= \frac{1}{\mu}\frac{\partial \mu}{\partial x}\tau_z + \mu\frac{\partial^2 w}{\partial x\,\partial y}, \quad \frac{\partial \tau_z}{\partial z} = \frac{1}{\mu}\frac{\partial \mu}{\partial z}\tau_z + \mu\frac{\partial^2 w}{\partial y\,\partial z}.\end{aligned} \tag{7.9}$$

From the compatibility conditions eqs. (4.57) and (4.58) we get for the pressure gradients in x- and z-direction at $y = 0$:

$$\frac{\partial p}{\partial x} = \mu\,\frac{\partial^2 u}{\partial y^2} + \frac{1}{\mu}\,\frac{\partial \mu}{\partial y}\,\tau_x, \quad \frac{\partial p}{\partial z} = \mu\,\frac{\partial^2 w}{\partial y^2} + \frac{1}{\mu}\frac{\partial \mu}{\partial y}\,\tau_z. \tag{7.10}$$

With eqs. (7.9) and (7.10) we can write eq. (7.7) now in the form

$$\frac{\partial^2 v}{\partial y^2}|_{y=0} = \\ = -\frac{1}{\mu}\left[\frac{\partial \tau_x}{\partial x} + \frac{\partial \tau_z}{\partial z} - \frac{1}{\mu}\left(\frac{\partial \mu}{\partial x}\,\tau_x + \frac{\partial \mu}{\partial z}\,\tau_z\right) + \frac{1}{\rho}\left(\frac{\partial \rho}{\partial x}\,\tau_x + \frac{\partial \rho}{\partial z}\,\tau_z\right)\right]_{y=0}. \tag{7.11}$$

Putting all terms accordingly into the expansion eq. (7.3) gives for the two surface-parallel components of the velocity vector $\underline{V}$ at point P in the neighborhood of P_0:

$$\begin{aligned} u|_P = & \left[\frac{1}{\mu}\,\tau_x\right]_{P_0} y + \left[\frac{1}{\mu}(\frac{\partial \tau_x}{\partial x} - \frac{1}{\mu}\frac{\partial \mu}{\partial x}\,\tau_x)\right]_{P_0} x\,y + \\ & + \left[\frac{1}{\mu}(\frac{\partial \tau_x}{\partial z} - \frac{1}{\mu}\frac{\partial \mu}{\partial z}\,\tau_x)\right]_{P_0} y\,z + \frac{1}{2}\left[\frac{1}{\mu}(\frac{\partial p}{\partial x} - \frac{1}{\mu}\frac{\partial \mu}{\partial y}\,\tau_x)\right]_{P_0} y^2, \end{aligned} \tag{7.12}$$

$$\begin{aligned} w|_P = & \left[\frac{1}{\mu}\,\tau_z\right]_{P_0} y + \left[\frac{1}{\mu}(\frac{\partial \tau_z}{\partial x} - \frac{1}{\mu}\frac{\partial \mu}{\partial x}\,\tau_z)\right]_{P_0} x\,y + \\ & + \left[\frac{1}{\mu}(\frac{\partial \tau_z}{\partial z} - \frac{1}{\mu}\frac{\partial \mu}{\partial z}\,\tau_z)\right]_{P_0} y\,z + \frac{1}{2}\left[\frac{1}{\mu}(\frac{\partial p}{\partial z} - \frac{1}{\mu}\frac{\partial \mu}{\partial y}\,\tau_z)\right]_{P_0} y^2, \end{aligned} \tag{7.13}$$

and for the surface-normal component:

$$v|_P = -\frac{1}{2}\left[\frac{1}{\mu}[\frac{\partial \tau_x}{\partial x} + \frac{\partial \tau_z}{\partial z} - \frac{1}{\mu}(\frac{\partial \mu}{\partial x}\tau_x + \frac{\partial \mu}{\partial z}\,\tau_z) + \frac{1}{\rho}(\frac{\partial \rho}{\partial x}\,\tau_x + \frac{\partial \rho}{\partial z}\,\tau_z)]\right]_{P_0} y^2. \tag{7.14}$$

7.2.3 Singular Points on Body Surfaces

Singular points are points in the skin-friction field, where the two wall shear-stress components vanish simultaneously:

$$\tau_x = \mu \frac{\partial u}{\partial y}|_{y=0} = 0, \; \tau_z = \mu \frac{\partial w}{\partial y}|_{y=0} = 0. \tag{7.15}$$

The flow patterns in the immediate vicinity of singular points can have very different appearances. Always fulfilled, of course, is the continuity equation. Parts of the flow enter the singularity and accordingly other parts leave it.

A systematic identification of singular points can be made with the help of the phase-plane analysis, see, e.g., the book of W. Kaplan [22]. That analysis does not permit to find solutions of eqs. (7.12) to (7.14), but to find possible geometrical configurations of the patterns of the skin-friction lines ("phase portraits" of the surface shear-stress vector [23])—or the streamlines—in the neighborhood of $P_0(x_0, y_0, z_0)$. It employs a vector and matrix arrangement of the first-order terms of the equations, see, e.g., [9, 24, 25].

For convenience the analysis is usually done for the incompressible case with constant viscosity μ. Then eq. (7.14) becomes with the compatibility condition eq. (3.30) (note the change of coordinate notation!):

$$v|_P = \frac{1}{2} \left[\frac{1}{\mu} \frac{\partial p}{\partial y} \right]_{P_0} y^2. \tag{7.16}$$

Taking into account eq. (7.15), we write now eqs. (7.12), (7.13), and (7.16) as

$$\frac{1}{y} \underline{V}(P) = \frac{1}{\mu} \mathbf{A}(P_0) \, \underline{X} + \mathbf{B}(P_0). \tag{7.17}$$

Here $\underline{X} = (x, y, z)^T$ is the location vector of point P and $\mathbf{A}$ a Jacobian matrix which contains the expansion terms of lowest order at point P_0:

$$\mathbf{A} = \begin{pmatrix} \dfrac{\partial \tau_x}{\partial x} & \dfrac{1}{2}\dfrac{\partial p}{\partial x} & \dfrac{\partial \tau_x}{\partial z} \\ 0 & \dfrac{1}{2}\dfrac{\partial p}{\partial y} & 0 \\ \dfrac{\partial \tau_z}{\partial x} & \dfrac{1}{2}\dfrac{\partial p}{\partial z} & \dfrac{\partial \tau_z}{\partial z} \end{pmatrix}_{P_0} . \tag{7.18}$$

The matrix $\mathbf{B}$ contains small terms of higher order.

The eigenvalues of matrix $\mathbf{A}$ read:

$$\lambda_{1,3} = \frac{1}{2} \left(\frac{\partial \tau_x}{\partial x} + \frac{\partial \tau_z}{\partial z} \right) \pm \frac{1}{2} \sqrt{\left(\frac{\partial \tau_x}{\partial x} - \frac{\partial \tau_z}{\partial z} \right)^2 + 4 \frac{\partial \tau_x}{\partial z} \frac{\partial \tau_z}{\partial x}}, \tag{7.19}$$

$$\lambda_2 = \frac{1}{2}\frac{\partial p}{\partial y}. \tag{7.20}$$

The matrix $\mathbf{A}$ is further analyzed by investigating the trace T, the Jacobian determinant J, and the discriminant $\triangle$ of it, see, e.g., [22].

The trace reads:

$$T = \frac{\partial \tau_x}{\partial x} + \frac{1}{2}\frac{\partial p}{\partial y} + \frac{\partial \tau_z}{\partial z}, \tag{7.21}$$

the Jacobian determinant:

$$J = \frac{\partial \tau_x}{\partial x}\,\frac{1}{2}\frac{\partial p}{\partial y}\,\frac{\partial \tau_z}{\partial z} - \frac{\partial \tau_z}{\partial x}\,\frac{1}{2}\frac{\partial p}{\partial y}\,\frac{\partial \tau_x}{\partial z}, \tag{7.22}$$

and the discriminant:

$$\triangle = T^2 - 4J. \tag{7.23}$$

The combination and signs of these three parameters govern the pattern of the skin-friction lines in the immediate neighborhood of a singular point $P_0(x_0, y_0, z_0)$.

We collect the resulting singular points in a manner similar to that used in [22] and [9], Fig. 7.3. These are the basic singular points; other singular points or combinations of them (merged points), which we do not discuss here, are also possible.

We distinguish two classes of singular points, saddles S ($J < 0$) and nodes ($J > 0$, $\triangle > 0$), both for attaching flow: $T > 0$, right-hand side of Fig. 7.3 and separating/detaching flow: $T < 0$, left-hand side of Fig. 7.3. In topological rules foci F ($J > 0$, $\triangle < 0$) are counted as nodes, as well as centers ($J > 0$, $T = 0$) which are also considered as foci.

A node is the common point of an infinite number of skin-friction lines which are directed away from it for attaching flow and towards it for separating/detaching flow. At that point all skin-friction lines except one are tangential to a given skin-friction line. The star node is an exception.

An infinite number of skin-friction lines is also associated with a focus although without a common tangent line. No skin-friction line is associated with the singular point of a center.

For both attaching and separating flow two single skin-friction lines each towards and away from it are associated with a saddle. All other skin-friction lines in the neighborhood of a saddle are deflected from that point in the directions of the single skin-friction lines.

Away from the body surface the patterns of streamlines around singular points can be very different. We do not discuss that here and instead refer the reader to, e.g., [1, 9, 21, 24, 25].

In a plane normal to the surface through a singular point, the point changes its character. The surface as boundary now itself is a singular line,

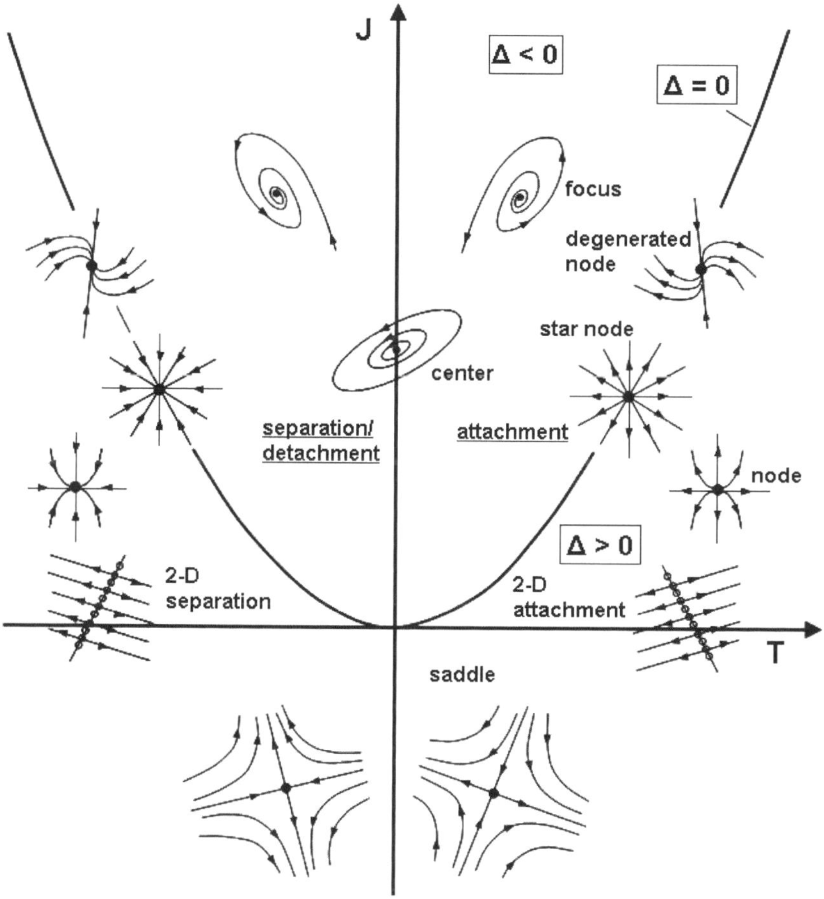

Fig. 7.3. Patterns of skin-friction lines in the neighborhood of $P_0(x_0, y_0, z_0)$: basic singular points in the chart of trace T and Jacobian determinant J, with the discriminant $\triangle$ of the Jacobian matrix $\mathbf{A}$ as parameter. This illustration follows those given in [22] and [9].

and singular points on the surface become, for instance, half-nodes N' or half-saddles S', J.C.R. Hunt et al. [26].

When considering separation patterns, structurally unstable singular points are of interest, see, e.g., [27]. If small changes of flow and/or geometrical parameters change a phase portrait, structural instability is given. We do not treat separated flows here, but we note Kaplan's statement that in Fig. 7.3 only the phase portraits lying in the quadrant $J > 0$ and $T < 0$ are stable.

7.3 Topological Rules

Peake and Tobak discuss in [1], Section 2.7, the "topography of streamlines in two-dimensional sections of three-dimensional flows". They give topological rules, two of which are of interest in view of the topic of this book.

General assumptions are that the flow is steady, the body is simply connected, that the velocity field and the skin-friction field past the body are continuous, and that the body is immersed in a uniform upstream flow field.

Rule 1, due to A. Davey [28] and also M.J. Lighthill [18], concerns the velocity field or skin-friction field on a three-dimensional body. It says that on the body surface the number of nodes N (foci are counted as nodes) is larger than the number of saddles S by two:

$$\sum N - \sum S = 2. \tag{7.24}$$

Rule 2, due to J.C.R. Hunt et al. [26], concerns skin-friction lines and streamlines in a two-dimensional plane cutting a three-dimensional body. The sum of nodes N plus one half of the number of half-nodes N' is one less than the sum of saddles S plus one half of the number of half-saddles S':

$$(\sum N + \frac{1}{2}\sum N') - (\sum S + \frac{1}{2}\sum S') = -1. \tag{7.25}$$

We illustrate these two rules with the flow past an axisymmetric body. Fig. 7.4 a) shows the streamlines of the inviscid flow on the body surface with the forward (attachment) and the rearward (detachment) stagnation point. Both are nodes, Rule 1 is fulfilled because saddles are absent. In Fig. 7.4 b) the body is cut by a two-dimensional plane through its axis. The nodes of Fig. 7.4 a) now become half-saddles. Rule 2 is fulfilled.

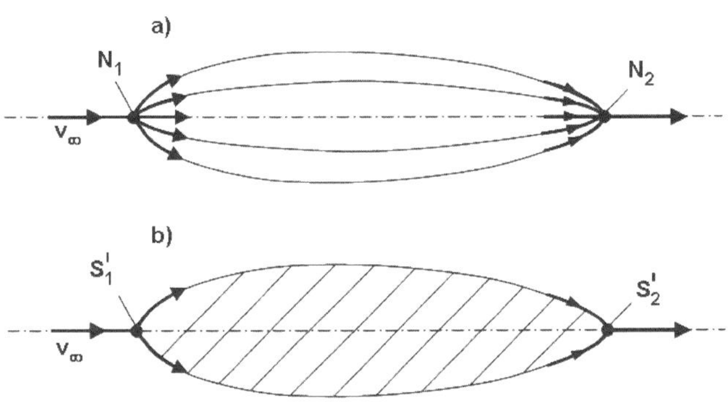

Fig. 7.4. Schematic of steady inviscid flow past an axisymmetric body. a) Streamlines on the surface: two nodes N. b) Streamlines seen in the two-dimensional plane cutting the body through its axis: two half-saddles S'.

If the flow near the surface is viscous, we see separation at some aft location of the body, Fig. 7.5. We assume a steady separation region.[8] The forward and the rearward stagnation point are half-saddles. Although inviscid flow attaches at the forward stagnation point, S'_1, the flow there is viscous, i.e. the viscous layer or boundary layer at that point has a finite thickness, Section 8.1.

Note that the rearward stagnation point, S'_2, now is an attachment point, too. The shear layers emanating from S'_3 and S'_4 are merging and split in S_1 and then move partly, with a wake-like appearance, towards S'_2. Actually it is the circumferential separation line which shows up in the cutting plane as the two half-saddles S'_3 and S'_4. The separation region is a toroid, which shows up as two centers (foci), F_1 and F_2. The separation region is closed by a saddle, S_1. Rule 2 is fulfilled.

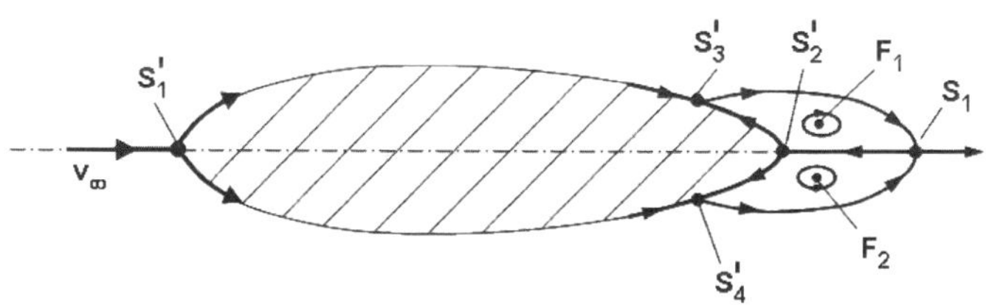

Fig. 7.5. Schematic of steady viscous flow past an axisymmetric body. Skin-friction lines and streamlines seen in the two-dimensional plane cutting the body through its axis: one saddle S, four half saddles S', two foci F, counted as nodes.

Quarter-saddles were introduced in [30] in order to treat the topology of the flow past a delta wing, independent of the type of leading edge, sharp or rounded. If the lower side of it—or of a fuselage—is flat or nearly flat and inclined against the free-stream, two primary attachment lines appear. Between them the flow is fully or nearly two-dimensional, see Section 10.4. This results in a favorable onset flow of, for instance, an engine air inlet or an aerodynamic control surface, Fig. 7.6 [31].

Consider in this regard Fig. 7.7. The skin-friction lines on the lower side of the wing leave the primary attachment lines like sketched in Fig. 7.6. In a plane two-dimensional cut A-A, then only the traces of the outward flow are seen and the primary attachment lines appear as quarter saddles S''.[9]

[8] In reality this is given only for very small Reynolds numbers, see, for instance, the beautiful pictures in [29].

[9] Note that at the attachment lines—the two primary lines, and also (!) at the two secondary lines and the tertiary line—*inviscid flow* attaches. However, the attachment-line flows themselves are viscous, i.e. the viscous layers or boundary layers at those lines have finite thicknesses, Section 7.4.

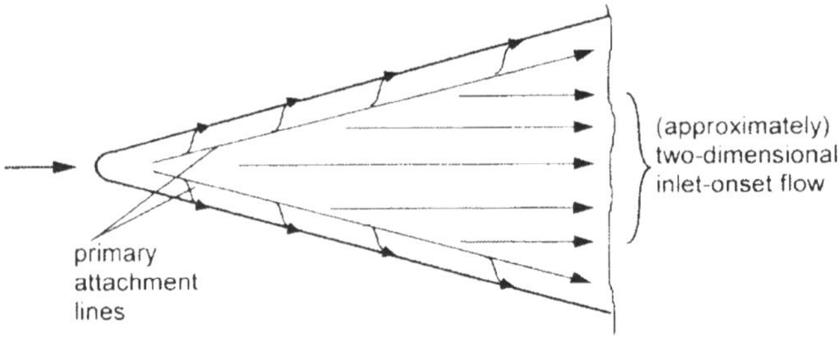

Fig. 7.6. Sketch of the skin-friction line pattern at the flat lower side of a forebody at angle of attack [31].

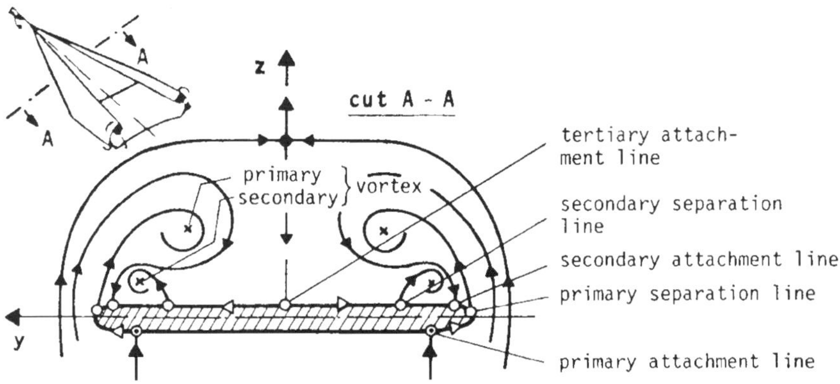

Fig. 7.7. Sketch of steady viscous flow past a delta wing with primary and secondary lee-side vortices [30]. Skin-friction lines and streamlines seen in the plane two-dimensional cut A-A, i.e. the Poincaré surface [32]: one saddle S ($\bullet$), seven half saddles S' ($\circ$), two quarter-saddles S'' ($\odot$), four foci F ($\times$), counted as nodes.

Rule 2 changes then into **Rule 2'**:

$$(\sum N + \frac{1}{2}\sum N') - (\sum S + \frac{1}{2}\sum S' + \frac{1}{4}\sum S'') = -1, \tag{7.26}$$

which in our case results in

$$4 - (1 + \frac{1}{2}7 + \frac{1}{4}2) = -1. \tag{7.27}$$

If the lower side of the body or wing has a convex shape—like shown in Fig. 7.8—the attachment line lies at the lower apex of it. Then, instead of the two quarter-saddles present at the lower side of the delta wing in Fig. 7.7, one half-saddle is present and **Rule 2** is fulfilled.

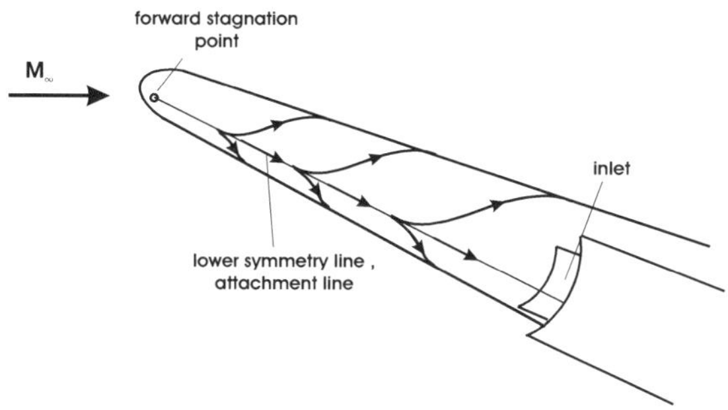

Fig. 7.8. Sketch of the skin-friction line pattern at the lower side of a circular forebody at angle of attack [31].

7.4 Singular Lines

7.4.1 Introduction

In the two preceding sections we treated singular points in a rather abstract way, obtaining and discussing only the phase portraits of the singular points. In this section we investigate the topological properties of attachment and separation lines, but now we take into account also certain flow properties. We believe that to bring in some concepts attributed to these lines—more or less familiar to the reader—will foster the general understanding of the different aspects of three-dimensional attached viscous flows.

Singular lines in the frame of this section are primary attachment and separation lines. Primary attachment lines as a rule are attachment lines also of the inviscid flow past a body. An attachment line may have its origin in a singular point, Fig. 7.4 a). This can be a node—forward (primary) stagnation point—or a saddle. However, like open type separation, also open type attachment is possible, Sub-Section 7.1.3.

At an ordinary airplane configuration with or without swept wings, only one primary stagnation point is present. That is located at the nose of the fuselage.[10] If the airplane has a forward swept wing, three primary attachment points are found.

We note that only in a singular point—attachment point—a streamline impinges on the body surface. Along an attachment line this does not happen. A streamline never becomes a skin-friction line. This holds for both viscous and inviscid flow. Likewise only in a singular point—separation point—a streamline leaves the body surface. Along a separation line this does not happen. A skin-friction line never becomes a streamline which leaves the body surface.

[10] We neglect possible forward stagnation points at the propulsion units and at antennas and the like.

From the sketch of the delta wing with lee-side vortices, Fig. 7.7, we have learned that attachment lines embedded in a separation region can also be attachment lines of inviscid flow. These secondary and tertiary attachment lines, however, would not be present, if the flow past that wing would not be separated in cross-flow direction. In that case only the primary attachment lines are present. We do not discuss secondary and higher attachment lines (see in this regard Sub-Section 7.1.3). However, most of our results apply also to them.

Separation lines are present only in viscous flow past a finite body, either as squeeze-off or as flow-off separation lines. As primary separation line we denote the, in the dominating flow direction first appearing squeeze-off separation line. Such a line may have, for instance, its origin in a saddle on the body surface. This is Lighthill's definition, page 138. If we have the case of open separation—Wang's finding—it does not begin in a singular point.

In the so-called separation region behind the primary (squeeze-off) separation line, secondary, see, e.g., Fig. 7.7, and even tertiary separation lines can be present, Sub-Section 7.1.4. If the flow is inviscid, we speak about detachment of the flow from the surface of the finite body.

The above holds quite in general for every configuration and the flow past it. The results of our investigations in principle apply to any kind of attachment and separation lines.

Attachment and separation/detachment lines appear in two canonical forms as shown in Fig. 7.9. Typical for an attachment line (a) is that an infinite number of skin-friction lines diverges from it. This holds also for the surface streamlines of the related inviscid flow. However, the inviscid and the viscous attachment line generally do not coincide with each other. The reason is that the skin-friction lines of three-dimensional attached viscous flow are more strongly curved than the surface streamlines of the related external inviscid flow, Section 4.4. The two attachment lines coincide only if they lie, for instance, on the surface generator of an infinite swept wing (ISW).[11]

Typical for a squeeze-off separation line (b) is that an infinite number of skin-friction lines converges towards it. This holds also for a detachment line of inviscid flow.

We look now at some general properties of attachment and squeeze-off separation/detachment lines and single out the following five items:

1. Relative maximum of the surface pressure At a curved inviscid attachment line, the surface pressure has a relative maximum in direction normal to the inviscid attachment line, 1 - - - 1 in Fig. 7.9 a). In general, the p_{max}-line lies close to the attachment line (also to the viscous one), and only on it, if the ISW situation is given.

[11] This surface generator is a geodesic, the boundary layer is a quasi-two-dimensional one, Appendix A.2.3.

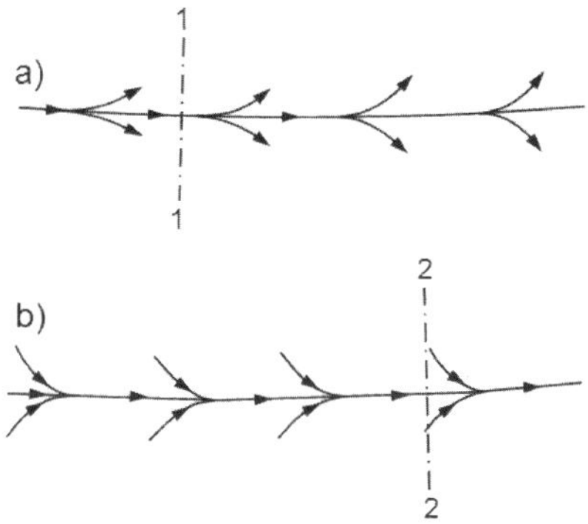

Fig. 7.9. Schematic of general singular lines of both inviscid and viscous flow: a) attachment line, b) squeeze-off separation/detachment line.

2. Points-of-inflection line On the convex sides of the singular lines shown in Fig. 7.9, the stream lines or the skin-friction lines have a point of inflection. A points-of-inflection line is present at every curved attachment or separation line. In general it lies close to the respective attachment or separation line. In the ISW situation it disappears.
3. Characteristic thickness of the viscous layer At an attachment line the characteristic thickness of the viscous layer $\triangle_c$ has a relative minimum in the direction orthogonal to it, 1 - · - 1 in Fig. 7.9 a). At an separation line $\triangle_c$ has a relative maximum in the direction orthogonal to it, 2 - · - 2 in Fig. 7.9 b).
4. Relative minimum of $|\tau_w|$ Along attachment and separation lines the skin-friction is non-zero. Normal to an viscous attachment line, 1 - · - 1 in Fig. 7.9 a), or to a separation line, 2 - · - 2 in Fig. 7.9 b), the absolute value $|\tau_w|$ of the skin-friction vector has a relative minimum. The minimum lies close to the respective line.
5. Extrema of the thermal state of the surface Along attachment and separation lines the temperature T_w and the heat flux in the gas at the wall q_{gw} are non-zero. Normal to an attachment line, 1 - · - 1 in Fig. 7.9 a), a relative maximum, and normal to a separation line, 2 - · - 2 in Fig. 7.9 b), a relative minimum exist. These extrema lie close to the respective line.

We note that at attachment lines the distances of the extremum lines or the points-of-inflection lines to the attachment lines as such in general are very small. The reason for this is that an attachment line usually is only weakly curved. In the ISW situation, both the p_{max}-line and the points-of-inflection line lie on it.

At separation lines, the situation is different. At squeeze-off separation lines the distances between the separation line and the extremum lines of the skin-friction vector and the thermal state of the surface are not necessarily very small. This holds also for points-of-inflection lines. The reason for this is that squeeze-off separation in general is not related directly to curvature maxima of the surface. This does not hold for flow-off separation lines at sharp edges, for instance wing trailing edges, Fig. 7.1 a), or sharp leading edges of highly swept wings, Fig. 7.1 b). There the distances can be very small.[12]

In the following sub-sections we give proofs of the listed properties. The results in principle are valid for incompressible and compressible, laminar and turbulent flow. However, for convenience, we sometimes simplify the presentation. We do not treat detachment lines as separate topic. What applies for them of the above five items is more or less self-evident.

Illustrative results are discussed here, other results can be found in Chapter 10. In many of the examples the coordinate notation is like that shown in Fig. 2.6 for general surface-oriented non-orthogonal curvilinear coordinates. Then the surface-tangential coordinates are x^1 and x^2 instead of x and z, the surface-normal coordinate is x^3 instead of y.

We treat first attachment lines and then separation/detachment lines. The results hold for both the respective lines which begin at singular points and the open-type lines.

7.4.2 Attachment Lines

Examples of primary attachment lines are given in Fig. 7.6 and in Fig. 7.8. For inviscid flow and for viscous flow they have the same form. In the sketch shown in Fig. 7.8 the straight attachment line lies in the lower symmetry line of the body. The two primary attachment lines shown in Fig. 7.6 may be straight or slightly curved.

Fig. 7.10 shows the computed inviscid surface streamlines around the leading-edge attachment line of a swept wing at large angle of attack. In this case the attachment line lies at a small distance away from the leading edge at the lower side of the wing. The line is slightly curved. This is the general situation at the leading edge of a lifting wing.

Consider now the surface element in Fig. 7.11 with the attachment line and the origin point P_0 on it. The general surface-oriented non-orthogonal curvilinear coordinate system is the ξ^i-system. In P_0 originate the surface-oriented locally orthogonal system $\overline{x}^i$ and the local Cartesian system x^i both with their axes tangential to the surface and rectilinear to it in P_0. These two systems are locally aligned with the attachment line.

The following considerations are made for convenience in the Cartesian system x^i. This is based on the assumptions that both the attachment line and the body surface are only slightly curved and that the interesting phenomena lie in the immediate neighborhood of the attachment line. We change to the

[12] For a detailed discussion of attachment lines see [33].

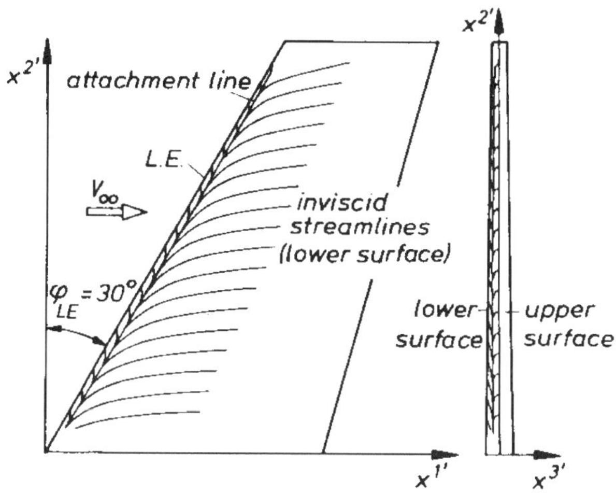

Fig. 7.10. Pattern of the inviscid surface streamlines at the leading-edge attachment line of the ONERA M6 wing [33]. Incompressible flow, $\alpha = 15°$, panel method. View towards the lower side of the wing.

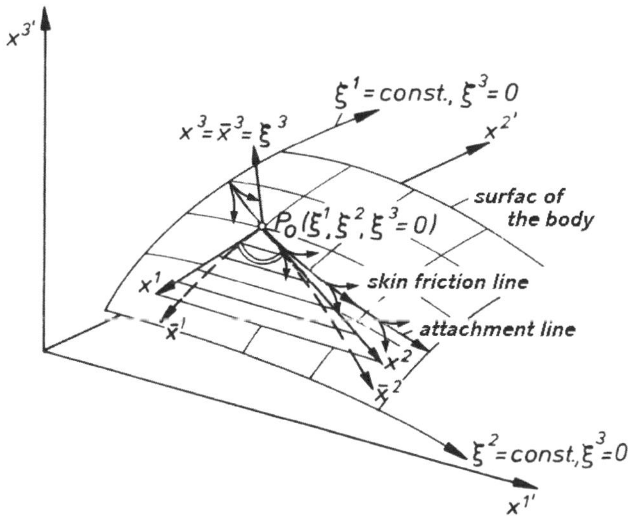

Fig. 7.11. Schematic of coordinate systems at an attachment line: general surface-oriented non-orthogonal curvilinear system ξ^i (i = 1,2,3). With their origins in point P_0 on the attachment line: the surface-oriented locally orthogonal system $\overline{x}^i$, and the local Cartesian system x^i, tangential to the surface in P_0.

nomenclature used throughout this chapter with $x^2 \to x$, $x^1 \to z$, $x^3 \to y$, Fig. 7.12. The velocity components are accordingly u and w in x- and z-direction, and v normal to the surface in y-direction.

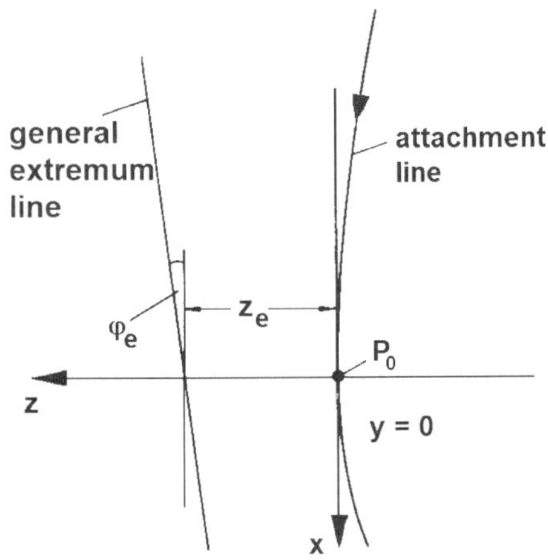

Fig. 7.12. The local Cartesian coordinate system from Fig. 7.11 with the attachment line and a general relative extremum line in its vicinity. View from above towards the surface $x^3 = 0$, and change of coordinate nomenclature: $x^1 \to z$, $x^2 \to x$, $x^3 \to y$.

In Fig. 7.12 also a line designated with "general extremum line" is shown. This line is the place-holder for the relative extremum lines, the p_{max}-line, the $|\tau_w|_{min}$-line etc. of the above five items which are discussed in this section. The extrema lie in direction normal, or approximately normal to the attachment line. These extrema are relative extrema. The absolute pressure maximum of a primary attachment line, for instance, lies in the primary attachment point.

We consider first, in terms of the listed items, the attachment line of inviscid flow and then that of viscous flow.

Attachment Lines, Inviscid Flow, Item 1: Relative Maximum of the Surface Pressure. We assume that the attachment line begins at a singular (primary attachment) point which must not necessarily be so, see above. We look first at that point. There the surface pressure has a maximun which is an absolute extremum. The relation for the stagnation-point pressure coefficient $c_{p_{stag}}$ for incompressible flow is with $p_{stag} = p_{t_\infty} = p_\infty + q_\infty$:

$$c_{p_{stag}} = \frac{p_{stag} - p_\infty}{q_\infty} = 1. \tag{7.28}$$

Here p_∞ is the free-stream pressure and $q_\infty = \rho_\infty v_\infty^2/2$ the dynamic pressure.

At a stagnation point in compressible subsonic flow with isentropic compression we have

$$c_{p_{stag}} = \frac{2}{\gamma M_\infty^2}\left[\left(1 + \frac{\gamma - 1}{2} M_\infty^2\right)^{\gamma/(\gamma-1)} - 1\right]. \tag{7.29}$$

This relation holds for $0 < M_\infty < 1$ and perfect gas. We note that $c_{p_{stag}}$ for compressible flow is larger than that for incompressible flow with $c_{p_{stag}} = 1$. At higher flight Mach numbers, a bow shock ahead of the body leads to a total-pressure loss. That and high-temperature real-gas effects then must be taken into account [31].

At the attachment line the pressure drops from the absolute maximum at the primary attachment point to smaller values. In direction normal to that line the relative maximum exists. To study this, we perform a Taylor expansion of the surface-tangential inviscid velocity components u_e and w_e around the origin point P_0 in Fig. 7.12. The general extremum line in that figure is now the p_{max}-line, z_e and φ_e now read z_M and φ_M.

In P_0 on the curved attachment line we observe the following flow properties. The velocity component u_e, tangential to the attachment line, is positive, the component w_e, normal to it, is zero. Of the first derivatives of the velocity components, $\partial w_e/\partial z$ is positive, $\partial u_e/\partial z$ is negative. $\partial u_e/\partial x$ and $\partial w_e/\partial x$ are smaller than the two z-derivatives and in general non-zero. In the ISW situation (x-direction), for instance, they would be identically zero. The pressure has a relative maximum in the vicinity of or at the attachment line (ISW). We assume that all derivatives, $\partial w_e/\partial z$ etc., are constant in the considered domain.

The expansion yields for a point $P \neq P_0$ in the plane $y = 0$:

$$u_e(x, z, y = 0) = [u_e]_{P_0} + \left[\frac{\partial u_e}{\partial x}\right]_{P_0} x + \left[\frac{\partial u_e}{\partial z}\right]_{P_0} z + ..., \tag{7.30}$$

and

$$w_e(x, z, y = 0) = \left[\frac{\partial w_e}{\partial x}\right]_{P_0} x + \left[\frac{\partial w_e}{\partial z}\right]_{P_0} z + ... \tag{7.31}$$

The momentum equation for the z-direction reads in the plane $y = 0$ (compare with eq. (4.16)):

$$\rho_e u_e \frac{\partial w_e}{\partial x} + \rho_e w_e \frac{\partial w_e}{\partial z} = -\frac{\partial p}{\partial z}. \tag{7.32}$$

We introduce eqs. (7.30) and (7.31) into this equation. Neglecting quadratic terms and second-order derivatives we obtain:

$$\left[(\frac{\partial w_e}{\partial z})^2 + \frac{\partial u_e}{\partial z} \frac{\partial w_e}{\partial x} \right]_{P_0} z + \left[\frac{\partial w_e}{\partial x} (\frac{\partial u_e}{\partial x} + \frac{\partial w_e}{\partial z}) \right]_{P_0} x = -\frac{1}{\rho_e} \frac{\partial p}{\partial z} - \left[u_e \frac{\partial w_e}{\partial x} \right]_{P_0}. \tag{7.33}$$

Because only linear terms are kept, this equation holds only in the immediate vicinity of the attachment line.

From eq. (7.33) we find with $\partial p/\partial z = 0$ the point P on the p_{max}-line:

$$z\,(\frac{\partial p}{\partial z} = 0) = z_M + x\, tan\, \varphi_M, \tag{7.34}$$

with

$$z_M = z(x = 0) = -\frac{u_e \frac{\partial w_e}{\partial x}}{(\frac{\partial w_e}{\partial z})^2 + \frac{\partial u_e}{\partial z} \frac{\partial w_e}{\partial x}} |_{P_0} \tag{7.35}$$

and

$$tan\, \varphi_M = \frac{dz}{dx} = -\frac{\frac{\partial w_e}{\partial x} (\frac{\partial u_e}{\partial x} + \frac{\partial w_e}{\partial z})}{(\frac{\partial w_e}{\partial z})^2 + \frac{\partial u_e}{\partial z} \frac{\partial w_e}{\partial x}} |_{P_0}. \tag{7.36}$$

This procedure can be repeated at any point on the curved attachment line in order to find the p_{max}-line. The relations show that the relative pressure maximum at a curved attachment line in general lies close to it, but not on it, because $\partial w_e/\partial x \neq 0$. On which side of the attachment line the maximum lies, depends on the sign of it and on the sign of the denominator. That, however, in general is positive, because $\partial w_e/\partial z$ usually is large compared to the other derivatives.

In the ISW situation the pressure maximum lies on the attachment line (x-direction). In that case $\partial/\partial x = 0$, and hence $z_M = 0$ and also φ_M. This also means that $\partial p/\partial x = 0$ and further

$$P_0: \quad p = p_{max}, \quad \frac{\partial p}{\partial z} = 0. \tag{7.37}$$

The condition that for $\partial p/\partial z = 0$ a maximum is given, is $\partial^2 p/\partial z^2 < 0$. We differentiate eq. (7.32) with respect to z and obtain:

$$\frac{\partial^2 p}{\partial z^2} = -\rho_e \{ (\frac{\partial w_e}{\partial z})^2 + \frac{\partial u_e}{\partial z} \frac{\partial w_e}{\partial x} \}_{P_0} < 0. \tag{7.38}$$

At the attachment line $\partial w_e/\partial z$ is large and we can assume $(\partial w_e/\partial z)^2 > \partial u_e/\partial z \cdot \partial w_e/\partial x$. The pressure therefore indeed has a relative maximum at a location close to the curved attachment line or for the ISW on the attachment line.

The curvature radius of the attachment line in P_0 is found from the relation for the curve in the plane $y = 0$:

$$R = \frac{[1 + (\frac{dx}{dz})^2]^{3/2}}{\frac{d^2 x}{dz^2}}. \tag{7.39}$$

From the general relation

$$\frac{dx}{dz} = \frac{u_e}{w_e} \tag{7.40}$$

follows

$$\frac{d^2x}{dz^2} = (\frac{\partial}{\partial z} + \frac{dx}{dz}\frac{\partial}{\partial x})(\frac{u_e}{w_e}) = \frac{1}{w_e^3}[w_e^2\frac{\partial u_e}{\partial z} - u_e\, w_e(\frac{\partial w_e}{\partial z} - \frac{\partial u_e}{\partial x}) - u_e^2\frac{\partial w_e}{\partial x}] \tag{7.41}$$

and finally

$$\frac{1}{R} = \frac{w_e^2\frac{\partial u_e}{\partial z} - u_e\, w_e(\frac{\partial w_e}{\partial z} - \frac{\partial u_e}{\partial x}) - u_e^2\frac{\partial w_e}{\partial x}}{(u_e^2 + w_e^2)^{3/2}}. \tag{7.42}$$

At P_0, because of $w_e = 0$ there, we get:

$$\frac{1}{R_0} = -\frac{\frac{\partial w_e}{\partial x}}{u_e}|_{P_0}. \tag{7.43}$$

With this the right-hand side of eq. (7.33) can be written:

$$-\frac{1}{\rho_e}\frac{\partial p}{\partial z} + \frac{u_e^2}{R_0}|_{P_0} \tag{7.44}$$

and for $x = z = 0$ eq. (7.33) is reduced to this right-hand side, resulting in the well-known balance between centrifugal force and pressure force:

$$\frac{\partial p}{\partial z} = \rho_e\frac{u_e^2}{R_0}. \tag{7.45}$$

As example we discuss in Fig. 7.13 the wall pressure along the attachment line of the wing-like ellipsoid shown in Fig. 7.2.

The attachment line lies on the lower side of the wing. The absolute pressure maximum is found at the stagnation point with $c_p = 1$. Along the attachment line c_p is nearly constant up to about two thirds of the span, then a gradual drop (flow acceleration along the attachment line) is seen to smaller values. The relative p_{max}-line lies close to the attachment line. The distance is very small, it becomes larger only towards the wing tip [33].

In chord direction we see a strong acceleration (pressure drop) towards the upper side of the wing and a weaker one on the lower side. Note that this pressure distribution is the result of an exact theory [17]. The indication of separation regards the boundary-layer solution which was made with this pressure distribution. It was performed with a finite-difference scheme for three-dimensional incompressible and laminar flow [33]. Given the small leading edge radii of the wing, the results are valid for reference Reynolds numbers $Re_{ref} \gtrapprox O(10^6)$.

The corresponding viscous attachment line lies close to the inviscid one. Therefore the result for the latter applies also to it.

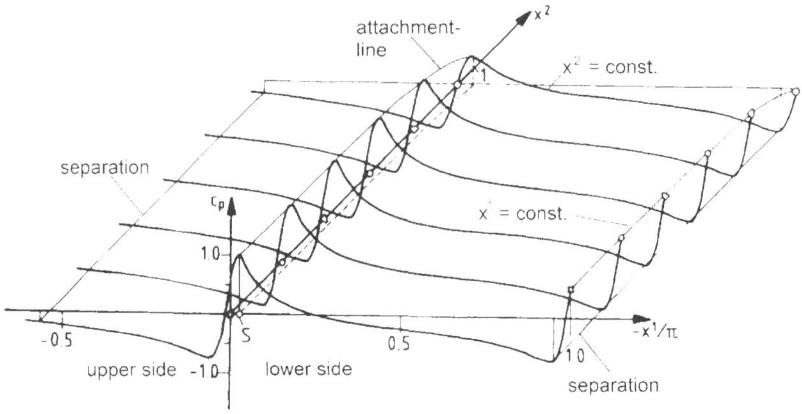

Fig. 7.13. Wing-like ellipsoid (Fig. 7.2) at $\alpha = 5°$, inviscid incompressible flow [33]. Wall pressure c_p in the surface-coordinate parameter map, view from above. Chord-wise coordinate: upper wing side $0 \leqq -x^1/\pi \leqq -0.5$, lower wing side $0 \leqq -x^1/\pi \leqq 1$; half-span coordinate: $0 \leqq x^2 \leqq 1$; S denotes the stagnation point.

Attachment Lines, Inviscid Flow, Item 2: Points-of-Inflection Line. If an attachment line is curved, the streamlines diverging from the convex side of it, Fig. 7.14, each have a point of inflection—zero curvature—which lies close to the attachment line [35]. Together they form a points-of-inflection line.

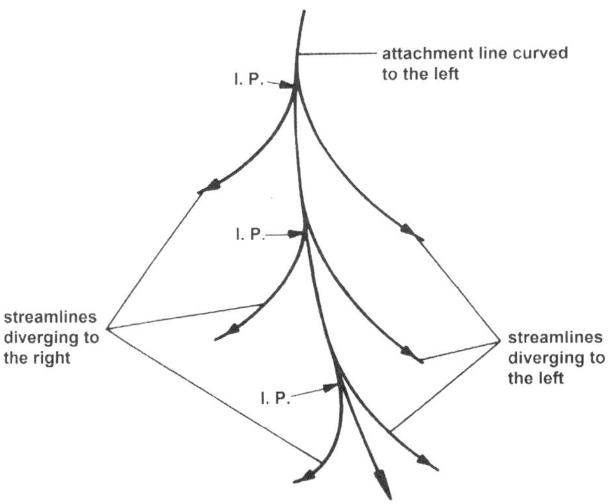

Fig. 7.14. Pattern of an inviscid attachment line and the diverging streamlines. The points of inflection of the streamlines which diverge to the right are marked with I. P..

The point of inflection is defined in the following with the condition $1/R = 0$. From eq. (7.42) we obtain

$$w_e^2 \frac{\partial u_e}{\partial z} - u_e\, w_e (\frac{\partial w_e}{\partial z} - \frac{\partial u_e}{\partial x}) - u_e^2 \frac{\partial w_e}{\partial x} = 0. \tag{7.46}$$

We proceed as with the p_{max}-line and look for the condition $1/R = 0$ at the point P in the neighborhood of P_0.[13] The general extremum line in Fig. 7.12 is now the points-of-inflection line (zero curvature), z_e and φ_e now read z_{ip} and φ_{ip}. With the Taylor expansions eqs. (7.30) and (7.31) introduced into eq. (7.46), we find:

$$z\,(\frac{1}{R} = 0) = z_{ip} + x\, tan\, \varphi_{ip}, \tag{7.47}$$

with

$$z_{ip} = z(x = 0) = - \frac{u_e \frac{\partial w_e}{\partial x}}{(\frac{\partial w_e}{\partial z})^2 - \frac{\partial u_e}{\partial x}\frac{\partial w_e}{\partial z} + 2\frac{\partial u_e}{\partial z}\frac{\partial w_e}{\partial x}}|_{P_0} \tag{7.48}$$

and

$$tan\, \varphi_{ip} = \frac{dz}{dx} = - \frac{\frac{\partial w_e}{\partial x}(\frac{\partial u_e}{\partial x} + \frac{\partial w_e}{\partial z})}{(\frac{\partial w_e}{\partial z})^2 - \frac{\partial u_e}{\partial x}\frac{\partial w_e}{\partial z} + 2\frac{\partial u_e}{\partial z}\frac{\partial w_e}{\partial x}}|_{P_0}. \tag{7.49}$$

This procedure again can be repeated at any point along the attachment line in order to find the points-of-inflection line, see also [35]. The result shows that the points-of-inflection line at a curved attachment line lies close to it. It obviously lies on the convex side. From this we can deduce that $\partial w_e/\partial x < 0$, if the denominator of eq. (7.48) is larger zero.

The points-of-inflection line only lies on the attachment line, if the ISW situation is given (x-direction). In this case $\partial/\partial x = 0$, and hence $z_{ip} = 0$ and also φ_{ip}, but the line itself has disappeared.

Attachment Lines, Viscous Flow, Item 3: Characteristic Thickness of the Viscous Layer. Attached viscous flow is characterized by a non-zero thickness throughout. This holds also for singular points, in particular for the forward (primary) stagnation point. This is not evident at the first look, but was proven a long time ago, see Section 8.1.

At a primary attachment line, originating from the primary stagnation point, first the flow is laminar, at some location then it becomes turbulent. A special case is given at the leading edge of a swept wing. There the leading-edge flow may be laminar or partly or fully turbulent due to leading edge/attachment-line contamination, Sub-Section 9.3.2, page 215 f.

Away from the stagnation point, the thickness of the viscous layer increases. If the flow is not too strongly three-dimensional, we can define for

[13] Also now only linear terms are kept. In [34] and [35] quadratic terms are regarded, too. That leads to a different formulation.

qualitative considerations, but also for approximate quantitative investigations, a characteristic thickness $\triangle_c$. It governs the wall shear stress and the heat flux in the gas at the wall [31]. Away from singular points and lines in laminar flow this thickness is the boundary-layer thickness (typically defined as 99 per cent thickness [36]), $\triangle_c = \delta_{lam}$, in turbulent flow it is the thickness of the viscous sub-layer, $\triangle_c = \delta_{vs}$, Appendix B.3.1. With the help of the reference-temperature concept, Mach number and wall-temperature influence can be taken into account, Appendix B.2.

From results of computations of three-dimensional boundary layers by means of integral methods it was observed in the early 1980s that ahead of and at separation lines a typical bulging—in terms of the δ (boundary-layer thickness) and the δ_1 (displacement thickness) surfaces—of the boundary layer contour occurs [37, 38]. This is seen in particular in cases of open type separation. The effect is due to the convergence of the two boundary-layers streams which squeeze each other off the surface. The convergence of the skin-friction lines is the visual indicator.

At attachment lines the opposite effect was observed, an indentation of the boundary layer contour [6]. It is due to the divergence of the flow there, visually indicated by the pattern of the skin-friction lines.

These observations later allowed an explanation of hot-spot and cold-spot situations in high-speed aerodynamics [39]. Along attachment lines the heat flux in the gas at the wall in lateral direction has a relative maximum on or near that line, see below. A direct proof of that phenomenon is not known, but it can be presumed that it is due to a relative minimum of the characteristic thickness occurring in direction normal to the attachment line, Fig. 7.15. At separation lines, an opposite phenomenon is observed, see next sub-section.

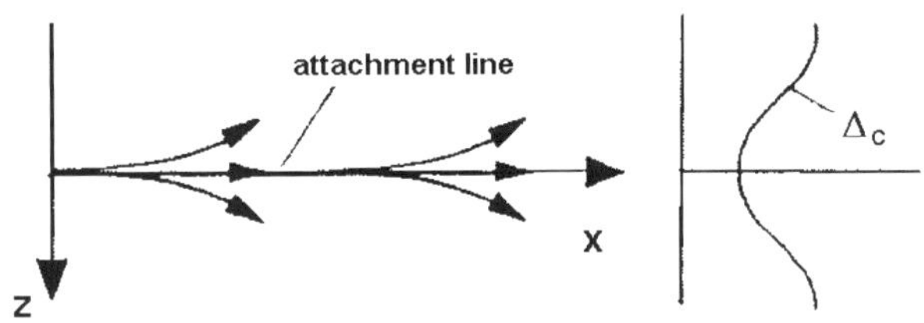

Fig. 7.15. Pattern of skin-friction lines at an attachment line and the lateral minimum of the characteristic boundary-layer thickness $\triangle_c$ (schematically) [39].

The minimum of $\triangle_c$ can be explained with an argument similar to that of Lighthill to explain why the skin-friction lines must leave the body surface in the vicinity of a separation line [18].

We follow the line of discussion given by Peake and Tobak [1]. We assume a plane surface ($y = 0$), Fig. 7.16. The attachment line lies along the x-axis.

The first hatched frame in Fig. 7.16 is assumed to have small width w and height h. The mass flow m through it then can be approximated by

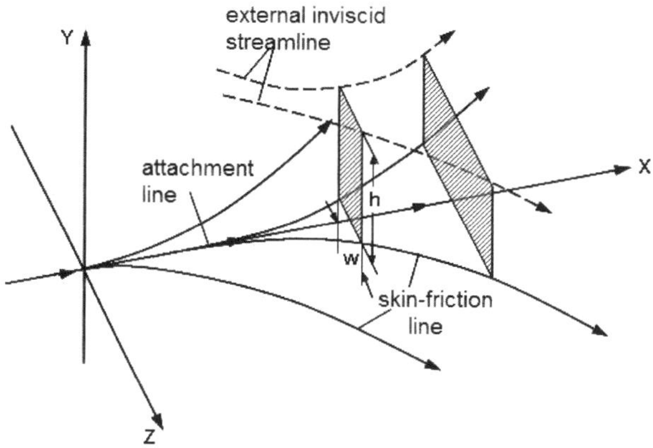

Fig. 7.16. Flow approaching an attachment line: inviscid streamlines and skin-friction lines.

$$m \approx \rho \frac{\overline{u}_e}{2} w\, h, \tag{7.50}$$

where ρ is the density and $\overline{u}_e$ the coplanar mean velocity with a linear profile.

The resultant mean skin friction $\overline{\tau}_w$ is

$$\overline{\tau}_w \approx \mu \frac{\overline{u}_e}{h/2}. \tag{7.51}$$

From this we obtain with $\nu = \mu/\rho$

$$m \approx \frac{h^2 w \tau_w}{4\nu} \tag{7.52}$$

and get finally for m = constant the proportionality

$$h \propto \left(\frac{\nu}{w\overline{\tau}_w}\right)^2. \tag{7.53}$$

In the second hatched frame hence the height h has decreased, because there the width w has increased and also $\overline{\tau}_w$. The width w increases because of the divergence of both the external inviscid streamlines and the skin-friction lines. $\overline{\tau}_w$ increases, because the lateral range with $|\tau_w| > |\tau_w|_{attachment\ line}$ increases. The result is that the characteristic boundary-layer thickness $\triangle_c \propto h$ decreases as the flow approaches the surface and diverges to the two sides of the attachment line, with the boundary-layer thicknesses increasing. At the attachment line $\triangle_c$ has a relative minimum normal to it, i.e. in z-direction.

As an example we discuss in Fig. 7.17 the distribution of the dimensionless three-dimensional displacement thickness δ_1 along the attachment line of the wing-like ellipsoid, Fig. 7.2. It was obtained with the finite-difference method of D. Schwamborn, [33], discussed in conjunction with Fig. 7.13. Because the

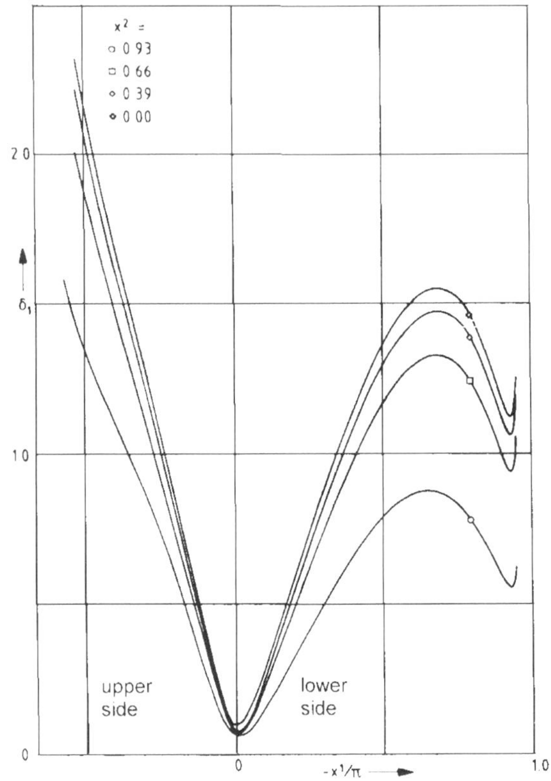

Fig. 7.17. Wing-like ellipsoid (Fig. 7.2) at $\alpha = 5°$, incompressible laminar flow [33]. Indentation of the boundary layer contour along the attachment line in terms of the dimensionless displacement thickness δ_1 as function of x^1 at four span-wise locations x^2. For the coordinates see Fig. 7.13.

flow is laminar, the characteristic thickness $\triangle_c$ would be the boundary-layer thickness δ which, however, is not available from [33].

The three-dimensional displacement thickness δ_1, Section 5.2,[14] is non-dimensionalized with the reference length $\overline{L}_{ref}$—the overline denotes dimensional quantities—and stretched with the square root of the reference Reynolds number $Re_{ref} = \overline{\rho}_{ref}\overline{u}_{ref}\overline{L}_{ref}/\overline{\mu}_{ref}$: $\delta_1 = \overline{\delta}_1 Re_{ref}/\overline{L}_{ref}$.

At the forward stagnation point, $x^2 = 0$, the displacement thickness is positive. It has there its absolute minimum.[15] Along the attachment line δ_1 then decreases first a little and after that stays nearly constant. The relative minimum across the attachment line—the indentation of the δ_1 contour—is

[14] The indentation is a property of the three-dimensional displacement thickness δ_1. It can not be found considering only the components $\delta_{1_{x^1}}$ and $\delta_{1_{x^2}}$.

[15] The figure is somewhat misleading. The minimum should lie at a small distance to the right of $-x^1/\pi = 0$.

well demonstrated for all locations x^2. If we accept δ_1 as the proxy of $\triangle_c$, Fig. 7.17 demonstrates the observed phenomenon and the argument similar to Lighthill's argument.

Attachment Lines, Viscous Flow, Item 4: Relative Minimum of $|\tau_w|$. At the attachment line on the body surface the pattern of the skin-friction lines is similar to the pattern of the surface streamlines of the inviscid flow. In our case both attachment lines have their origin in the (forward) stagnation point, where the external inviscid flow velocity and the skin friction are zero, Section 8.1. Along the attachment line the skin friction is non-zero. An approximate analytic tool for the attachment line on a swept cylinder is, for instance, the generalized reference-temperature formulation, Appendix B.3.3.

Both attachment lines are curved in the same sense, the viscous one stronger than the inviscid one (see in this regard the discussion on page 27). As long as the curvature is small, the distance between the two attachment lines is small. This also holds for the respective streamlines and skin-friction lines which diverge from the attachment lines.

The skin friction has a relative minimum in direction normal or approximately normal to the attachment line. The reason for that is that in the diverging flow, which is accelerated away from the attachment line, the skin friction increases.

The investigation follows now the same line as for the inviscid case with the coordinate system defined in Fig. 7.12. The general extremum line in that figure is now the line, on which the absolute value of the skin-friction vector $|\tau_w|$ has a relative minimum in direction normal to the attachment line. Now z_e and φ_e read z_m and φ_m.

However, instead of the surface-tangential velocity components u_e and w_e in the origin point P_0, we consider now the wall shear-stress components τ_x and τ_z, as defined in the eqs. (4.18) and (4.19).

We observe first the shear-stress components and their derivatives in point P_0 of the curved viscous attachment line. By definition the tangential component τ_x is positive, the normal component τ_z is zero. Of the first derivatives of the shear-stress components, $\partial\tau_z/\partial z$ is positive, $\partial\tau_x/\partial z$ is negative. $\partial\tau_x/\partial x$ and $\partial\tau_z/\partial x$ are smaller than the two z-derivatives and in general non-zero. In the ISW situation they would be identical zero. The absolute value of the skin-friction vector has a relative minimum in the vicinity of the attachment line. We assume that the derivatives are constant in the considered domain.

The shear-stress components in P are found by a Taylor expansion around the origin point P_0:

$$\tau_x(x, z, y = 0) = [\tau_x]_{P_0} + \left[\frac{\partial \tau_x}{\partial x}\right]_{P_0} x + \left[\frac{\partial \tau_x}{\partial z}\right]_{P_0} z + ..., \tag{7.54}$$

and

$$\tau_z(x, z, y = 0) = \left[\frac{\partial \tau_z}{\partial x}\right]_{P_0} x + \left[\frac{\partial \tau_z}{\partial z}\right]_{P_0} z + ... \tag{7.55}$$

The minimum of the wall shear stress with respect to the attachment line is defined by

$$\frac{\partial |\tau_w|}{\partial z} = \frac{\tau_x \frac{\partial \tau_x}{\partial z} + \tau_z \frac{\partial \tau_z}{\partial z}}{|\tau_w|} = 0. \tag{7.56}$$

Introducing eqs. (7.54) and (7.55) into this equation and neglecting second-order terms yields the point P on the $|\tau|_{min}$-line:

$$z\,(\frac{\partial |\tau_w|}{\partial z} = 0) = z_m + x \tan \varphi_m, \tag{7.57}$$

with

$$z_m = z(x = 0) = -\frac{\tau_x \frac{\partial \tau_x}{\partial z}}{(\frac{\partial \tau_x}{\partial z})^2 + (\frac{\partial \tau_z}{\partial z})^2}|_{P_0} \tag{7.58}$$

and

$$\tan \varphi_m = \frac{dz}{dx} = -\frac{\frac{\partial \tau_x}{\partial x}\frac{\partial \tau_x}{\partial z} + \frac{\partial \tau_z}{\partial x}\frac{\partial \tau_z}{\partial z}}{(\frac{\partial \tau_x}{\partial z})^2 + (\frac{\partial \tau_z}{\partial z})^2}|_{P_0}. \tag{7.59}$$

This procedure can be repeated at any point on the attachment line in order to find the $|\tau_w|_{min}$-line.

We summarize:

- The relative minimum of the absolute value of the skin-friction vector, the $|\tau_w|_{min}$-line, lies close to the attachment line, but not on it, because $\partial \tau_x / \partial z \neq 0$.
- The $|\tau_w|_{min}$-line is not a skin-friction line.
- On which side of the attachment line the minimum-line lies, depends on the sign of the product $\tau_x \, \partial \tau_x / \partial z$.
- At the attachment line of an ISW (x-direction) with a symmetric airfoil at zero angle of attack, we have $\partial \tau_x / \partial z|_{P_0} = 0$ and also $\partial / \partial x = 0$. In this case $z_m = 0$, $\varphi_z = 0$ and the $|\tau_w|_{min}$-line lies on the attachment line.

In two-dimensional flow and plane-of-symmetry flow in z-direction we have $\tau_x = 0$, $\partial / \partial x = 0$, but $\partial \tau_z / \partial z \neq 0$. Therefore $z_m = 0$, $\varphi_m = 0$, and attachment point and $|\tau_w|_{min}$ ($\tau_z = 0$ in P_0) fall together.

We do not show that the extremum indeed is a minimum. For the proof the reader is referred to [34].

As example we discuss in Fig. 7.18—like in Fig. 7.17 the displacement thickness δ_1—the absolute value of the dimensionless wall shear stress $|\tau_w|$

along the attachment line of the wing-like ellipsoid shown in Fig. 7.2. $|\tau_w|$ is non-dimensionalized in the following way: $|\tau_w| = |\overline{\tau}_w|\sqrt{Re_{ref}}/[\overline{\rho}_{ref}(\overline{u}_{ref})^2]$ [33].

The absolute minimum is found at the stagnation point with $|\tau_w| = 0$. Along the attachment line $|\tau_w|$ is non-zero and nearly constant up to about two thirds of the span. Then a gradual increase is seen which is due to the acceleration of the flow (pressure in tip direction. Also the relative $|\tau_w|_{min}$-line lies close to the attachment line. The distance again is very small, it becomes larger only towards the wing tip [33].

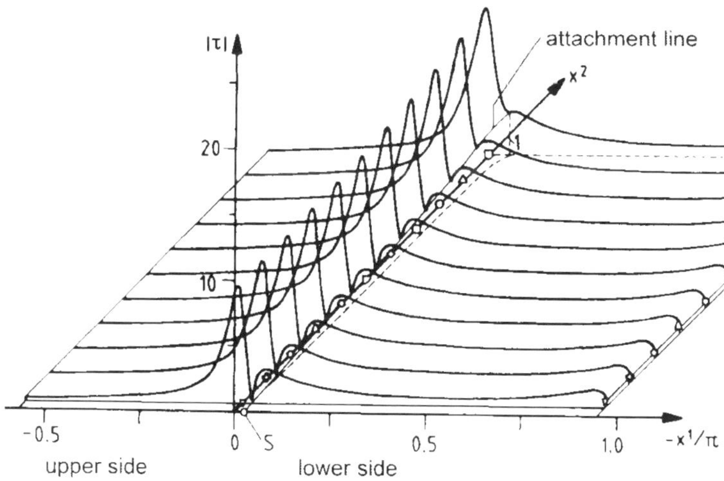

Fig. 7.18. Wing-like ellipsoid (Fig. 7.2) at $\alpha = 5°$, incompressible laminar flow [33]. Dimensionless wall shear stress $|\tau|$ ($\equiv |\tau_w|$) in the surface-coordinate parameter map, view from above. For the coordinates see Fig. 7.13.

In chord direction we see a strong rise of $|\tau_w|$—due to the acceleration (pressure drop) in that direction and despite the rise of the characteristic thickness in the same direction—towards the upper side of the wing and a weaker one on the lower side.

Attachment lines, viscous flow, item 2: Points-of-inflection line The skin-friction lines leaving the viscous attachment line on its convex side exhibit a point of inflection close to the attachment line, too. The picture is similar to that one for the inviscid streamlines, Fig. 7.14. We keep again only linear terms. z_e and φ_e now read z_{ipv} and φ_{ipv}.

The curvature radius of the viscous attachment line in P_0 is found now to

$$\frac{1}{R} = \frac{\tau_z^2\frac{\partial \tau_x}{\partial z} - \tau_x\,\tau_z\left(\frac{\partial \tau_z}{\partial z} - \frac{\partial \tau_x}{\partial x}\right) - \tau_x^2\frac{\partial \tau_z}{\partial x}}{|\tau|^3}\Big|_{P_0}. \tag{7.60}$$

At P_0, because of $\tau_z = 0$ there, we get

$$\frac{1}{R_0} = -\frac{\frac{\partial \tau_z}{\partial x}}{\tau_x}|_{P_0}. \tag{7.61}$$

We proceed as in the inviscid case and find:

$$z\,(\frac{1}{R} = 0) = z_{ipv} + x\,tan\,\varphi_{ipv}, \tag{7.62}$$

with

$$z_{ipv} = z(x = 0) = -\frac{\tau_x \frac{\partial \tau_z}{\partial x}}{(\frac{\partial \tau_z}{\partial z})^2 - \frac{\partial \tau_x}{\partial x}\frac{\partial \tau_z}{\partial z} + 2\frac{\partial \tau_x}{\partial z}\frac{\partial \tau_z}{\partial x}}|_{P_0} \tag{7.63}$$

and

$$tan\,\varphi_{ipv} = \frac{dz}{dx} = -\frac{\frac{\partial \tau_z}{\partial x}(\frac{\partial \tau_x}{\partial x} + \frac{\partial \tau_z}{\partial z})}{(\frac{\partial \tau_z}{\partial z})^2 - \frac{\partial \tau_x}{\partial x}\frac{\partial \tau_z}{\partial z} + 2\frac{\partial \tau_x}{\partial z}\frac{\partial \tau_z}{\partial x}}|_{P_0}. \tag{7.64}$$

We do not discuss these relations. The line of argumentation goes like the one which we had for the points-of-inflection line in the inviscid case.

Attachment Lines, Viscous Flow, Item 5: Extrema of the Thermal State of the Surface The thermal state of the surface encompasses both the wall temperature T_w and the heat flux in the gas at the wall q_{gw}.[16] The observation is that at a forward stagnation point q_{gw} and T_w have absolute maxima. At attachment lines, we see a relative maximum in direction normal to it.

For the stagnation point, several theories exist which explain that observation. They were developed in view of the mentioned high-speed vehicles, see, e.g., [31]. For the attachment line at a swept cylinder, like for the stagnation point, the generalized reference-enthalpy formulations are suitable approximate analytic tools, Appendix B.3.3.

To study in detail the relative maximum normal to the attachment line, we have no theory at hand. We instead discuss the classical approximation for the heat flux in the gas at the wall

$$q_{gw} \sim k_w \frac{T_r - T_w}{\triangle_c}, \tag{7.65}$$

where k_w is the thermal conductivity of the gas at the wall, T_r the recovery temperature, T_w the wall temperature, and $\triangle_c$ the characteristic thickness of the viscous layer.

[16] Usually on speaks simply of the wall heat flux q_w. We use here the more general term, because we have in the background also attached viscous flow past vehicles flying with supersonic and hypersonic speed. At flight Mach numbers higher than $M_\infty \approx 3$, the outer surfaces of such vehicles are radiation cooled. Then q_{gw} is the heat flux in the gas, whereas $q_w \neq q_{gw}$ is the actual heat flux into or out of the wall material [39]. If radiation cooling is absent, we have simply $q_{gw} = q_w$.

Above we have seen that $\triangle_c$ has a relative minimum at the attachment line. Therefore we can deduce that q_{gw} has a relative maximum there.[17] This argument holds also for the forward stagnation point, where q_{gw} has an absolute maximum.

In the general case, the wall temperature T_w will attain a maximum, too, see the hypersonic speed example in Section 10.4. There it is also shown that a minimum of the thermal state of the surface in direction normal to a separation line occurs, as is discussed in the next sub-section. In analogy to the skin-friction situation, one can expect that the extremum line lies close to the actual attachment line.

7.4.3 Separation Lines

Like in the case of attachment lines, we consider now the surface element in Fig. 7.19 with the separation line and the origin point P_0 on it. Again the general surface-oriented non-orthogonal curvilinear coordinate system is the ξ^i-system. In P_0 originate the surface-oriented locally orthogonal system $\overline{x}^i$ and the local Cartesian system x^i with their axes tangential to the surface and rectilinear to it in P_0. The latter systems are locally aligned with the attachment line.

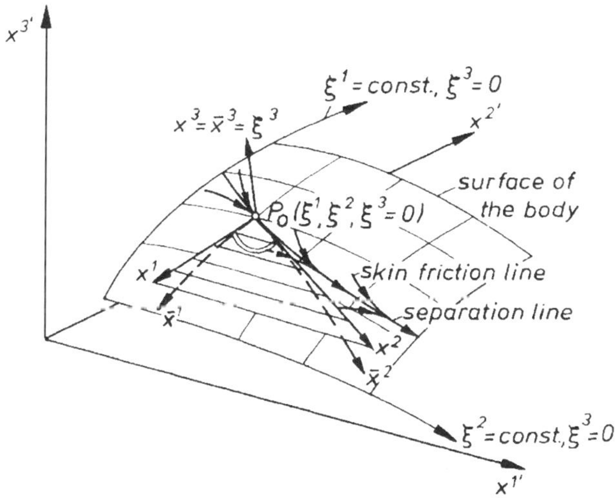

Fig. 7.19. Schematic of coordinate systems at a separation line [35]: general surface-oriented non-orthogonal curvilinear system ξ^i (i = 1,2,3). With their origins in point P_0 on the separation line: the surface-oriented locally orthogonal system $\overline{x}^i$, and the local Cartesian system x^i, tangential to the surface in P_0.

[17] A deduction in this way for the absolute value of the skin-friction vector is not possible, because we cannot make a simple guess for the behavior of the absolute value of the external inviscid velocity vector. With q_{gw} it is no problem, because $T_r - T_w$ can be taken as approximately constant in the considered domain.

For the separation line the investigations are made for convenience also in the Cartesian system shown in Fig. 7.12. Again the assumptions hold that both the separation line and the body surface are only slightly curved and that the interesting phenomena lie in the immediate neighborhood of the separation line. The used nomenclature then is the same as for the attachment lines. We consider the listed items in Sub-Section 7.4.1, as far as they apply for attachment lines.

Separation Lines, Item 3: Characteristic Thickness of the Viscous Layer. As discussed for attachment lines, ahead of and at separation lines, the characteristic thickness has a $\triangle_c$-maximum across the line, Fig. 7.20.

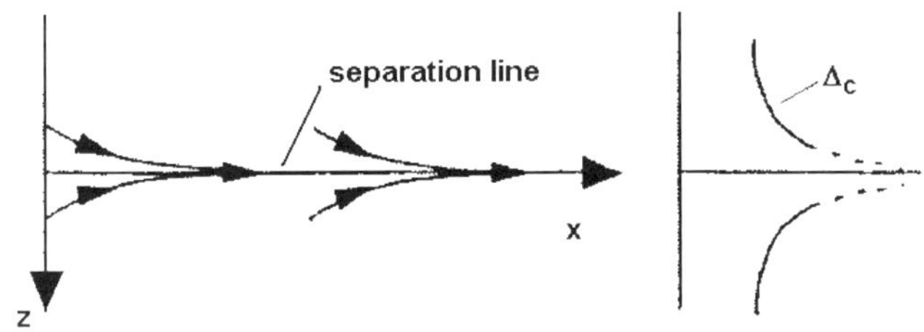

Fig. 7.20. Pattern of skin-friction lines at a separation line and the lateral maximum of the characteristic boundary-layer thickness $\triangle_c$ (schematically) [39].

$\triangle_c$, however, cannot be expressed in terms of the boundary-layer properties δ_{lam} or δ_{vs}. In the case of open type separation, the convergence of the skin-friction lines happens already upstream of the separation line as such. Then, and this also holds for the boundary layer at a detachment line, $\triangle_c$ may be expressed in terms of the above boundary-layer properties.

The relative maximum of $\triangle_c$ can be explained with the argument of Lighthill. He did show, why the skin-friction lines must leave the body surface in the vicinity of a separation line [18].

We do not give the derivation here. The reader is asked to redraw—maybe only in his imagination—Fig. 7.16, such that the situation at a separation line is described. Then the discussion is like above for the characteristic thickness of an attachment line. Also the work of Peake and Tobak, [1], can be consulted. The result finally is that the characteristic boundary-layer thickness $\triangle_c \propto h$ increases towards the separation line. There it has a maximum normal to it.

As example we discuss in Figs. 7.21 and 7.22 the incompressible laminar flow past a 1:6 ellipsoid at an angle of attack. Given are the patterns of the inviscid wall streamlines, the skin-friction lines, the primary separation line, and the point-wise plotted $|\tau_w|$-minimum line in the surface-coordinate parameter map. Shown in the second figure are the displacement thickness δ_1 and the energy-loss thickness δ_{3t} (disregarded in the following discussion) at locations x^1 = constant. The data were found with the three-dimensional boundary-layer integral method of the second author of this book [40]. The method was extended by B. Aupoix and used for the study in the MBB-Version [41]. Because the flow is laminar, the characteristic thickness $\triangle_c$ is the boundary-layer thickness δ.

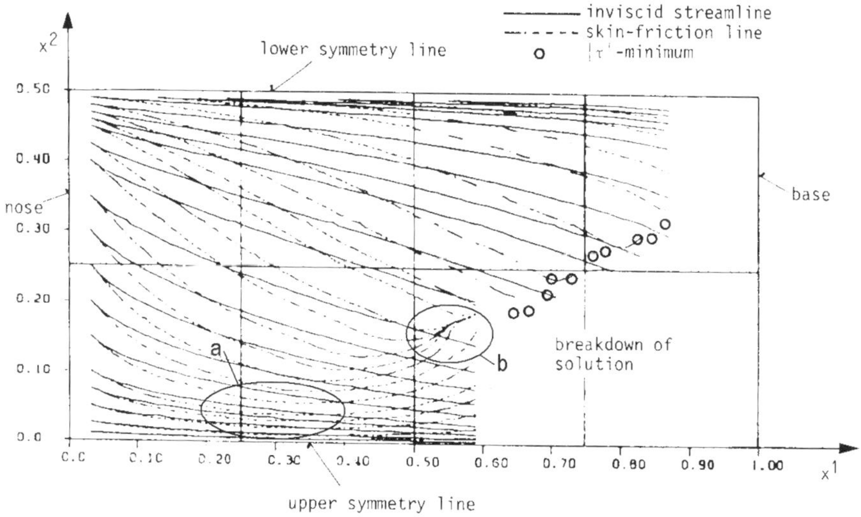

Fig. 7.21. Incompressible laminar flow past a 1:6 ellipsoid at $\alpha = 5°$ angle of attack, $L = 2.4$ m, $Re = 7.2 \cdot 10^6$ [38]. Surface-coordinate parameter map (right half of ellipsoid): pattern of inviscid wall streamlines, skin-friction lines, and point-wise $|\tau_w|$-minimum location.

We see a typical open-type separation pattern in Fig. 7.21. At $x^1 \approx 0.3$ the skin-friction lines are turning away from the upper symmetry line[18] (location a in the lower part of the parameter map). Then the two boundary-layer streams from the lower and the upper side converge toward each other at $x^1 \approx 0.5$, $x^2 \approx 0.15$ (location b). Finally, along the separation line (indicated by the breakdown of the solution), they squeeze each other off the surface.

The contours of the three-dimensional displacement thickness δ_1, Fig. 7.22, show for $0.31 \lessapprox x^1 \lessapprox 0.51$ the typical bulging ahead of the separation location. For $x^1 \gtrapprox 0.55$ the boundary-layer solution begins to break down in the separation region above the separation line. Below it, it is sustained up to the immediate vicinity of the separation line.

If we accept δ_1 as proxy of $\triangle_c$, Fig. 7.22 demonstrates the crosswise maximum of the latter. Of course, the boundary-layer solution does not permit to determine the maximum of $\triangle_c$ along the separation line.

Separation Lines, Item 4: Relative Minimum of $|\tau_w|$. In order to find the $|\tau|_{min}$-line, we proceed in exactly the same way as for item 4 of

[18] Note that the external inviscid streamlines still converge to that line.

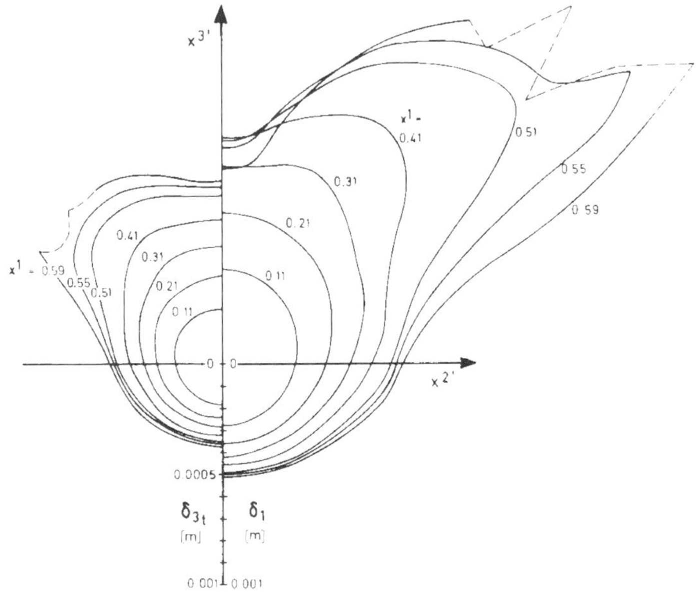

Fig. 7.22. Incompressible laminar flow past a 1:6 ellipsoid at $\alpha = 5°$ angle of attack, $L = 2.4$ m, $Re = 7.2 \cdot 10^6$ [38]. Displacement thickness δ_1 and energy-loss thickness δ_{3t} at locations x^1 = constant.

attachment lines.[19] We do not repeat the derivation, however—for a more convenient discussion—show again the result. In point P on the general extremum line of Fig. 7.12, now thought as $|\tau_w|_{min}$-line, we have:

$$z\,(\frac{\partial|\tau_w|}{\partial z} = 0) = z_m + x\,tan\,\varphi_m, \tag{7.66}$$

with

$$z_m = z(x=0) = -\frac{\tau_x \frac{\partial \tau_x}{\partial z}}{(\frac{\partial \tau_x}{\partial z})^2 + (\frac{\partial \tau_z}{\partial z})^2}|_{P_0} \tag{7.67}$$

and

$$tan\,\varphi_m = \frac{dz}{dx} = -\frac{\frac{\partial \tau_x}{\partial x}\frac{\partial \tau_x}{\partial z} + \frac{\partial \tau_z}{\partial x}\frac{\partial \tau_z}{\partial z}}{(\frac{\partial \tau_x}{\partial z})^2 + (\frac{\partial \tau_z}{\partial z})^2}|_{P_0}. \tag{7.68}$$

We observe the shear-stress components and their derivatives in point P_0 of the curved separation line. The tangential component τ_x is positive, the normal component τ_z is zero. Of the first derivatives of the shear-stress components, $\partial\tau_z/\partial z$ is positive, $\partial\tau_x/\partial z$ is negative. $\partial\tau_x/\partial x$ and $\partial\tau_z/\partial x$ are smaller than the two z-derivatives and in general non-zero.

[19] A very detailed discussion can be found in [34] and [35].

The relations show that the minimum of the absolute value of the skin-friction vector at a curved separation line in general lies close to it, but not on it.

We summarize, see also [34]:

- The relative minimun of the absolute value of the skin-friction vector lies close to the separation line, but not on it, because $\partial \tau_x / \partial z \neq 0$.
- The $|\tau_w|_{min}$-line is not a skin-friction line.
- On which side of the separation line the minimun-line lies, depends on the sign of the product $\tau_x \, \partial \tau_x / \partial z$.
- In the ISW case of separation with x parallel to the generator and $\partial / \partial x = 0$, the distance z_m remains non-zero. The angle φ_m becomes zero, showing that the $|\tau_w|_{min}$-line lies parallel to the separation line.

In two-dimensional flow and plane-of-symmetry flow in z-direction we have $\tau_x = 0$, $\partial / \partial x = 0$, but $\partial \tau_z / \partial z \neq 0$. Therefore $z_m = 0$, $\varphi_m = 0$, and separation point and $|\tau_w|_{min}$ ($\tau_z = 0$ in P_0) fall together.

In experiments it is difficult to determine the separation line, except if, for instance, oil-flow visualization is used. The $|\tau|_{min}$-line, however, can be obtained point-wise experimentally [42, 43]. Then, assuming that the separation line is approximately parallel to that line, we obtain from eq. (7.54) with $x = 0$ the distance between the two lines:

$$z_A = \frac{\tau_z|_m}{\frac{\partial \tau_z}{\partial z}|_m}. \tag{7.69}$$

An example of the location of the $|\tau|_{min}$-line was given in Fig. 7.21. For that ellipsoid at larger angle of attack, we show in Fig. 7.23 the separation line, the $|\tau|_{min}$-line, both from experiment, and the separation line, found from the $|\tau|_{min}$-line with eq. (7.69). Given the uncertainties of the oil-flow visualization technique—no other method was available at that time—the agreement of the measured and the computed separation line is satisfactory.

Whether open type separation occurs at the large angle of attack, is not clear from [43]. Noteworthy is the influence of laminar-turbulent transition. It leads to an upward shift of the separation line, beginning at $\tilde{x}^1 \approx 0.15$, completed at $\tilde{x}^1 \approx 0.25$. It demonstrates, how much more an adverse pressure gradient can be negotiated by turbulent flow compared to laminar flow.

Details in this regard can be found in Fig. 7.24. Experimentally determined transition and separation regions are indicated. At zero and small angles of attack laminar flow is present up to nearly half of the length of the ellipsoid. The authors note that harmonic perturbations, which can be interpreted as Tollmien-Schlichting waves, are present already at $x/2a = 0.14$. Actual transition sets in only at $x/2a \approx 0.43$. This is due to the combined effect of the rather small Reynolds number, the small stream-wise turbulence level of the wind tunnel, $Tu_x = 0.1$ to 0.2 per cent, see Section 9.4, and

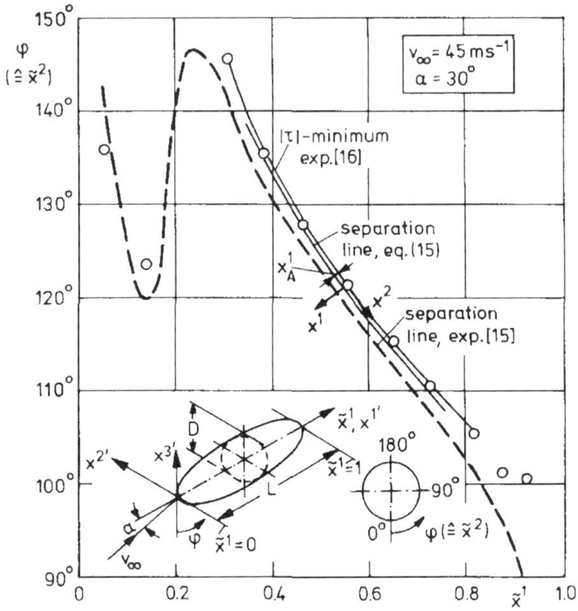

Fig. 7.23. Incompressible laminar flow past a 1:6 ellipsoid (right half of ellipsoid) at $\alpha = 30°$ angle of attack, $L = 3.4$ m, $Re = 7.2 \cdot 10^6$ [34]. Surface-coordinate parameter map: separation line from oil-flow pattern - - - [43] (exp.[15] in the figure), $|\tau_w|$-minimum line from surface hot-film experiment –o– [44], [42] (exp.[16] in the figure), and separation line determined with eq. (7.69) — (eq.(15) in the figure). The distance between the separation line and the $|\tau|$-minimum line is denoted by x^1_A ($\equiv z_A$ in eq. (7.69)).

accelerated flow up to $x/2a = 0.5$. The transition zone extends up to $x/2a \approx 0.7$. For this is responsible the initially only weak stream-wise deceleration of the flow. The fully turbulent flow separates at $x/2a \approx 0.95$.

At larger angles of attack the picture changes completely. Now a strong cross-flow exists. At $\alpha = 30°$ the flow initially separates while being still laminar. At the flank of the body transition sets in early, the transition zone is small, fully turbulent flow is present soon.[20] As a consequence, the separation line is pushed upwards, Fig. 7.23, compared to the laminar-flow situation.

Separation Lines, Item 2: Points-of-Inflection Line. Again we proceed in the same way as for item 2 of attachment lines. We show only the result:

$$z\,(\frac{1}{R} - 0) = z_{ipv} + x\; tan\, \varphi_{ipv}, \tag{7.70}$$

[20] In Section 10.5 results are discussed of a numerical study of laminar-turbulent transition on the ellipsoid for the $\alpha = 10°$ case.

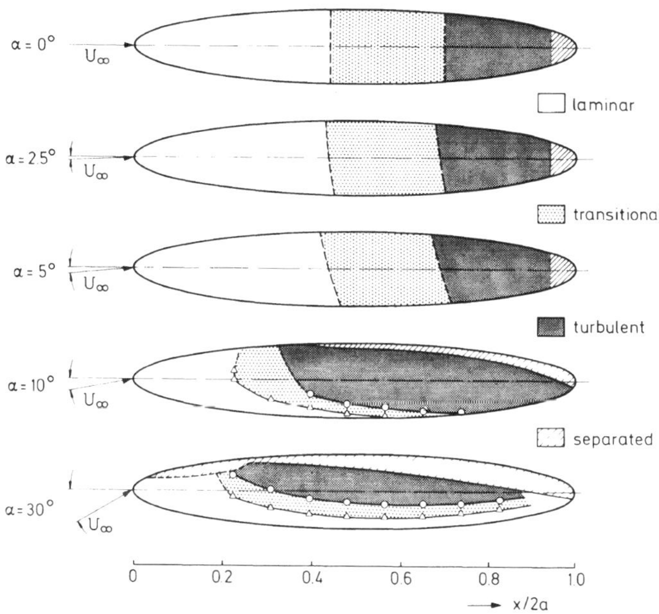

Fig. 7.24. Experimentally found laminar-turbulent transition and separation regions at the 1:6 ellipsoid of Fig. 7.23 for different angles of attack [43]. The axial coordinate $x/2a$ ($2a \equiv L$) is equivalent to $\tilde{x}^1$ in Fig. 7.23.

with

$$z_{ipv} = z(x=0) = -\frac{\tau_x \frac{\partial \tau_z}{\partial x}}{(\frac{\partial \tau_z}{\partial z})^2 - \frac{\partial \tau_x}{\partial x}\frac{\partial \tau_z}{\partial z} + 2\frac{\partial \tau_x}{\partial z}\frac{\partial \tau_z}{\partial x}}|_{P_0} \tag{7.71}$$

and

$$tan\, \varphi_{ipv} = \frac{dz}{dx} = -\frac{\frac{\partial \tau_z}{\partial x}(\frac{\partial \tau_x}{\partial x} + \frac{\partial \tau_z}{\partial z})}{(\frac{\partial \tau_z}{\partial z})^2 - \frac{\partial \tau_x}{\partial x}\frac{\partial \tau_z}{\partial z} + 2\frac{\partial \tau_x}{\partial z}\frac{\partial \tau_z}{\partial x}}|_{P_0}. \tag{7.72}$$

Again we do not discuss these relations. The line of argumentation goes like that one which we had for the points-of-inflection line in the inviscid and the viscous case of attachment lines.

Separation Lines, Item 5: Extrema of the Thermal State of the Surface. Whereas at attachment lines in high-speed flow a relative maximum of the thermal state of the surface can be found, see item 5 of attachment lines, we find a relative minimum at separation lines. Also for this case we do not have a theory at hand. Looking at eq. (7.65), we see that the properties of the characteristic thickness Δ_c again permit to make a statement. As we have found in the case of separation lines—item 3—Δ_c has a relative maximum there. Therefore we can deduce that q_{gw} has a relative minimum at the separation line. The example discussed in Section 10.4 gives proof of this result.

7.5 Attachment and Separation of Three-Dimensional Viscous Flow—More Results and Indicators

We begin with the separation topic. In two-dimensional viscous flow the obvious criterium for separation is the vanishing of the wall shear stress τ_w. Experimental evidence of separation usually is given by changes of the wall-pressure distribution compared to that of the unseparated case. A pressure plateau may be formed, or at the aft of the two-dimensional body the recompression is severely suppressed.

However, it should be noted that at any body or wing of finite length the boundary layers leaving the surface at the end of the body—squeeze-off or flow-off separation—inhibit the recompression to the inviscid value. As a consequence, the pressure drag appears, Section 6.1. The mentioned changes of the pressure field of course appear also in numerical solutions for laminar flow on Navier-Stokes level. For turbulent flow solutions on RANS-level turbulence modelling may be insufficient for their appearance.

We can make some more statements about two-dimensional separation. At the separation point actually τ_w does not vanish, but is changing only its sign. If in flow direction both the functions $\tau_w(x)$ and $p_w(x)$ are known, we can determine the finite angle λ, under which a streamline in the separation point leaves the surface.[21] This was shown by K. Oswatitsch for two-dimensional or axisymmetric flow [21].

With eqs. (7.12) and (7.14) reduced to two-dimensional incompressible flow with constant viscosity, we find at the separation point P_0, where $\tau_w = 0$, with

$$tan\lambda = \frac{y}{x} = \frac{v}{u}. \tag{7.73}$$

After some manipulations we obtain

$$tan\lambda = -3\,\frac{\frac{\partial \tau_w}{\partial x}}{\frac{\partial p}{\partial x}}\Big|_{P_0}. \tag{7.74}$$

Note that the minus sign appears here because we use for τ_w the notation given in eq. (7.8). The pressure gradient $\partial p/\partial x$ is positive because it is the adverse gradient which leads to separation (of course, it is not that of an external inviscid flow used in a boundary-layer approach). The gradient $\partial \tau_w/\partial x$ is negative, hence λ is positive, as is to be expected.

Along a three-dimensional separation line a streamline line does not leave the body surface, as was also shown in [21]. That happens only at singular points.

We assume that locally the direction of the separating skin-friction line is the same as that of the separation line itself, Fig. 7.25. The angle λ_2, under which the streamline is assumed to leave the surface, is again

---

[21] We keep the coordinate convention of the preceding sections.

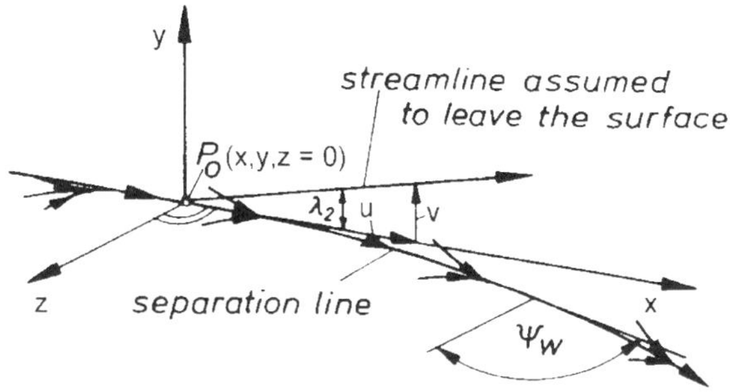

Fig. 7.25. Sketch of a streamline assumed to leave the body surface along a separation line [35].

$$tan\lambda_2 = \frac{y}{x} = \frac{v}{u}. \tag{7.75}$$

From eqs. (7.12) and (7.14) we obtain with $z = 0$:

$$tan\lambda_2 = -\frac{1}{2}\frac{(\frac{\partial \tau_x}{\partial x} + \frac{\partial \tau_z}{\partial z})\, y}{\tau_x + \frac{\partial \tau_x}{\partial x}\, x + \frac{1}{2}\frac{\partial p}{\partial x}\, y}|_{P_0}. \tag{7.76}$$

With $tan\lambda_2 \rightarrow y/x$ for $y \rightarrow 0$, $x \rightarrow 0$, this equation can be rearranged to yield:

$$tan\lambda_2 = -\frac{3\frac{\partial \tau_x}{\partial x} + \frac{\partial \tau_z}{\partial z} + 2\frac{\tau_x}{x}}{\frac{\partial p}{\partial x}}|_{P_0}. \tag{7.77}$$

From this equation no meaningful result for λ_2 can be obtained unless we have $\tau_x|_{P_0} = 0$. This would mean, because in P_0 by definition $\tau_z = 0$, that P_0 is a singular point. The result is that indeed no streamline can leave the surface. That is possible only in the singular point P_0. In the two-dimensional case $\partial \tau_z/\partial z \equiv 0$, and we are back to the result of eq. (7.74).

We consider now the—dividing—stream surface between the two boundary-layer streams, which squeeze each other off the surface. Its intersection with the body surface is the separation line, Fig. 7.26. With the help of eq. (7.13), again for incompressible flow with constant viscosity, it is possible to estimate the angle λ_1 of the separating surface with respect to the z-direction at the point P_0 on the separation line. At that point the separating surface is defined by $\tau_z = 0$. For $z \geqq 0$, if the elevation angle of the flow in the surface is small, we have $w \approx 0$.

Thus we find for the angle λ_1 with $w = 0$ from eq. (7.13):

$$tan\lambda_1 = -2\,\frac{\frac{\partial \tau_z}{\partial z}}{\frac{\partial p}{\partial z}}|_{P_0}. \tag{7.78}$$

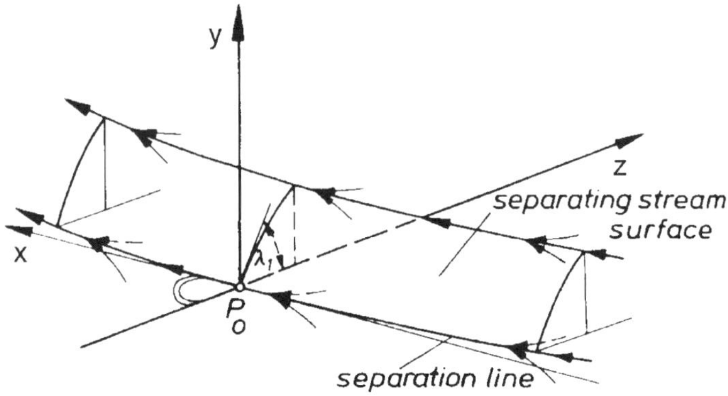

Fig. 7.26. Sketch of a separation line and the separating stream surface leaving the body surface [35].

Our result so far is that streamlines do not depart upwards from a separation line. The separating stream surface, which emanates from it, is formed by the two boundary-layer streams squeezing each other off the body surface. The question now is, what is the situation at attachment lines?

We look first at the attachment point. We consider again the two-dimensional or the axisymmetric case. At the primary attachment point we have an absolute pressure maximum, $\partial p/\partial x = 0$, and an absolute skin-friction minimum, $\partial \tau_x/\partial x = 0$. Away from the attachment point, we have a negative pressure gradient and a positive skin-friction gradient.

The relation for the angle in the separation point, eq. (7.74), holds also for an attachment point. With $\partial p/\partial x$ negative, and $\partial \tau_x/\partial x$ positive in its vicinity, when approaching the attachment point, $tan\lambda$ approaches infinity. The result is that the attaching streamline in that point impinges on the surface at a right angle. This is the same result which potential theory gives for the attachment point. Generalizing this we state that both in inviscid and viscous flow the streamline impinges on the primary attachment point at a right angle.

With the reasoning which we used for the separation line, we can show that no streamline impinges on the attachment line along it. From eq. (7.78) we deduce also that the attachment stream surface stands at a right angle to the surface, because $\partial p/\partial z \to$ o for $z \to 0$, whereas $\partial \tau_z/\partial z$ is finite.

In closing this chapter, we ask how to recognize three-dimensional attachment or separation lines on the body surface. Visually the respective patterns are easily to recognize, see the examples in Chapter 10. But there is no simple criterion like in two-dimensional flow with $\tau_w = 0$.

In [6] the following indicators were proposed to detect separation in computed data:

1. Local convergence of skin-friction lines.
2. Occurrence of a $|\tau_w|$-minimum line.
3. Bulging of the boundary-layer thickness and the displacement thickness.

For attachment lines we note accordingly:

1. Local divergence of skin-friction lines.
2. Occurrence of a $|p_w|$-maximum line.
3. Occurrence of a $|\tau_w|$-minimum line.
4. Indentation of the boundary-layer thickness and the displacement thickness.

7.6 Problems

Problem 7.1. Show that eqs. (7.19) and (7.20) are correct.

Problem 7.2. Sketch the flow in the cross-section of Fig. 7.8 and show that the topological rule 2 is fulfilled.

Problem 7.3. Fig. C.9 in Appendix C.4 shows a small separation bubble at the suction side of the wing at about four per cent chord length. Show that the topological rule 2 is fulfilled.

Problem 7.4. Assume a curved inviscid attachment line. Why do inflection points appear on streamlines on one side of that attachment line?

Problem 7.5. When do general extrema lines and attachment/spearation lines coincide?

Problem 7.6. What is the prerequisite for plane-of-symmetry flow? Give a verbal definition of that prerequisite.

Problem 7.7. In Sub-Section 7.4.1 five items associated with attachment and separation lines are discussed. Give a summary of these items.

Problem 7.8. Which are the topological rules 1 and 2 including the extension to quarter-saddles?

References

1. Peake, D.J., Tobak, M.: Three-Dimensional Interaction and Vortical Flows with Emphasis on High Speeds. AGARDograph 252 (1980)
2. Sears, W.R.: The Boundary Layer of Yawed Cylinders. J. Aeronatical Sc. 15, 49–52 (1948)

3. Maltby, R.L.: Flow Visualization in Wind Tunnels Using Indicators. AGARDograph 70 (1962)
4. Eichelbrenner, E.A.: Three-Dimensional Boundary Layers. Annual Review of Fluid Mechanics 5, 339–360 (1973)
5. Hirschel, E.H.: On the Creation of Vorticity and Entropy in the Solution of the Euler Equations for Lifting Wings. MBB-LKE122-AERO-MT-716, Ottobrunn, Germany (1985)
6. Hirschel, E.H.: Evaluation of Results of Boundary-Layer Calculations with Regard to Design Aerodynamics. AGARD R-741, 5-1–5-29 (1986)
7. Moffat, H.K., Tsinober, A. (eds.): Topological Fluid Mechanics. Proc. IUTAM Symp., Cambridge, GB, 1989. Cambridge University Press (1990)
8. Délery, J.: Robert Legendre and Henri Werlé: Towards the Elucidation of Three-Dimensional Separation. Annual Review of Fluid Mechanics 33, 129–154 (2001)
9. Dallmann, U.: Topological Structures of Three-Dimensional Flow Separations. DLR Rep. 221-82 A 07 (1983)
10. Dallmann, U.: On the Formation of Three-Dimensional Vortex Flow Structures. DLR Rep. 221-85 A 13 (1985)
11. Eberle, A., Rizzi, A., Hirschel, E.H.: Numerical Solutions of the Euler Equations for Steady Flow Problems. NNFM, vol. 34. Vieweg, Braunschweig Wiesbaden (1992)
12. Hirschel, E.H.: Vortex Flows: Some General Properties, and Modelling, Configurational and Manipulation Aspects. AIAA-Paper 96-2514 (1996)
13. Schlichting, H., Truckenbrodt, E.: Aerodynamics of the Aeroplane, 2nd edn. (revised). McGraw Hill Higher Education, New York (1979)
14. Wang, K.C.: Boundary Layer Over a Blunt Body at High Incidence with an Open Type of Separation. Proc. Royal Soc., London A340, 33–55 (1974)
15. Hirschel, E.H., Fornasier, L.: Flowfield and Vorticity Distribution Near Wing Trailing Edges. AIAA-Paper 84-0421 (1984)
16. Schwamborn, D.: Boundary Layers on Finite Wings and Related Bodies with Consideration of the Attachment-Line Region. In: Viviand, H. (ed.) Proc. 4th GAMM-Conference on Numerical Methods in Fluid Mechanics, Paris, France, October 7-9, 1981. NNFM, vol. 5, pp. 291–300. Vieweg, Braunschweig Wiesbaden (1982)
17. Zahm, A.F.: Flow and Force Equations for a Body Revolving in a Fluid. NACA Rep. No. 323 (1930)
18. Lighthill, M.J.: Attachment and Separation in Three-Dimensional Flow. In: Rosenhead, L. (ed.) Laminar Boundary Layers, pp. 72–82. Oxford University Press (1963)
19. Lugt, H.J.: Introduction to Vortex Theory. Vortex Flow Press, Potomac (1996)
20. Poincaré, H.: Les points singuliers des équations différentielles. C. R. Acad. Sci. 94, 416–418 (1882), Oeuvres Complètes 1, 3–5 (1882)
21. Oswatitsch, K.: Die Ablösebedingungen von Grenzschichten. In: Görtler, H. (ed.) Proc. IUTAM Symposium on Boundary Layer Research, Freiburg, Germany, 1957, pp. 357–367. Springer, Heidelberg (1958); Also: The Conditions for the Separation of Boundary Layers. In: Schneider, W., Platzer, M. (eds.) Contributions to the Development of Gasdynamics, pp. 6–18. Vieweg, Braunschweig Wiesbaden, Germany (1980)
22. Kaplan, W.: Ordinary Differential Equations. Addison-Wesley Publishing Company, Reading (1958)

23. Andronov, A.A., Leontovich, E.A., Gordon, I.I., Maier, A.G.: Qualitative Theory of Second-Order Dynamic Systems. J. Wiley, New York (1973)
24. Hornung, H., Perry, A.E.: Some Aspects of Three-Dimensional Separation, Part I: Streamsurface Bifurcations. Z. Flugwiss. und Weltraumforsch (ZFW) 8, 77–87 (1984)
25. Bakker, P.G., de Winkel, M.E.M.: On the Topology of Three-Dimensional Separated Flow Structures and Local Solutions of the Navier-Stokes Equations. In: Moffat, H.K., Tsinober, A. (eds.) Topological Fluid Mechanics. Proc. IUTAM Symp., 1989, pp. 384–394. Cambridge University Press, Cambridge (1990)
26. Hunt, J.C.R., Abell, C.J., Peterka, J.A., Woo, H.: Kinematical Studies of the Flows Around Free or Surface-Mounted Obstacles; Applying Topology to Flow Visualization. J. Fluid Mechanics 86, 179–200 (1978)
27. Tobak, M., Peake, D.J.: Topology of Three-Dimensional Separated Flows. Annual Review of Fluid Mechanics 14, 61–85 (1982)
28. Davey, A.: Boundary-Layer Flow at a Saddle Point of Attachment. J. Fluid Mechanics 10, 593–610 (1961)
29. Van Dyke, M.: An Album of Fluid Motion. The Parabolic Press, Stanford (1982)
30. Hirschel, E.H.: Viscous Effects. Space Course 1991, RWTH Aachen, Germany, pp. 12-1–12-35 (1991)
31. Hirschel, E.H., Weiland, C.: Selected Aerothermodynamic Design Problems of Hypersonic Flight Vehicles, AIAA, Reston, Va. Progress in Astronautics and Aeronautics, vol. 229. Springer, Heidelberg (2009)
32. Dallmann, U., Hilgenstock, A., Riedelbauch, S., Schulte-Werning, B., Vollmers, H.: On the Footprints of Three-Dimensional Separated Vortex Flows Around Blunt Bodies. Attempts of Defining and Analyzing Complex Vortex Structures. AGARD-CP-494, 9-1–9-13 (1991)
33. Schwamborn, D.: Laminare Grenzschichten in der Nähe der Anlegelinie an Flügeln und flügelähnlichen Körpern mit Anstellung (Laminar Boundary Layers in the Vicinity of the Attachment Line at Wings and Wing-Like Bodies at Angle of Attack). Doctoral thesis, RWTH Aachen, Germany, also DFVLR-FB 81–31 (1981)
34. Hirschel, E.H., Kordulla, W.: Local Properties of Three-Dimensional Separation Lines. Z. Flugwiss. und Weltraumforsch (ZFW) 4, 295–307 (1980)
35. Hirschel, E.H., Kordulla, W.: Shear Flow in Surface-Oriented Coordinates. NNFM, vol. 4. Vieweg, Braunschweig Wiesbaden (1981)
36. Schlichting, H., Gersten, K.: Boundary Layer Theory, 8th edn. Springer, Heidelberg (2000)
37. Stock, H.-W.: Laminar Boundary Layers on Inclined Ellipsoids of Revolution. Z. Flugwiss. und Weltraumforsch (ZFW) 4, 217–224 (1980)
38. Hirschel, E.H.: Three-Dimensional Boundary-Layer Calculations in Design Aerodynamics. In: Fernholz, H.H., Krause, E. (eds.) Three-Dimensional Turbulent Boundary Layers. Proc. IUTAM Symp., Germany, pp. 353–365. Springer, Heidelberg (1982)
39. Hirschel, E.H.: Basics of Aerothermodynamics, AIAA, Reston, Va. Progress in Astronautics and Aeronautics, vol. 204. Springer, New York (2004)
40. Cousteix, J.: Analyse théorique et moyens de prévision de la couche limite turbulente tridimensionelle. Doctoral thesis, University of Paris VI, Paris, France (1974); Also: Theoretical Analysis and Prediction Methods for a Three-Dimensional Turbulent Boundary-Layer. ESA TT-238 (1976)

41. Hirschel, E.H.: Das Verfahren von Cousteix-Aupoix zur Berechnung von turbulenten, dreidimensionalen Grenzschichten. MBB-UFE122-AERO-MT-484, Ottobrunn, Germany (1983)
42. Kreplin, H.-P., Vollmers, H., Meier, H.U.: Experimental Determination of Wall Shear Stress Vectors on an Inclined Spheroid. In: Proc. DEA-Meeting on Viscous and Interacting Flow-Field Effects, Annapolis MD, USA, pp. 315–332. AFFDL-TR-80-3088 (1980)
43. Meier, H.U., Kreplin, H.-P.: Experimental Investigation of the Boundary Layer Transition and Separation on a Body of Revolution. Z. Flugwiss. und Weltraumforsch (ZFW) 4, 65–71 (1980)
44. Meier, H.U., Kreplin, H.-P.: Experimentelle Untersuchung von Ablösephänomenen an einem rotationssymmetrischen Körper. In: DGLR Symposium "Strömungen mit Ablösung", München, Germany, September 19 & 20. DGLR Reprint 79-071 (1979)

8

Quasi-One-Dimensional and Quasi-Two-Dimensional Flows

In this chapter we consider quasi-one-dimensional and quasi-two-dimensional attached viscous flow cases, being of particular interest in airplane design. The topics encompass stagnation point flow, flow in symmetry planes and infinite-swept-wing flow, the latter in different forms, as employed in design and research work.

After a short introduction into each topic, the governing equations are given in boundary-layer formulation. These are based on the dimensionless first-order boundary-layer equations in contravariant formulation, Appendix A.2.4. The flow is assumed to be steady, as well as laminar or turbulent and incompressible or compressible.

The coordinate convention is that given for the surface element in Fig. 2.6 which is embedded in the Cartesian reference system $x^{i'}$ $(i' = 1,2,3)$. The surface-tangential coordinates are x^1 and x^2, the surface-normal coordinate is x^3. The surface-tangential velocity components are v^1 and v^2, the surface-normal component is v^3. Metric properties can be found in Appendix C.

The boundary conditions at the wall are the usual no-slip condition $v^1(x^3= 0) = 0$, $v^2(x^3= 0) = 0$, $v^3(x^3= 0) = 0$, and the temperature or temperature-gradient condition. Suction and surface-normal blowing, $v^3(x^3= 0) \neq 0$ in principle can be prescribed.

We do not discuss the relations for the displacement thickness etc. which can be derived from the general formulations given in Chapter 5, but see also [1].

At the end of the chapter two-dimensional and axisymmetric flow cases are considered shortly as well as the Mangler effect and the reverse Mangler effect.

We follow partly closely the representations of the topics given in [1]. The list of references in this chapter is not complete, since no review is intended.

8.1 Stagnation Point Flow

The—quasi-one-dimensional—flow at the stagnation point has been object of theoretical investigations since the early times of viscous-flow theory. K. Hiemenz in 1911 gave an exact solution of the Navier-Stokes equations for

E.H. Hirschel, J. Cousteix, and W. Kordulla, *Three-Dimensional Attached Viscous Flow*,
DOI: 10.1007/978-3-642-41378-0_8,

plane stagnation-point flow [2]. This solution was improved in 1935 by L. Howarth [3]. The latter then was one of the first authors to present—in 1951—results for incompressible flow at a general three-dimensional nodal stagnation point [4]. A solution for axisymmetric stagnation-point flow was published first—in 1936—by F. Homann [5]. For work on stagnation-point flow in the decades after 1945 see, e.g., [6].

The problem of heat transfer at stagnation points of blunt bodies at hypersonic flight in particular spawned much work in the 1950s, e.g., [7]–[10], see also [11]. A solution for compressible boundary layers with heat and mass transfer was given in 1967 by P.A. Libby which was applicable also to nodal and saddle points of attachment [12].

The above and other not cited investigations concern stagnation-point flow at general curved surfaces in the frame of first-order boundary-layer theory. Second-order theory was applied in 1977 for instance by H.D. Papenfuss to nodal attachment points with strong suction and blowing [13]. Three-dimensional stagnation points then were treated in the frame of second-order boundary-layer computations for general bodies in the second half of the 1980s/first half of the 1990s by F. Monnoyer [14].

Concluding we note that the boundary layer at the stagnation point has a positive thickness. We note further that at a forward stagnation point the flow is laminar. This is in contrast to stagnation points embedded in a separation domain. There at a stagnation point the flow can be turbulent.

The Stagnation-Point Flow in Boundary-Layer Formulation. Consider Fig. 8.1 which shows a forward stagnation point on a curved surface. The stagnation point S in the figure is a general nodal point. At the stagnation point of an axisymmetric body at zero angle of attack, it would be a star node. Point S is located at the origin of the surface-oriented non-orthogonal curvilinear coordinate system $S = S(x^i = 0)$. The coordinates x^1, x^2 are the surface coordinates, x^3 is the rectilinear surface-normal coordinate.

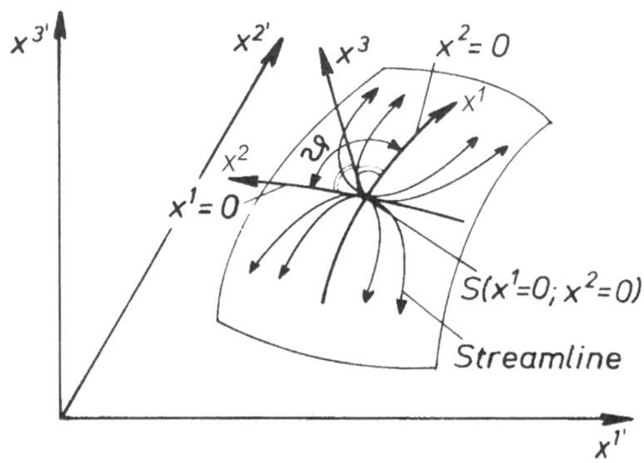

Fig. 8.1. Stagnation point in a surface-oriented non-orthogonal curvilinear coordinate system [1]. The stagnation-point streamline impinges on the body surface at $S(x^i = 0\ (i = 1,2,3))$ in negative x^3-direction, see also Fig. 7.4.

At the stagnation point S the pressure has an absolute maximum. Its first derivatives in x^1- and in x^2-direction are zero. The second derivatives are negative. This is the condition for the maximum. Along the stagnation-point streamline the tangential velocity components v^1 and v^2 are zero, their gradients in x^1- and in x^2-direction are positive, and since the x^3-coordinate points in the direction against the free stream, v^3 is negative. In the stagnation point the skin-friction components τ^1 and τ^2 are zero, their surface-tangential gradients are positive, too.

To obtain the quasi-one-dimensional formulation of the boundary-layer equations for the stagnation point, one has to assume that at least in the vicinity of S the coordinates x^1 and x^2 coincide with external inviscid streamlines. These streamlines furthermore must not be too strongly curved close to S so that stream surfaces normal to the body surface can be defined with sufficient accuracy by the external streamlines. These conditions are fulfilled automatically, if the body, and the flow past it, exhibits two planes of symmetry.

The assumption of external inviscid streamlines x^1 and x^2 with weak curvature at S assures symmetry of flow in the boundary layer along the direction normal to the surface at S (the * defines a physical quantity, Appendix C):

$$x^1 = 0,\ x^2 = 0,\ x^3 \geqq 0: \ \frac{\partial v^{*1}}{\partial x^2} = 0,\ \frac{\partial v^{*2}}{\partial x^1} = 0. \tag{8.1}$$

We assume further symmetry for the thermodynamic variables temperature T, density ρ and hence also pressure p:

$$x^1 = 0,\ x^2 = 0,\ x^3 \geqq 0: \ \frac{\partial T}{\partial x^1} = \frac{\partial T}{\partial x^2} = \frac{\partial \rho}{\partial x^1} = \frac{\partial \rho}{\partial x^2} = \frac{\partial p}{\partial x^1} = \frac{\partial p}{\partial x^2} = 0. \tag{8.2}$$

If the wall is being heated or cooled or if, in supersonic/hypersonic flow past an inclined blunt body, the point-of-attachment streamline does not coincide with the maximum-entropy streamline behind the curved bow shock surface, see Sub-Section 6.4.1, the conditions given in eq. (8.2) may not hold.

In general, both sets of conditions given in eqs. (8.1) and (8.2) are necessary to derive the stagnation-point equations. Actually only the weaker condition of a symmetric pressure distribution would be sufficient: $\partial p/\partial x^1 = 0$, $\partial p/\partial x^2 = 0$. But because symmetric temperature and density profiles are present at quasi-two-dimensional dividing stream surfaces, Section 8.2, and since in most cases the stagnation point is part of such surfaces, it is suitable to employ the conditions given in eq. (8.2).

The conditions in eqs. (8.1) and (8.2) together with the definition of a three-dimensional stagnation point

$$x^1 = 0,\ x^2 = 0,\ x^3 \geqq 0: \ v^{*1} = 0,\ v^{*2} = 0,\ \frac{\partial v^{*1}}{\partial x^1} \neq 0,\ \frac{\partial v^{*2}}{\partial x^2} \neq 0 \tag{8.3}$$

cause the momentum equations (A.43) and (A.44) to vanish. Not vanishing is the continuity equation (A.42) which becomes:

$$\rho(\frac{\partial v^1}{\partial x^1} + \frac{\partial v^2}{\partial x^2}) + \frac{\partial \rho v^3}{\partial x^3} = 0, \tag{8.4}$$

The approach now is to differentiate equation (A.43) with respect to x^1 and (A.44) with respect to x^2. The two derivatives in the bracket of eq. (8.4) become the new dependent variables A^1 and A^2 instead of v^1 and v^2:

$$A^1(x^3) = \frac{\partial v^1(x^3)}{\partial x^1} = \frac{\partial}{\partial x^1}\left[\frac{v^{*1}(x^3)}{\sqrt{a_{11}}}\right], \; A^2(x^3) = \frac{\partial v^2(x^3)}{\partial x^2} = \frac{\partial}{\partial x^2}\left[\frac{v^{*2}(x^3)}{\sqrt{a_{22}}}\right], \tag{8.5}$$

and we obtain—for the stagnation point—the "new" continuity equation:

$$\rho(A^1 + A^2) + \frac{\partial \rho v^3}{\partial x^3} = 0, \tag{8.6}$$

together with the two momentum equations—assuming that the velocity components v^1 and v^2 are continuous at S:

$$\rho(A^1)^2 + \rho v^3 \frac{\partial A^1}{\partial x^3} - \frac{\partial}{\partial x^3}(\mu \frac{\partial A^1}{\partial x^3}) = \rho_e(A_e^1)^2, \tag{8.7}$$

$$\rho(A^2)^2 + \rho v^3 \frac{\partial A^2}{\partial x^3} - \frac{\partial}{\partial x^3}(\mu \frac{\partial A^2}{\partial x^3}) = \rho_e(A_e^2)^2. \tag{8.8}$$

The pressure-gradient terms have been replaced by the convective terms at the outer edge of the boundary layer

$$k_{16}\frac{\partial^2 p}{(\partial x^1)^2} + k_{17}\frac{\partial^2 p}{\partial x^1 \partial x^2} = \rho_e(A^1)^2, \tag{8.9}$$

$$k_{26}\frac{\partial^2 p}{\partial x^1 \partial x^2} + k_{27}\frac{\partial^2 p}{(\partial x^2)^2} = \rho_e(A^2)^2. \tag{8.10}$$

For orthogonal coordinates the cross derivatives vanish, because the respective metric factors are $k_{17} = k_{26} = 0$.

The energy equation (A.45) reduces to

$$c_p \rho v^3 \frac{\partial T}{\partial x^3} = \frac{1}{Pr}\frac{\partial}{\partial x^3}(k \frac{\partial T}{\partial x^3}). \tag{8.11}$$

These equations are equivalent to those derived in the investigations cited above once the appropriate assumptions and transformations are introduced. The special cases of two-dimensional or axisymmetric flow are easily obtained.

We finally note the results for the positive non-zero thickness of the viscous layer at the stagnation point in the case of plane, [2], and axisymmetric, [5], incompressible flow. The 99 per cent thickness δ is found there to be

$$\delta = c\sqrt{\frac{\nu}{A^1}}, \tag{8.12}$$

where ν is the kinematic viscosity, and—in our notation—A^1 the surface-tangential velocity gradient at the stagnation point. For plane stagnation-point flow the result of Hiemenz is $c = 2.4$, and for axisymmetric flow that of Homann $c = 2.8$.

8.2 Flow in Symmetry Planes

Three-dimensional viscous flow is called quasi-two-dimensional, if it depends on two coordinates only. If it depends on one coordinate only, like in the case of the stagnation-point flow, Section 8.1, it is called quasi-one-dimensional, because only one coordinate appears in the governing equations, namely that one normal to the surface.

The quasi-two-dimensional situation is given, for example, in planes of symmetry of flow past bodies, in main-flow direction past infinite swept cylinders and wings, but also in planes of symmetry of channels and other configurations. An often studied example of a plane-of symmetry flow is the quasi-one-dimensional flow locally at the attachment line of an infinite swept cylinder or wing [15]–[18]. This attachment-line flow separates the flow on the upper surface of the wing from that one on the lower side, thus establishing a dividing stream surface. The concept has been extended to finite attachment lines, where the quasi-two-dimensional situation is given [19]–[23].

Quasi-two-dimensional boundary-layer solutions represent an essential ingredient for three-dimensional boundary-layer predictions, if initial data are needed on at least one, but mostly two surfaces, in order to start the integration of the field equations.

If real configurations are considered such as lifting finite wings or fuselages at angle of attack, the plane-of-symmetry concept, like the infinite-swept-wing concept, is fairly easy to apply. However, in most cases these concepts are only approximately valid. The dividing stream surface at the leading edge of a swept wing, for instance, in general is not a plane-of-symmetry flow [24], see also Section 8.5. In many applications, however, the quasi-two-dimensional approach is sufficiently accurate.

The Plane-of-Symmetry Flow in Boundary-Layer Formulation. A plane-of-symmetry flow exists only, in general, if the geometry of the body under consideration exhibits a plane of symmetry and if the vector of the oncoming flow lies within that plane. Due to the symmetry we can adopt a surface-oriented coordinate system which locally is orthogonal in the plane of symmetry, Fig. 8.2, but not necessarily elsewhere.

A coordinate system of this kind is easily established for the attachment line at the leading edge of a wing with symmetrical airfoil at zero angle of attack. The surface coordinates then are orthogonal at $x^2 = 0$ as shown in the

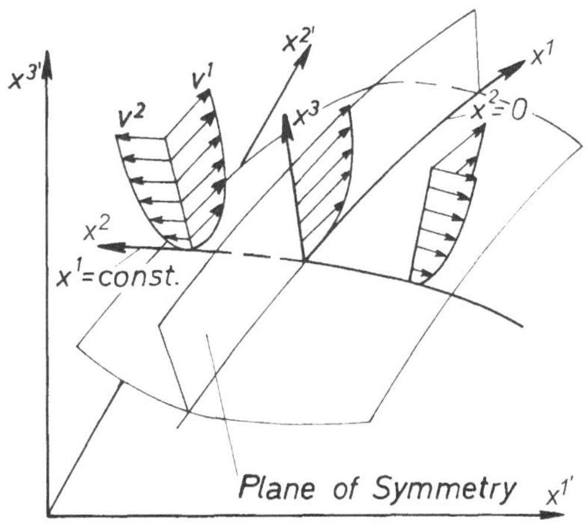

Fig. 8.2. Plane-of-symmetry flow in the locally orthogonal coordinate system in the plane of symmetry at $x^2 = 0$ [1].

figure. If the wing is at an angle of attack, the attachment line moves away from the leading edge and the surface of symmetry exists, at least locally, only approximately. In the case of a symmetric body at angle of attack with a stream-wise plane of symmetry, the usually employed surface coordinates are orthogonal there.

If the coordinates are orthogonal, the metric factors k_{mn} in the equations in Appendix A.2.4 reduce because the off-diagonal term $a_{12} = a_{21}$ of the surface metric tensor vanishes.

Plane-of-symmetry flows are characterized by the variation of the dependent variables across the plane of symmetry: one variable is symmetric of the second kind (odd symmetry) while the rest of the variables is symmetric of the first kind (even symmetry). Referring to Fig. 8.2 this means that $v^{*2} = 0$ in the plane of symmetry $x^2 = 0$, but away from it $v^{*2}(x^1, x^2, x^3) = -v^{*2}(x^1, -x^2, x^3)$ and therefore $\partial v^{*2}/\partial x^2 \neq 0$ at $x^2 = 0$.

The other variables are finite at $x^2 = 0$, but their first derivatives with respect to x^2 vanish due to even symmetry. Since geometrical symmetry is assumed, the metric coefficients $a_{\alpha\beta}$ $(\alpha, \beta = 1{,}2)$ are even symmetric as well. If the coordinate system is orthogonal also in the vicinity of the line of symmetry, we have $\partial a_{12}/\partial x^2 = 0$ for $x^2 = 0$. In general, however, this is not true, and a_{12} is of odd symmetry. The metric factors, which are needed in the following equations of motion, reduce at $x^2 = 0$ to:

$$
\begin{aligned}
k_{01}^2 = a = a_{11}a_{22},\ k_{11} = \frac{1}{2a_{11}}\frac{\partial a_{11}}{\partial x^1},\ k_{16} = -\frac{1}{a_{11}}, \\
k_{22} = \frac{1}{a_{22}}\frac{\partial a_{22}}{\partial x^1},\ k_{27} = -\frac{1}{a_{22}},\ k_{41} = a_{11}.
\end{aligned}
\tag{8.13}
$$

With the symmetry conditions and some of the other metric factors vanishing (see Problem 8.4) we derive from eqs. (A.42) to (A.45) the set of equations for plane-of-symmetry flow.

The continuity equation becomes:

$$\frac{\partial}{\partial x^1}(k_{01}\rho v^1) + k_{01}\rho A^2 + k_{01}\frac{\partial}{\partial x^3}(\rho v^3) = 0, \tag{8.14}$$

where A^2 is defined by eq. (8.5) applied in the plane-of-symmetry:

$$x^2 = 0: \; A^2(x^3) = \frac{\partial v^2(x^3)}{\partial x^2}|_{x^2=0} = \frac{1}{\sqrt{a_{22}}}\frac{\partial v^{*2}(x^3)}{\partial x^2}|_{x^2=0}. \tag{8.15}$$

The momentum equation for the x^1-direction becomes:

$$\begin{aligned}\rho[v^1\frac{\partial v^1}{\partial x^1} + v^3\frac{\partial v^1}{\partial x^3} + k_{11}(v^1)^2] = \\ = k_{16}\frac{\partial p}{\partial x^1} + \frac{\partial \tau^1}{\partial x^3} = \rho_e[v_e^1\frac{\partial v_e^1}{\partial x^1} + k_{11}(v_e^1)^2] + \frac{\partial \tau^1}{\partial x^3}.\end{aligned} \tag{8.16}$$

The momentum equation for the x^2-direction becomes meaningless and has to be differentiated with respect to x^2 in order to yield an equation for the determination of A^2 at $x^2 = 0$:

$$\begin{aligned}\rho[v^1\frac{\partial A^2}{\partial x^1} + (A^2)^2 + v^3\frac{\partial A^2}{\partial x^3} + \frac{\partial k_{21}}{\partial x^2}(v^1)^2 + k_{22}v^1A^2] = \\ = \frac{\partial k_{26}}{\partial x^2}\frac{\partial p}{\partial x^1} + k_{27}\frac{\partial^2 p}{(\partial x^2)^2} + \frac{\partial^2\tau^2}{\partial x^3\partial x^2} = \\ = \rho_e[v_e^1\frac{\partial A_e^2}{\partial x^1} + (A_e^2)^2 + \frac{\partial k_{21}}{\partial x^2}(v_e^1)^2 + k_{22}v_e^1A_e^2] + \frac{\partial^2\tau^2}{\partial x^3\partial x^2},\end{aligned} \tag{8.17}$$

where, see eq. (A.47)

$$\begin{aligned}\frac{\partial^2\tau^2}{\partial x^3\partial x^2} = \frac{\partial^2}{\partial x^3\partial x^2}(\mu\frac{\partial v^2}{\partial x^3} - \frac{1}{\sqrt{a_{22}}} < \tilde{\rho} v^{*2'} v^{*3'} >) = \\ = \frac{\partial}{\partial x^3}(\mu\frac{\partial A^2}{\partial x^3}) - \frac{\partial^2}{\partial x^3\partial x^2}(\frac{1}{\sqrt{a_{22}}} < \tilde{\rho} v^{*2'} v^{*3'} >).\end{aligned} \tag{8.18}$$

The two differentiated metric factors are

$$\frac{\partial k_{21}}{\partial x^2} = \frac{1}{2a_{11}a_{22}}[a_{11}(2\frac{\partial^2 a_{12}}{\partial x^1\partial x^2} - \frac{\partial^2 a_{11}}{(\partial x^2)^2}) - \frac{\partial a_{12}}{\partial x^2}\frac{\partial a_{11}}{\partial x^1}], \quad \frac{\partial k_{26}}{\partial x^2} = \frac{1}{a_{11}a_{22}}\frac{\partial a_{12}}{\partial x^2}. \tag{8.19}$$

The energy equation reduces to

$$c_p \rho (v^1 \frac{\partial T}{\partial x^1} + v^3 \frac{\partial T}{\partial x^3}) = - \frac{1}{Pr_{ref}} \frac{\partial q^3}{\partial x^3} + E_{ref} \left\{ v^1 \frac{\partial p}{\partial x^1} + k_{41} \tau^1 \frac{\partial v^1}{\partial x^3} \right\}. \quad (8.20)$$

For the integration of eqs. (8.14) to (8.20) initial and boundary conditions have to be supplied. The initial conditions are obtained with the stagnation-point equations, Section 8.1.[1] For the case that a coordinate singularity at the body nose exists, see below. The boundary conditions at the wall are the usual ones, but now we have instead of the condition for $v^2(x^3 = 0) = 0$ the condition $A^2(x^3 = 0) \equiv 0$.

The distribution of the external variable $A^2(x^1)$ is often not given or difficult to obtain from the external inviscid velocity field. An alternative is to obtain it from the pressure field. At the outer edge of the boundary layer, $x^3 = \delta$, eq. (8.17) reduces to:

$$\frac{\partial k_{26}}{\partial x^2} \frac{\partial p}{\partial x^1} + k_{27} \frac{\partial^2 p}{(\partial x^2)^2} = \rho_e [v_e^1 \frac{\partial A_e^2}{\partial x^1} + (A_e^2)^2 + \frac{\partial k_{21}}{\partial x^2} (v_e^1)^2 + k_{22} v_e^1 A_e^2]. \quad (8.21)$$

This ordinary differential equation for A_e^2 can be integrated if the other variables are known. The initial value for the integration of eq. (8.21) is given at the stagnation point with

$$A_e^2 \, (x^1 = 0, \, x^2 = 0) = \pm \left[\frac{1}{\rho_e} \, k_{27} \frac{\partial^2 p}{(\partial x^2)^2} \right]^{1/2}_{x^1=0, \, x^2=0} . \quad (8.22)$$

Because the pressure has a maximum at the stagnation point and $k_{27} < 0$, expression (8.22) is real. A_e^2 is positive if a nodal point of attachment (convex surface) is present. It is negative for a saddle point of attachment if the surface is concave with respect to the x^2-direction [25].

With the initial value given, eq. (8.21) can be integrated with, for instance, a finite-difference scheme. In the case of an axisymmetric body at angle of attack the stagnation point lies away from the nose on the lower symmetry line. Along that line, the integration poses no problem. Along the upper symmetry line, with a cross-section oriented coordinate system, Appendix C.3, a coordinate singularity occurs at the body nose. In such a case, this singularity needs a special consideration, see, e.g., [26]. This holds also for the integration of the above given boundary-layer equations.

In the case of symmetric leading edge flow, initial data can be supplied with the approaches for the different infinite-swept-wing cases, Section 8.5.

The equations presented above pose a quasi-two-dimensional initial value/ boundary condition problem. This holds for both diverging and converging

[1] Depending on the coordinate system employed at the stagnation point, appropriate transformations of the variables into the symmetry-plane coordinates are needed, see, e.g., [1].

flow in the symmetry plane. When considering the domain of dependence properties of three-dimensional boundary-layer flow, Section 4.4, for the case of converging flow a contradiction seems to appear. The reason is that in this case only the property of the inviscid flow, via the second cross-wise derivative of the pressure of the external inviscid flow field is taken into account, but not the properties of the converging boundary layers [27]. It appears that this problem has not been solved satisfactorily. In [28] different cases are considered, but no "reasonably complete physical picture" has been found, see also [20].

8.3 The Infinite Swept Wing

The essential feature of the flow past infinite swept wings—or more in general, past finite swept cylinders—is the independence of the flow properties from the span-wise direction, which is a "natural" coordinate in this case. Because the isobars of the external inviscid flow are parallel to both the leading and the trailing edge of the wing, the span-wise derivatives of the flow, as well of the geometrical variables are identically zero. The span-wise coordinate disappears from the governing boundary-layer equations and quasi-two-dimensional formulations result.

The boundary layer, however, is three-dimensional because of the sweep of the wing and of the variation of the flow across the isobars. The assumption of irrotational inviscid flow yields an inviscid span-wise velocity component which remains constant in the direction normal to the isobars.

It is obvious from its definition that the infinite-swept-wing (ISW) flow can only be an approximation to the flow past finite-span and, where applicable, tapered wings. The approximation improves, if the wing shape results from a parallel-isobar design. Then the isobars for a large part are parallel or nearly parallel to the lines of constant chord. In this case the concept of the infinite swept wing can be applied to the mid span region of the wing. Close to the root and the tip of the wing isobars and lines of constant chord deviate markedly from each other, the concept is not applicable.

At those locations, however, infinite-swept-wing solutions are sometimes used to initiate a fully three-dimensional boundary-layer prediction. Then the concept of the "locally infinite swept wing (LISW)", Section 8.4, may lead to a more realistic approach. It allows for a chordwise change of the inviscid span-wise velocity component at the edge of the boundary layer. The correct external boundary conditions are used, but the span-wise derivatives in the governing equations are neglected. Problems may arise with the initialization of the computation. This topic is treated in Section 8.5.

The infinite-swept-wing approach goes back to L. Prandtl [29]. He treated the laminar flow past an infinite swept cylinder—see the short survey of the development of the field in the present book, Section 1.3. Overviews of early work can be found in [30, 31], and from the work environment of the authors of this book in [16]–[18], [32]–[36]. Infinite-swept-wing boundary-layer solutions are still a useful tool for laminar-turbulent transition investigations regarding laminar flow control.

We present three different formulations of the infinite swept wing, beginning with the simplest one. The most complex one, the locally infinite swept wing, is treated in Section 8.4. The related initial data problem is discussed in Section 8.5.

The infinite-swept wing equations have in common—with one small exception for the LISW case—that the span-wise coordinate x^2 disappears, i.e. all derivatives in that direction are zero: $\partial/\partial x^2 \equiv 0$. However, the x^2-momentum equation and the span-wise velocity component v^2 do not vanish.

Details of the metric properties of the respective coordinate systems are given in Appendix C.4.

The Infinite Swept Wing Flow with Leading-Edge Oriented Orthogonal Curvilinear Coordinates in Boundary-Layer Formulation. The leading-edge oriented orthogonal curvilinear coordinate system is shown in Fig. 8.3. The x^1-coordinate lies normal to the leading edge, and the rectilinear x^2-coordinate in span-direction. Due to the orthogonality of the coordinates the angle ϑ between the coordinates x^1 and x^2 is $\pi/2$ everywhere. Hence also everywhere $a_{12} = a_{21} \equiv 0$. The x^1-coordinate is curved only in chord direction orthogonal to the leading edge. The components of the metric tensor of the surface are $a_{11} = 1$, $a_{22} = 1$. Most of the metric factors k_{mn} vanish, the remaining ones are constant: $k_{01} = 1$, $k_{16} = k_{27} = -1$, $k_{41} = k_{43} = 1$.

The first-order boundary-layer equations eqs. (A.42) to (A.45) become very simple:

– continuity equation:

$$\frac{\partial}{\partial x^1}(\rho v^1) + \frac{\partial}{\partial x^3}(\rho v^3) = 0. \tag{8.23}$$

– x^1-momentum equation:

$$\rho[v^1 \frac{\partial v^1}{\partial x^1} + v^3 \frac{\partial v^1}{\partial x^3}] = -\frac{\partial p}{\partial x^1} + \frac{\partial \tau^1}{\partial x^3} = \rho_e v_e^1 \frac{\partial v_e^1}{\partial x^1} + \frac{\partial \tau^1}{\partial x^3}. \tag{8.24}$$

– x^2-momentum equation:

$$\rho[v^1 \frac{\partial v^2}{\partial x^1} + v^3 \frac{\partial v^2}{\partial x^3}] = \frac{\partial \tau^2}{\partial x^3}. \tag{8.25}$$

– energy equation for calorically and thermally perfect gas:

$$c_p\rho(v^1 \frac{\partial T}{\partial x^1} + v^3 \frac{\partial T}{\partial x^3}) = -\frac{1}{Pr_{ref}} \frac{\partial q^3}{\partial x^3} + E_{ref} \left\{ v^1 \frac{\partial p}{\partial x^1} + \tau^1 \frac{\partial v^1}{\partial x^3} + \tau^2 \frac{\partial v^2}{\partial x^3} \right\}. \tag{8.26}$$

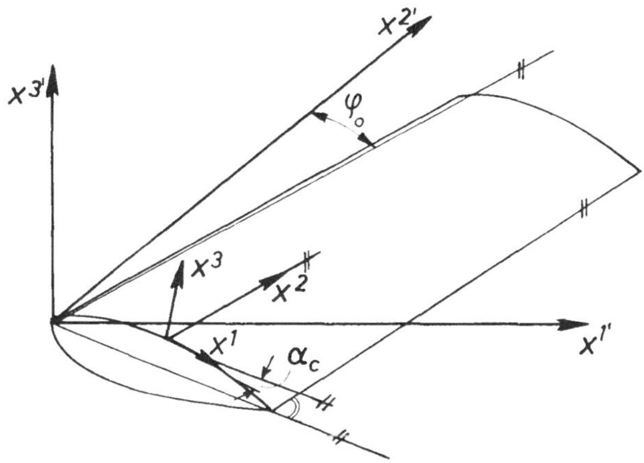

Fig. 8.3. Infinite swept wing with surface-oriented orthogonal curvilinear coordinate system [1]. φ_0 is the sweep angle of the leading edge, $\alpha_c = \alpha_c(x^1)$ the contour angle of the airfoil. The x^2-direction is parallel, the x^1-x^3-plane normal to the leading and the trailing edge.

For laminar incompressible flow the "independence principle" holds, i.e. the eqs. (8.23) and (8.24) depend only on the coordinates x^1 and x^3 and are decoupled from eq. (8.25) for the x^2-coordinate.

The boundary conditions at the outer edge of the boundary layer are for $x^1 \geqq 0$, $x^2 = \text{const.}$

$$v^1 = v^1_e(x^1),\ v^2 = v^2_e = \sin\varphi_0 = const.,\ T = T_e(x^1), \tag{8.27}$$

where we have $v^2_e = \sin\varphi_0$, instead of $v^2_e = \sin\varphi_0 u_\infty$, because all velocities are non-dimensionalized with $v_{ref} = u_\infty$, with u_∞ pointing in x^1-direction. The wall boundary conditions are the usual ones.

The Infinite Swept Wing Flow with Leading-Edge Oriented Non-orthogonal Curvilinear Coordinates in Boundary-Layer Formulation. Consider the leading-edge oriented non-orthogonal curvilinear coordinates in Fig. 8.4. The coordinate in chord direction is the x^1-coordinate, that in span-direction the x^2-coordinate. All derivatives $\partial/\partial x^2$ are zero by definition. The angle ϑ between the coordinates x^1 and x^2 is the same as that between lines of constant chord and constant span. At back-swept wings it is therefore equal or less than $\pi/2$ on the upper surface and equal or larger $\pi/2$ on the lower surface (if at the lower side x^1 is counted negative). Hence $a_{12} = a_{21} \neq 0$ for $|x^1| > 0$.

At the leading edge, not necessarily coincident with the attachment line, $a_{12} = a_{21} = 0$, because there the surface coordinates are orthogonal to each other. Besides that, both are functions (only) of x^1 owing to the change in the contour. For a planar wing we have $a_{12} = a_{21} =$ constant. The components a_{11} and a_{22} of the metric tensor are in first-order theory equal to one for

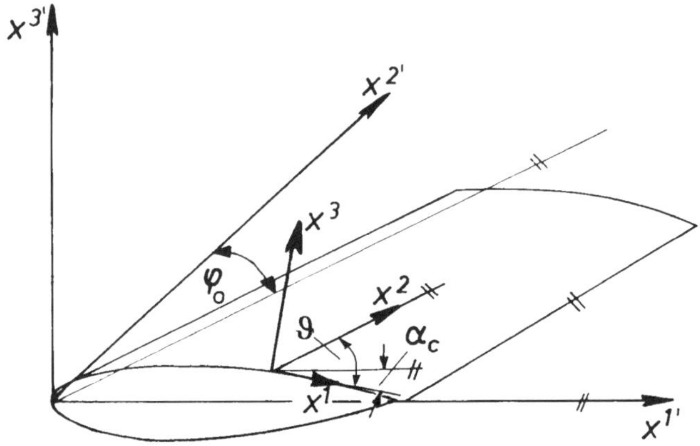

Fig. 8.4. Infinite swept wing with surface-oriented non-orthogonal curvilinear coordinate system [1]. φ_0 is the sweep angle of the leading edge, $\alpha_c = \alpha_c(x^1)$ the contour angle of the airfoil. The x^2-direction is parallel to the leading edge, the x^1-x^3-plane is a plane of constant span and parallel to the $x^{1'}$-$x^{3'}$-plane.

the infinite-swept-wing coordinate system. Thus the metric factors k_{mn}, eqs. (A.49) to (A.52), are finite except for k_{12}, k_{13}, k_{22} and k_{23} which vanish due to the above assumptions.

The boundary-layer equations eqs. (A.42) to (A.45) are simplified to:

– continuity equation:

$$\frac{\partial}{\partial x^1}(k_{01}\rho v^1) + k_{01}\frac{\partial}{\partial x^3}(\rho v^3) = 0. \tag{8.28}$$

– x^1-momentum equation:

$$\rho[v^1\frac{\partial v^1}{\partial x^1} + v^3\frac{\partial v^1}{\partial x^3} + k_{11}(v^1)^2] = k_{16}\frac{\partial p}{\partial x^1} + \frac{\partial \tau^1}{\partial x^3} = \rho_e[v_e^1\frac{\partial v_e^1}{\partial x^1} + k_{11}(v_e^1)^2] + \frac{\partial \tau^1}{\partial x^3}. \tag{8.29}$$

– x^2-momentum equation:

$$\rho[v^1\frac{\partial v^2}{\partial x^1} + v^3\frac{\partial v^2}{\partial x^3} + k_{21}(v^1)^2] = k_{26}\frac{\partial p}{\partial x^1} + \frac{\partial \tau^2}{\partial x^3} = \rho_e[v_e^1\frac{\partial v_e^2}{\partial x^1} + k_{21}(v_e^1)^2] + \frac{\partial \tau^2}{\partial x^3}. \tag{8.30}$$

– energy equation for calorically and thermally perfect gas:

$$\begin{aligned} c_p\rho(v^1\frac{\partial T}{\partial x^1} + v^3\frac{\partial T}{\partial x^3}) &= -\frac{1}{Pr_{ref}}\frac{\partial q^3}{\partial x^3} + \\ + E_{ref}\left\{v^1\frac{\partial p}{\partial x^1} + k_{41}\tau^1\frac{\partial v^1}{\partial x^3} + k_{42}\tau^1\frac{\partial v^2}{\partial x^3} + k_{43}\tau^2\frac{\partial v^2}{\partial x^3}\right\}&. \end{aligned} \tag{8.31}$$

The boundary conditions at the outer edge of the boundary layer for $x^1 \geqq 0$, $x^2 =$ const. are either given

$$v^1 = v_e^1(x^1),\ v^2 = v_e^2 = \sin\varphi_0 = const.,\ T = T_e(x^1), \tag{8.32}$$

or, regarding the velocity components, computed from those for the orthogonal case:

$$v_e^1(x^1) = v_e^1(x^1)|_{orthog.} \cdot [cos^2\alpha_c(x^1)\ cos^2\varphi_0 + sin^2\alpha_c(x^1)]^{-1/2} \tag{8.33}$$

and

$$v_e^2(x^1) = sin\,\varphi_0 \cdot [1 - cos\,\alpha_c(x^1) v_e^1(x^1)]. \tag{8.34}$$

8.4 The Locally Infinite Swept Wing

We have argued above that the infinite-swept-wing concept can be applied with sufficient accuracy only to those regions of finite wings, where the isobars are at least nearly parallel to the wing's leading edge and to each other. In such cases the sweep of the equivalent infinite swept wing can be approximated by the sweep of the leading edge, and the external boundary conditions can be chosen in the way described in the preceding section.

If, however, a swept wing is tapered, and/or if the isobars diverge or converge markedly on portions of the wing's surface, it is necessary to drop the infinite-swept-wing concept in favor of the locally-infinite-swept-wing (LISW) concept [35]. This concept uses only locally the assumption of negligible span-wise flow variation and the span-wise velocity component v_e^2 is no longer related to one sweep angle as before. Instead the external velocity components v_e^1 and v_e^2, as predicted by inviscid flow methods or measured, are used as local outer boundary conditions.

The locally-infinite-swept-wing concept also allows for a span-wise variation of the metric coefficients. These, in general, do not vanish, although the span-wise flow variation may be negligible along lines of constant chord on a tapered wing. The LISW concept permits a good approximation if a parallel isobar design for a particular case results in nearly constant pressure along lines of constant-percent chord, which serve as span-wise coordinates.

The Locally Infinite Swept Wing Flow in Boundary-Layer Formulation. Consider now a location $x^2 =$ const. of the coordinate system shown in Fig. C.6 of Appendix C. The boundary-layer equations for the locally-infinite-swept-wing at that location are defined by the following conditions:

$$\frac{\partial}{\partial x^2} \equiv 0, \tag{8.35}$$

but, and that is important to note

$$\frac{\partial v^{*\alpha}}{\partial x^2} \equiv 0, \ \textit{not} \ \frac{\partial v^{\alpha}}{\partial x^2} \equiv 0, \tag{8.36}$$

with $\alpha = 1{,}2$. Further $\partial p/\partial x^2$ is not necessarily zero, because such an inviscid flow in general would not be irrotational.

The LISW equations read:

– continuity equation:

$$\frac{\partial}{\partial x^1}(k_{01}\rho v^1) + k_{01}^+ \rho v^2 + k_{01}\frac{\partial}{\partial x^3}(\rho v^3) = 0, \tag{8.37}$$

where

$$k_{01}^+ = \frac{\partial k_{01}}{\partial x^2} - \frac{k_{01}}{2a_{22}}\frac{\partial a_{22}}{\partial x^2}. \tag{8.38}$$

– x^1-momentum equation:

$$\begin{aligned} \rho[v^1\frac{\partial v^1}{\partial x^1} + v^3\frac{\partial v^1}{\partial x^3} + k_{11}(v^1)^2 + k_{12}^+ v^1 v^2 + k_{13}(v^2)^2] = \\ = \rho_e[v_e^1\frac{\partial v_e^1}{\partial x^1} + k_{11}(v_e^1)^2 + k_{12}^+ v_e^1 v_e^2 + k_{13}(v_e^2)^2] + \frac{\partial \tau^1}{\partial x^3}, \end{aligned} \tag{8.39}$$

where

$$k_{12}^+ = k_{12} - \frac{1}{2a_{11}}\frac{\partial a_{11}}{\partial x^2}. \tag{8.40}$$

– x^2-momentum equation:

$$\begin{aligned} \rho[v^1\frac{\partial v^2}{\partial x^1} + v^3\frac{\partial v^2}{\partial x^3} + k_{21}(v^1)^2 + k_{22}v^1 v^2 + k_{23}^+(v^2)^2] = \\ = \rho_e[v_e^1\frac{\partial v_e^2}{\partial x^1} + k_{21}(v_e^1)^2 + k_{22}v_e^1 v_e^2 + k_{23}^+(v_e^2)^2] + \frac{\partial \tau^2}{\partial x^3}, \end{aligned} \tag{8.41}$$

where

$$k_{23}^+ = k_{23} - \frac{1}{2a_{22}}\frac{\partial a_{22}}{\partial x^2}. \tag{8.42}$$

– energy equation for calorically and thermally perfect gas:

$$\begin{aligned} c_p\rho(v^1\frac{\partial T}{\partial x^1} + v^3\frac{\partial T}{\partial x^3}) = -\frac{1}{Pr_{ref}}\frac{\partial q^3}{\partial x^3} + \\ + E_{ref}\left\{[v^1\frac{\partial p}{\partial x^1} + v^2\frac{\partial p}{\partial x^2}] + k_{41}\tau^1\frac{\partial v^1}{\partial x^3} + k_{42}\tau^1\frac{\partial v^2}{\partial x^3} + k_{43}\tau^2\frac{\partial v^2}{\partial x^3}\right\}. \end{aligned} \tag{8.43}$$

The boundary conditions at the outer edge of the boundary layer for $x^1 \geqq 0$, $x^2 =$ const. read

$$v^1 = v_e^1(x^1),\ v^2 = v_e^2(x^1),\ T = T_e(x^1). \tag{8.44}$$

Note that these conditions in the general case are functions of x^2.

8.5 Initial Data for Infinite-Swept-Wing Solutions

The prediction of quasi-two-dimensional boundary layers requires the specification of initial data at $x^1 = 0$. In the case of the plane-of-symmetry flow initial data are given by the three-dimensional, but quasi-one-dimensional stagnation-point flow solution, Section 8.1. The initial data for infinite-swept-wing flows are furnished by the corresponding attachment-line flow, a special form of a plane-of-symmetry flow [37, 38, 18, 35].

The attachment-line flow of a lifting wing in general is not a plane-of-symmetry flow. If the attachment line is curved, like indicated in Fig. 8.5, it can piecewise be approximated by lines of constant chord, where $v^1 = 0$ [35]. At such locations then either an infinite-swept-wing solution or a locally-infinite-swept-wing solution can be initiated. The governing equations are derived from those for the locally infinite swept wing, Section 8.4.

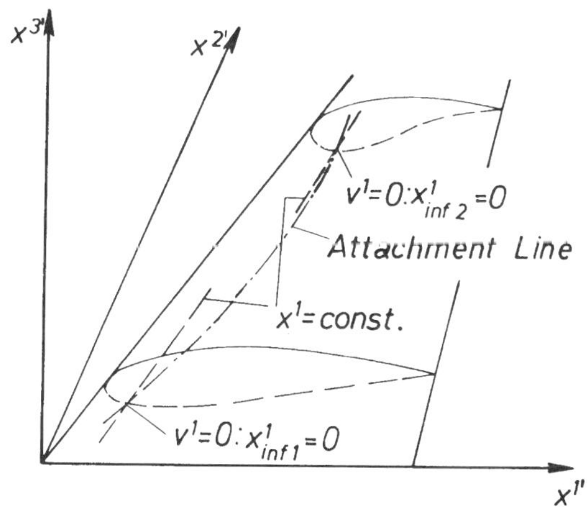

Fig. 8.5. Schematized approximation of the attachment line for a locally-infinite-swept-wing flow at two span stations (the index 'inf' stands for infinite-swept-wing assumption) [1]

Initial Solution for the Locally Infinite Swept Wing in Boundary-Layer Formulation. The boundary-layer equations at the attachment line are obtained from eqs. (8.37) to (8.43) by taking the limit for $x^1 \to 0$, with

$v^1(x^1 = 0, x^2 = const., x^3) = 0$. Because $v^1 = 0$, we take its derivative as unknown, as we did it for the stagnation point flow:

$$A^1(x^3) = \frac{\partial v^1(x^3)}{\partial x^1} = \frac{\partial}{\partial x^1}\left[\frac{v^{*1}(x^3)}{\sqrt{a_{11}}}\right]. \tag{8.45}$$

The LISW initial data equations read, with k^+_{01}, k^+_{12}, and k^+_{23} defined in Section 8.4:

- continuity equation:

$$k_{01}\rho A^1 + k^+_{01}\,\rho v^2 + k_{01}\frac{\partial}{\partial x^3}(\rho v^3) = 0. \tag{8.46}$$

- x^1-momentum equation after differentiation and introduction of flow symmetry with respect to x^1:

$$\begin{aligned}\rho[(A^1)^2 + v^3\frac{\partial A^1}{\partial x^3} + k^+_{12}A^1 v^2 + \frac{\partial k_{13}}{\partial x^1}(v^2)^2] = \\ = \rho_e[(A^1_e)^2 + k^+_{12}A^1_e v^2_e + \frac{\partial k_{13}}{\partial x^1}(v^2_e)^2] + \frac{\partial \tau^1}{\partial x^3}.\end{aligned} \tag{8.47}$$

- x^2-momentum equation:

$$\rho[v^3\frac{\partial v^2}{\partial x^3} + k^+_{23}(v^2)^2] = \rho_e[k^+_{23}(v^2_e)^2] + \frac{\partial \tau^2}{\partial x^3}. \tag{8.48}$$

- energy equation for calorically and thermally perfect gas:

$$c_p\rho v^3\frac{\partial T}{\partial x^3} = -\frac{1}{Pr_{ref}}\frac{\partial q^3}{\partial x^3} + E_{ref}[k_{43}\tau^2\frac{\partial v^2}{\partial x^3}]. \tag{8.49}$$

The boundary conditions at the outer edge of the boundary layer for $x^1 \geqq 0$, x^2 = const. read

$$A^1 = A^1_e(x^2),\ v^2 = v^2_e(x^2),\ T = T_e(x^2). \tag{8.50}$$

For the derivation of eq. (8.47) it has been assumed tacitly that $k_{13} = 0$ at $x^1 = 0$, otherwise the first momentum equation—eq. (8.39)—would not have vanished completely. This means that locally the contour is assumed to be cylindrical, which further implies that the attachment line locally is straight and lies on a generator of that cylinder. Basically this means that locally the attachment line is a geodesic, Appendix A.2.3.

Another point of view would result if one would deal from the beginning with the momentum equation for the x^1-direction after having it differentiated with respect to the x^1-direction. Then one has to assume $k_{13}v^2(\partial v^2/\partial x^1) = 0$ at $x^1 = 0$. If $k_{13} \neq 0$, then this would demand that $\partial v^2/\partial x^1 = 0$, that is, the flow is symmetrical with respect to the attachment line. However, for the general case this is not true.

In practice often an incompatibility arises between the above quasi-one-dimensional attachment-line solution and the locally-infinite-swept-wing solution which is supposed to continue it. If the flow or the surface is not sufficiently symmetrical, no continuing solution can be achieved at all [39].

8.6 Two-Dimensional and Axisymmetric Flow

The reader might wish to consider two-dimensional or axisymmetric flow. Since the first-order boundary-layer equations for these flows can be found in the literature, e.g., [40, 6], we do not reproduce them here. We note, however that the equations for two-dimensional flow can easily be derived from those for the flow past infinite swept wings with leading-edge oriented orthogonal curvilinear coordinates, Section 8.3. Because then $v^2 \equiv 0$, the x^2-momentum equation vanishes, like the v^2-term in the energy equation. For the metric coefficients see Appendix C.3. There also the metric coefficients for axisymmetric bodies are provided. By employing the latter for instance in the general boundary-layer equations given in Appendix A.2.4, the equations for axisymmetric flow are recovered, see also Appendix A.2.2. If second-order boundary-layer flow is to be described, the equations for that can be derived from those found, for instance, in [1, 14]. This also holds for all equations given in the present chapter.

8.7 The Mangler Effect

Connected to axisymmetric boundary-layer flow is the Mangler transformation, see, e.g., [40, 6]. In 1948 W. Mangler published a general relationship between two-dimensional and axisymmetric boundary layers [41]. It permits to reduce the calculation for an axisymmetric body to that for a cylindric body. The transformation holds for both incompressible and compressible laminar flow.

We do not present the transformation here, but point to an effect, the Mangler effect, which becomes evident, when discussing the transformation. The effect basically is that the boundary layer on a body with a cross-section that is increasing in flow direction, is thinner than that—with the same inviscid pressure field—on a flat or cylindrical surface. The boundary-layer thickness grows in flow direction, but because the perimeter of the body increases, the boundary layer is "stretched" laterally and the thickness grows less strongly.

Then of course, at a given point also the skin-friction is larger, as well as the heat flux in the gas at the wall. This holds not only for axisymmetric bodies, but for any body with increasing cross-section[2]. If one makes for such a body a guess of boundary-layer properties with, for instance, a handbook method, one should be aware of the Mangler effect.

If the cross-section of a body decreases, a situation typical for the after-body of a fuselage, the reverse Mangler effect happens. The boundary layer is thicker than that—with the same inviscid pressure field—on the flat or cylindrical surface. It is easy to understand what occurs: the boundary layer is thickened because the perimeter of the body decreases. Both skin friction and heat flux in the gas at the wall are reduced. Also here, if one makes a guess of boundary-layer properties with a handbook method for an afterbody or the like, one should be aware of the reverse Mangler effect.

With a fully three-dimensional boundary-layer solution or a Navier-Stokes/RANS solution both the Mangler effect and the reverse Mangler effect are covered automatically.

8.8 Problems

Problem 8.1. Derive the gradient of the external inviscid velocity $\partial u_e/\partial x$ in the stagnation point x_0 on the surface of a sphere, a circular cylinder, and an infinite swept circular cylinder. Use the coordinate convention from Chapter 4 with x and z being the surface-tangential coordinates and u_e and w_e the respective inviscid velocity components, Fig. 8.6. Assume incompressible flow and employ the potential-flow result $u_e = c_1 u_\infty \sin\psi$ with $c_1 = 1{,}5$ for the sphere and $c_1 = 2$ for the circular cylinder.

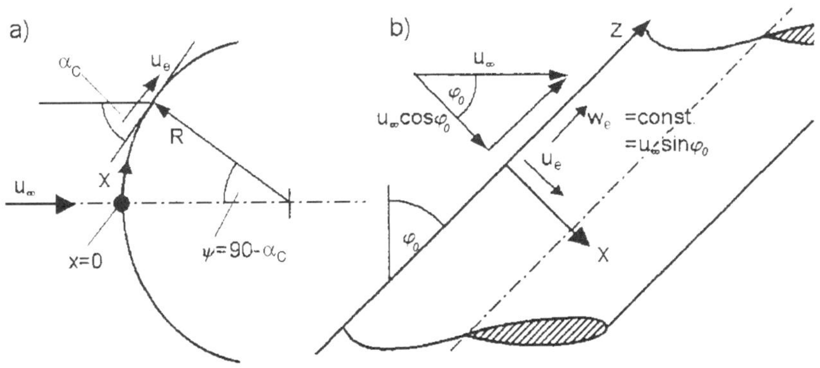

Fig. 8.6. Schematic and notation of flow a) past a sphere or circular cylinder (2-D), b) infinite swept circular cylinder [42]. The sweep angle is φ_0 and the contour angle α_c.

[2] On a facetted forebody, or on the flat part of a forebody which acts as a pre-compression surface of an air inlet, the Mangler effect is not present.

In Sub-Section 7.4.2 it was stated that the pressure attains an absolute extremum at the forward attachment point, i.e., an absolute maximum. This would mean that $\partial p/\partial x|_{x=0} = 0$. Show that this is true.

Problem 8.2. Why are assumptions like symmetry-plane flow, infinite-swept-wing flow, and locally-infinite-swept-wing flow needed for the computation of three-dimensional attached viscous flow with boundary-layer methods and not with Navier-Stokes/RANS methods?

Problem 8.3. Characterize the flow "field" at a three-dimensional stagnation point.

Problem 8.4. Show that for plane-of-symmetry flow the metric factors reduce considerably.

Problem 8.5. We investigate the incompressible laminar flow past a swept wing with the help of the boundary-layer equations. The free-stream velocity is u_∞, the sweep angle of the wing is φ_0. We choose the case with orthogonal coordinates, Fig. 8.3. In Appendix C.4 the metric tensor for this case was shown to be simply

$$\underline{\underline{a}} = \begin{pmatrix} 1 & 0 \\ 0 & 1 \end{pmatrix},$$

hence we have $v^1 = v^{*1}$, $v^2 = v^{*2}$.

1. Show that the span-wise component v_e^2 of the external inviscid velocity is constant on the wing and that
$$v_e^2 = u_\infty \sin \varphi_0.$$
2. If for an infinite swept wing flow the chordwise pressure gradient $\partial p/\partial x^1 = 0$, show that the external inviscid streamlines are straight lines.
3. Show that for laminar flow, the boundary layer is collateral, i.e., the solution is such that
$$v^2 = v^1 \frac{v_e^2}{v_e^1}.$$
4. Show that this solution is the same as the solution of flow over a two-dimensional flat plate.
5. Is this solution also valid for turbulent flow?

Problem 8.6. What are the boundary and the initial conditions for an infinite-swept-wing boundary-layer solution?

Problem 8.7. Assume that the external inviscid flow conditions are the same for two-dimensional and axisymmetric flow. What is the Mangler effect and what the inverse Mangler effect?

References

1. Hirschel, E.H., Kordulla, W.: Shear Flow in Surface-Oriented Coordinates. NNFM, vol. 4. Vieweg, Braunschweig Wiesbaden (1981)
2. Hiemenz, K.: Die Grenzschicht an einem in den gleichförmigen Flüssigkeitsstrom eingetauchten geraden Zylinder (The Boundary Layer at a Linear Cylinder Immerged in an Uniform Fluid Stream). Doctoral thesis, University Göttingen, Germany. Dingl. Polytech. J. 326, 321 (1911)
3. Howarth, L.: On the Calculation of the Steady Flow in the Boundary Layer Near the Surface of a Cylinder in a Stream. ARC-RM-1632 (1935)
4. Howarth, L.: The Boundary Layer in Three-Dimensional Flow—Part II: The Flow Near a Stagnation Point. Phil. Mag., Ser. 7 42(335), 1433–1440 (1951)
5. Homann, F.: Der Einfluß großer Zähigkeit bei der Strömung um den Zylinder und um die Kugel. ZAMM 16, 153–164 (1936)
6. Schlichting, H., Gersten, K.: Boundary Layer Theory, 8th edn. Springer, Heidelberg (2000)
7. Van Driest, E.R.: On Skin Friction and Heat Transfer Near the Stagnation Point. NACA Report, AL-2267 (1956)
8. Lees, L.: Laminar Heat Transfer over Blunt-Nosed Bodies at Hypersonic Flight Speeds. Jet Propulsion 26(4), 259–269 (1956)
9. Fay, J.A., Riddell, F.R.: Theory of Stagnation Point Heat Transfer in Dissociated Air. Journal of Aeronautical Sciences 25(2), 73–85 (1958)
10. Cohen, N.B.: Boundary Layer Similar Solutions and Correlation Equations for Laminar Heat Transfer Distribution in Equilibrium Air at Velocities up to 41,100 Feet per Second. NASA TR R-118 (1961)
11. Hirschel, E.H., Weiland, C.: Selected Aerothermodynamic Design Problems of Hypersonic Flight Vehicles, AIAA, Reston, Va. Progress in Astronautics and Aeronautics, vol. 229. Springer, Heidelberg (2009)
12. Libby, P.A.: Heat and Mass Transfer at a General Three-Dimensional Stagnation Point. AIAA J. 5, 507–517 (1967)
13. Papenfuss, H.D.: Die Grenzschichteffekte zweiter Ordnung am dreidimensionalen Staupunkt mit starkem Absaugen oder Ausblasen. ZFW 1, 87–96 (1977)
14. Monnoyer, F.: Calculation of Three-Dimensional Viscous Flow on General Configurations Using Second-Order Boundary-Layer Theory. ZFW 14, 95–108 (1990)
15. Rosenhead, L. (ed.): Laminar Boundary Layers. Oxford University Press (1963)
16. Beasley, J.A.: Calculation of the Laminar Boundary Layer and the Prediction of Transition on a Sheared Wing. ARC/R&M-3787 ARC-35316 RAE/TR-73156 (1973)
17. Hirschel, E.H.: The Influence of the Free-Stream Reynolds Number on Transition in the Boundary Layer on an Infinite Swept Wing. AGARD R-602, 1-1–1-11 (1973)
18. Cebeci, T.: Attachment-Line Flow on an Infinite Swept Wing. AIAA J. 12, 242–245 (1974)
19. Trella, M., Libby, P.A.: Similar Solutions for the Hypersonic Laminar Boundary Layer Near a Plane of Symmetry. AIAA J. 3, 75–83 (1965)
20. Wang, K.C.: Three-Dimensional Boundary Layer near the Plane of Symmetry of a Spheroid at Incidence. J. Fluid Mech. 43, 187–209 (1970)

21. Cebeci, T., Kaups, K., Ramsey, J.A.: A General Method for Calculating Three-Dimensional Compressible Laminar and Turbulent Boundary Layers on Arbitrary Wings. McDonnel Douglas Rep. J. 7267 (1976)
22. Hirschel, E.H., Schwamborn, D.: Ein Verfahren zur Berechnung von Grenzschichten in Strömungssymmetrieebenen. In: Haase, W. (ed.) Beiträge zu Transportphänomenen in der Strömungsmechanik und anverwandten Gebieten. DLR-FB 77-16, pp. 125–132 (1977)
23. Grundmann, R.: Dreidimensionale Grenzschichtberechnungen entlang Symmetrielinien auf Körpern. ZFW 5, 389–395 (1981)
24. Schwamborn, D.: Boundary Layers on Finite Wings and Related Bodies with Consideration of the Attachment-Line Region. In: Viviand, H. (ed.) Proc. 4th GAMM-Conference on Numerical Methods in Fluid Mechanics, Paris, France, October 7-9, 1981. NNFM, vol. 5, pp. 291–300. Vieweg, Braunschweig Wiesbaden (1982)
25. Lighthill, M.J.: Introduction. Boundary-Layer Theory. In: Rosenhead, L. (ed.) Laminar Boundary Layers, pp. 46–113. Oxford University Press (1963)
26. Cebeci, T., Chen, H.H., Kaups, K.: A Method for Removing the Coordinate Singularity on Bodies with Blunt Rounded Noses at Incidence. Computers and Fluids 28(4), 369–389 (1990)
27. Moore, F.K. (ed.): Theory of Laminar Flow. High Speed Aerodynamics and Jet Propulsion, vol. IV. Princeton University Press (1964)
28. Head, M.R., Prahlad, T.S.: The Boundary-Layer on a Plane of Symmetry. Aeronautical Quarterly 25, 293–304 (1974)
29. Prandtl, L.: Über Reibungsschichten bei dreidimensionalen Strömungen. Festschrift zum 60. Geburtstage von A. Betz, Göttingen, Germany, pp. 134–141 (1945)
30. Eichelbrenner, E.A.: Three-Dimensional Boundary Layers. Annual Review of Fluid Mechanics 5, 339–360 (1973)
31. Tani, I.: History of Boundary-Layer Theory. Annual Review of Fluid Mechanics 9, 87–111 (1977)
32. Treadgold, D.A., Beasley, J.A.: Some Examples of the Application of Methods for the Prediction of the Influence of Boundary-Layer Transition on Sheared Wings. AGARD R-602, 2-1–2-11 (1973)
33. Krause, E., Hirschel, E.H., Kordulla, W.: Fourth-Order "Mehrstellen" Integration for Three-Dimensional Turbulent Boundary Layers. In: Proc. AIAA Comp. Fluid Dynamics Conference, Palm Springs, Calif., pp. 77–92 (1973); also Computers and Fluids 4, 77–92 (1976)
34. Schmitt, V., Cousteix, J.: Etude de la couche limite tridimensionelle sur une aile en flèche. ONERA Rapport Technique No. 14/1713 AN (1975)
35. Hirschel, E.H., Jawtusch, V.: Nachrechnung des experimentell ermittelten Übergangs laminar-turbulent an einem gepfeilten Flügel. In: Maurer, F. (ed.): Beiträge zur Gasdynamik und Aerodynamik. DLR-FB 77-36, pp. 179–190 (1977)
36. Fannelop, T.K., Krogstad, P.A.: Three-Dimensional Turbulent Boundary Layers in External Flows: A Report on Euromech 60. J. Fluid Mech. 71, 815–826 (1975)
37. Nash, J.F., Patel, V.C.: Three-Dimensional Turbulent Boundary-Layers. ISBC Technical Books, Atlanta, GA (1972)
38. Adams Jr., J.C.: Numerical Calculation of the Subsonic and Transonic Turbulent Boundary Layer on an Infinite Yawed Airfoil. AEDC-TR-73-112 (1973)

39. Schwamborn, D.: Laminare Grenzschichten in der Nähe der Anlegelinie an Flügeln und flügelähnlichen Körpern mit Anstellung (Laminar Boundary Layers in the Vicinity of the Attachment Line at Wings and Wing-Like Bodies at Angle of Attack). Doctoral thesis, RWTH Aachen, Germany, also DFVLR-FB 81–31 (1981)
40. Cebeci, T., Cousteix, J.: Modeling and Computation of Boundary-Layer Flows, 2nd edn. Horizons Publ., Springer, Long Beach, Heidelberg (2005)
41. Mangler, W.: Zusammenhang zwischen ebenen und rotationssymmetrischen Grenzschichten in kompressiblen Flüssigkeiten. ZAMM 28, 97–103 (1948)
42. Hirschel, E.H.: Basics of Aerothermodynamics, AIAA, Reston, Va. Progress in Astronautics and Aeronautics, vol. 204. Springer, New York (2004)

9

Laminar-Turbulent Transition and Turbulence

The state of the boundary layer, laminar or turbulent, influences the drag, the performance of the wing, of stabilization and control devices etc., and the flight characteristics of an airplane. A given flight vehicle is considered as "transition insensitive", if the location of laminar-turbulent transition does not affect these items.

A flight vehicle can be "transition sensitive", if, for instance, for economical and ecological reasons the drag of the vehicle must be reduced. Usually, then the determination of the transition location and/or its control becomes a very big challenge.

Our aim is to provide the reader—if he is confronted with a transition-sensitive flow problem—with this chapter an overview of the important stability/instability and transition phenomena/mechanisms present in three-dimensional attached viscous flow. Several of the important topics are discussed for two-dimensional flow only. Our understanding is that the results hold for weakly three-dimensional flow, too. Not intended is a review of the vast literature of the field.

Because of the high complexity of laminar-turbulent transition, much room is given to this topic. Turbulence phenomena and models as well as general flow control issues are treated only superficially. The separate consideration of laminar-turbulent transition and turbulence is the rule today. But there are flow cases, where this is not permitted. Here we do not consider such cases. We point to [1] and [2], where the concept of flow-physics modelling was proposed. This concept asks for a concurrent treatment of both transition and turbulence where it is necessary.

9.1 Laminar-Turbulent Transition—An Introduction

When the state of the attached viscous flow changes from the laminar to the turbulent one, the skin friction rises, and the displacement property of the attached viscous flow changes as well as the separation behavior. The thermal state of the surface is affected, Sub-Section 2.3.3, regarding both thermal surface effects and thermal loads.

In many cases, the overall flow behavior does not depend much on the transition location. Then the flow is considered to be transition insensitive.

E.H. Hirschel, J. Cousteix, and W. Kordulla, *Three-Dimensional Attached Viscous Flow*,
DOI: 10.1007/978-3-642-41378-0_9,

Laminar-turbulent transition and in particular its location are of importance, if skin-friction drag including laminar flow control and/or if thermal loads are critical. In high-speed airbreathing flight, for instance, viscous drag, engine inlet performance, thermal-surface effects and thermal loads depend strongly on the location of laminar-turbulent transition, see, e.g., the case of the National Aerospace Plane (NASP) of the USA [3].

We give here an introduction to laminar-turbulent transition in the flight-speed domain up to low supersonic Mach numbers. For the higher Mach number domain see, e.g., [4].

The difficulties associated with the determination of the location and extent of the laminar-turbulent transition zone are still the insufficient understanding of several of the involved phenomena on the one hand, and the deficits of the simulation means on the other hand. This holds for both ground-facility and computational simulation, although for the former high Reynolds-number, low disturbance-level facilities are available today. These consist of several research facilities and also of large transonic wind-tunnel facilities for industrial purposes which permit—partly even from low subsonic up to low supersonic speeds—true and independent Mach number, Reynolds number and also dynamic pressure simulation [5]–[8].

In the classical ground-simulation facilities, including transonic facilities, basically the low attainable Reynolds numbers, the disturbance environment, which the tunnel poses for the boundary layer on the model,[1] and possibly also the thermal state of the model surface are the problems.[2]

If the Reynolds number (though lower than in flight) is large enough for laminar-turbulent transition to occur in a ground-simulation facility, the shape and location of the transition zone may be wrong due to the wrong disturbance environment and the wrong thermal state of the model surface.

If the Reynolds number is too low, artificial turbulence triggering must be employed (where shape and location of the transition zone must have been somehow guessed). In this case, the Reynolds number still must be large enough to sustain the artificially created turbulence. If that is the case, turbulent attached flow and strong interaction phenomena and separation can be simulated with sufficient confidence.

In the following sections we describe the different instability and transition phenomena and their dependencies on flow-field parameters and vehicle surface properties, including the thermal state of the surface. We consider further state-of-the-art transition criteria and shortly also transition control in Section 9.7. Generally we follow the presentation given in [4].

---

[1] The (unstable) boundary layer responds to disturbances which are present in flight or in the ground-simulation facility, Sub-Section 9.4.

[2] The thermal state of the surface is especially important in view of research activities in cryogenic and in hypersonic ground-simulation facilities and in view of flight experiments. In both ground-simulation facilities and flight experiments so far the thermal state of the surface often is either uncontrolled or not recorded.

Turbulence was treated already in Chapters 3 and 4. We reconsider it here rather briefly regarding turbulence models in Section 9.6 and with emphasis on control in Section 9.7.

9.2 Instability/Transition Phenomena and Criteria

Laminar-turbulent transition in three-dimensional attached viscous flow is a phenomenon with several possible instability and receptivity mechanisms, which depend on a multitude of flow, surface and environment parameters. In the frame of this book only an overview over the most important issues can be given. Newer introductions to the topic are found in, e.g., [9]–[11].

Possibly two basic transition scenarios can be distinguished, which, however, may overlap to a certain degree:[3]

1. regular transition,
2. forced or by-pass transition.

These two scenarios can be characterized as follows:

- Regular transition occurs if—once a boundary layer is unstable—low-intensity level disturbances, which fit the receptivity properties of the unstable boundary layer, undergo first linear, then non-linear amplification(s), until turbulent spots appear and actual transition to self-sustained turbulence happens. In [12] this is called "transition emanating from exponential instabilities".

  This scenario has been discussed in detail in the classical paper by M.V. Morkovin, [13], see also [14], who considers the (two-dimensional) laminar boundary layer as "linear and non-linear operator" which acts on small disturbances like free-stream vorticity, sound, entropy spots, but also high-frequency vibrations. This begins with linear amplification of Tollmien-Schlichting type disturbance waves [9], which can be modified by boundary-layer and surface properties like those which occur on real flight-vehicle configurations: pressure gradients, thermal state of the surface, three-dimensionality, surface roughness, waviness et cetera. Non-linear and three-dimensional effects, secondary instability and scale changes and finally turbulent spots and transition follow. In flow with adverse pressure gradient the transition scenario is different, no turbulent spots are observed [10].
- Forced transition is present, if large amplitude disturbances, caused, e.g., by surface irregularities, lead to turbulence without the boundary layer acting as convective exponential amplifier, like in the first scenario. Morkovin calls this "high-intensity bypass" transition. Leading-edge or more general,

[3] For a detailed discussion in a recent publication see [12].

attachment-line contamination, Sub-Section 9.3.2, also falls under this scenario.

If the Reynolds number is too small for regular transition to occur, boundary-layer tripping on wind-tunnel models is forced transition on purpose. However, forced transition can also be an—unwanted and hard to control—issue in ground-simulation facilities, if a large disturbance level is present in the test section. Indeed Tollmien-Schlichting instability originally could only be studied properly and verified in (a low-speed) experiment after such a wind-tunnel disturbance level was discovered and systematically reduced. This is the classical work of G.B. Schubauer and H.K. Skramstadt [15].

In supersonic/hypersonic wind tunnels the sound field radiated from the turbulent boundary layers of the tunnel wall was shown by J.M. Kendall, [16], to govern transition at $M = 4.5$, generally at medium and higher supersonic Mach numbers, see the discussion of L.M. Mack in [17].

In the following sub-sections we sketch basic concepts of stability and transition in two-dimensional flow, and kind and influence of the major involved phenomena. We put emphasis on regular transition. Special three-dimensional issues are treated as flight-vehicle effects in Section 9.3.

9.2.1 Some Basic Observations

We consider laminar-turbulent transition of the two-dimensional boundary layer over a flat plate as prototype of regular transition, and ask what can be observed macroscopically at its surface. We employ the usual coordinate convention for two-dimensional boundary layers: the coordinate x is the coordinates tangential to the body surface in stream-wise direction, the rectilinear coordinate normal to the surface is y. Accordingly u is the tangential velocity component and v is the component normal to the body surface.

We study the qualitative behavior of wall shear stress along the surface, Fig. 9.1.[4] We distinguish three branches of τ_w. The laminar branch (I)—the disturbance-reception branch—is sketched in accordance with $\tau_w \propto x^{-0.5}$ (Blasius boundary layer), and the turbulent branch (III) with $\tau_w \propto x^{-0.2}$ (viscous sub-layer), Appendix B.3.1. We call the distance between the location of primary instability x_{cr} (critical point) and the "upper" location of transition $x_{tr,u}$ the transitional branch (II). It consists of the instability sub-branch (IIa) between x_{cr} and the "lower" location of transition $x_{tr,l}$, and the transition sub-branch (IIb) between $x_{tr,l}$ and $x_{tr,u}$. The instability sub-branch overlaps with the laminar branch (see below).

Consider in Fig. 9.1 the point of primary instability x_{cr}. Upstream of x_{cr} the laminar boundary layer is stable, i.e., a small disturbance introduced into it will be damped out. At x_{cr} the boundary layer is neutrally

[4] The heat flux in the gas at the wall with fixed wall temperature shows a similar qualitative behavior.

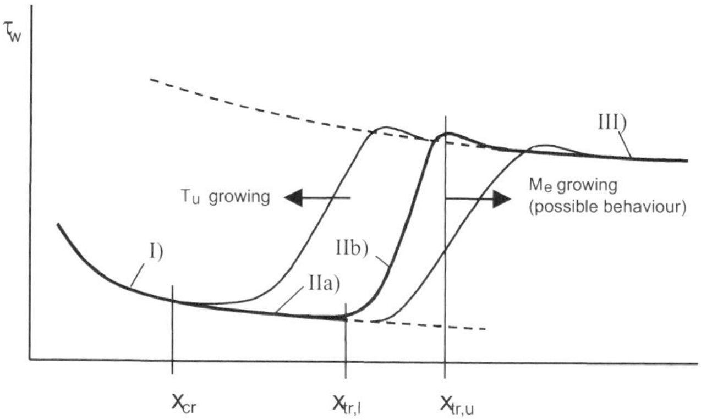

Fig. 9.1. Schematic of behavior of wall shear stress τ_w in flat-plate boundary-layer flow undergoing laminar-turbulent transition [4]: **I**) laminar branch, **II**) transitional branch with **IIa**) instability sub-branch and **IIb**) transition sub-branch, and **III**) turbulent branch of the boundary layer. x_{cr} is the location of primary instability, $x_{tr,l}$ the "lower" and $x_{tr,u}$ the "upper" location of transition.

stable, and downstream of it is unstable. Disturbances there trigger Tollmien-Schlichting waves (normal modes of the boundary layer) whose amplitudes grow rather slowly, however exponentially.[5] Secondary instability sets in after the Tollmien-Schlichting amplitudes have reached approximately 1 per cent of u_e, i.e. at amplitudes where non-linear effects are still rather small regarding the (primary) Tollmien-Schlichting waves. Finally turbulent spots appear and the net-production of turbulence begins (begin of sub-branch IIb). This location is the "lower" location of transition, $x_{tr,l}$. At the "upper" location of transition, $x_{tr,u}$, the boundary layer is fully turbulent.[6] This means that now the turbulent fluctuations transport fluid and momentum towards the

[5] Tollmien-Schlichting waves can propagate with the wave vector aligned with the main-flow direction ("normal" wave as a two-dimensional disturbance, wave angle $\psi = 0$) or lying at a finite angle to it ("oblique" wave as three-dimensional disturbance, wave angle $\psi \neq 0$). The most amplified Tollmien-Schlichting waves are usually in two-dimensional low-speed flows the two-dimensional waves, and in two-dimensional supersonic and hypersonic flows the oblique waves.

[6] In [12] the branches I and II are, less idealized, divided into five stages : 1) disturbance reception (branch I ahead of x_{cr} and part of sub-branch IIa), 2) linear growth of (unstable) disturbances (largest part of sub-branch IIa), 3) non-linear saturation (towards the end of sub-branch IIa), 4) secondary instability (towards the very end of sub-branch IIa), 5) breakdown (begin of sub-branch IIb). The term "breakdown" has found entry into the literature. It is a somewhat misleading term in so far as it suggests a sudden change of the (secondary unstable) flow into the turbulent state. Actually it means the "breakdown" of identifiable structures in the disturbance flow. In sub-branch IIb a true "transition" into turbulence occurs. Sub-branch IIb, i.e., the length $x_{tr,u}$ - $x_{tr,l}$, can be rather large, especially in high-speed flows.

surface such that the full time-averaged velocity profile shown in Fig. 9.4 b) develops. However, in reality the picture is not that simple [18]. It appears that this location lies still in the intermittency region $\triangle x^{\shortmid}_{tr}$ (see below), where the intermittency factor is approximately 0.5.

The length of the transition region, related to the location of primary instability, can be defined either as:

$$\triangle x_{tr} = x_{tr,l} - x_{cr}, \tag{9.1}$$

or as

$$\triangle x_{tr} = x_{tr,u} - x_{cr}. \tag{9.2}$$

In Fig. 9.1 some important features of the transition region are indicated:

- The time-averaged ("mean") flow properties practically do not deviate in the instability sub-branch (IIa) between x_{cr} and $x_{tr,l}$ from those of laminar flow (branch I). This is an important feature, because it permits the formulation of stability and especially transition criteria and models based on the properties of the laminar flow branch. Therefore the accurate knowledge of the laminar flow is of very large importance for practical instability and transition predictions, the latter still based on empirical or semi-empirical models and criteria.
- The transition sub-branch (IIb) (intermittency region), i.e. $\triangle x^{\shortmid}_{tr} = x_{tr,u}$ - $x_{tr,l}$, usually is very narrow.[7] It is characterized by the departure of τ_w from that of the laminar branch and by its joining with that of branch III. For boundary-layer edge flow Mach numbers $M_e \gtrapprox 4$ to 5 the (temporal) amplification rates of disturbances can decrease with increasing Mach number, therefore a growth of $\triangle x^{\shortmid}_{tr}$ is possible. In such cases transition criteria based on the properties of the laminar flow branch would become questionable. The picture in reality, however, is very complicated, as was shown first by Mack in 1965 [20]. We will come briefly back to that later.
- At the end of the transitional branch (II), $x_{tr,u}$, the wall shear stress overshoots somewhat that of the turbulent branch (III). This overshoot occurs also for the heat flux in the gas at the wall.
- With increasing disturbance level, for example increasing free-stream turbulence T_u in a ground-simulation facility, the transition sub-branch (IIa) will move upstream while becoming less narrow, see, e.g., [21]. When disturbances grow excessively and transition becomes forced transition, transition criteria based on the properties of the laminar flow branch (I) become questionable.

[7] If $\triangle x^{\shortmid}_{tr}$ is large compared to a characteristic body length—a problem especially appearing in turbo-machinery—, this must be taken into account in the turbulence model of an employed computation method, see, e.g., [19].

– In general it can be observed that boundary-layer mean flow properties, which destabilize the boundary layer, see Sub-Section 9.2.3, shorten the length of sub-branch IIa ($x_{tr,l}$ - x_{cr}) as well as that of sub-branch IIb ($x_{tr,u}$ - $x_{tr,l}$). The influence of an adverse stream-wise pressure gradient is strongest pronounced in this regard. If the mean flow properties have a stabilizing effect, the transition sub-branches IIa and IIb become longer.

9.2.2 Outline of Stability Theory

We sketch now some features of linear stability theory. This will give us insight into the basic dependencies of instability but also of transition phenomena [9, 12]. Of course, linear stability theory does not explain all of the many phenomena of regular transition which can be observed. It seems, however, that at a sufficiently low external disturbance level, e.g., in free flight, linear instability is the primary cause of regular transition [13, 17].

For the sake of simplicity we consider only the two-dimensional incompressible flat plate case (Tollmien-Schlichting instability). The basic approach and many of the formulations, e.g., concerning temporally and spatially amplified disturbances, however are the same for both incompressible and compressible flow [22, 23].

Tollmien-Schlichting theory is based on the introduction of split flow parameters $\widehat{q} = q + q'$ into the Navier-Stokes equations and their linearization (q denotes mean flow, and q' disturbance flow). It follows the assumption of parallel boundary-layer mean flow, i.e., $v \equiv 0$.[8] The consequence—following from the continuity equation—is $\partial u/\partial x \equiv 0$. Hence in this theory only a mean flow profile $u(y)$ is considered, without dependence on x. Therefore we speak about linear and local stability theory. The latter means that only locally, i.e., in locations x on the surface under consideration, which can be arbitrarily chosen, stability properties of the boundary layer are investigated.[9]

The disturbances q' are then formulated as sinusoidal disturbances:[10]

$$q'(x, y, t) = q'_A(y)e^{i(\alpha x - \omega t)}. \tag{9.3}$$

Here $q'_A(y)$ is the complex disturbance amplitude as function of y, and α and ω are parameters regarding the disturbance behavior in space and time.

The complex wave number α is with $i = \sqrt{-1}$:

---

[8] In reality this holds only for very large Reynolds numbers, see Section 4.5, page 95. Cases exist in which this assumption is critical, because information of the mean flow is lost, which is of importance for the stability/instability behavior of the boundary layer ("non-parallel effects" [12]).

[9] Non-local stability theory and methods, see Sub-Section 9.5.1, are based on parabolized stability equations (PSE). With them the stability properties are investigated taking into account the whole boundary-layer domain of interest.

[10] This form results, because the linearized equations (see below) are such that the coefficients depend on y only.

$$\alpha = \alpha_r + i\alpha_i, \tag{9.4}$$

with

$$\alpha_r = \frac{2\pi}{\lambda_x}, \tag{9.5}$$

λ_x being the length of the disturbance wave propagating in x-direction.

The complex circular frequency ω reads

$$\omega = \omega_r + i\omega_i, \tag{9.6}$$

with

$$\omega = 2\pi f, \tag{9.7}$$

f being the complex frequency of the wave.

The complex phase velocity c is:

$$c = c_r + ic_i = \frac{\omega}{\alpha}. \tag{9.8}$$

Temporal amplification of an amplitude A is found, with α real-valued, by

$$\frac{1}{A}\frac{dA}{dt} = \frac{d}{dt}(lnA) = \omega_i = \alpha_r c_i, \tag{9.9}$$

and spatial amplification by

$$\frac{1}{A}\frac{dA}{dx} = \frac{d}{dx}(lnA) = -\alpha_i. \tag{9.10}$$

We see that a disturbance is amplified, if $\omega_i > 0$, or $\alpha_i < 0$. It is damped, if $\omega_i < 0$, or $\alpha_i > 0$, and neutral, if $\omega_i = 0$, or $\alpha_i = 0$. A spatially growing disturbance, for instance, would be given, if $\omega_i = 0$, $\alpha_i < 0$.

The total amplification rate in the case of temporal amplification follows from eq. (9.9) with

$$\frac{A(t)}{A_0} = e^{\int_{t_0}^{t} \omega_i dt}, \tag{9.11}$$

and in the case of spatial amplification from eq. (9.10) with

$$\frac{A(x)}{A_0} = e^{\int_{x_0}^{x} (-\alpha_i) dx}. \tag{9.12}$$

If we assume ω_i or $-\alpha_i$ to be constant in the respective integration intervals, we observe for the amplified cases from these equations the unlimited exponential growth of the amplitude A which is typical for linear stability theory with

$$\frac{A(t)}{A_0} = e^{\omega_i t}, \tag{9.13}$$

and

$$\frac{A(x)}{A_0} = e^{-\alpha_i x} \tag{9.14}$$

respectively.

Introducing a disturbance stream function $\Psi'(x, y, t)$, with $\Phi(y)$ as the complex amplitude

$$\Psi'(x, y, t) = \Phi(y) e^{i(\alpha x - \omega t)} \tag{9.15}$$

into the linearized and parallelized Navier-Stokes equations finally the Orr-Sommerfeld equation is found:[11]

$$(u - c)(\Phi_{yy} - \alpha^2 \Phi) - u_{yy}\Phi = -\frac{1}{\alpha Re_\delta}(\Phi_{yyyy} - 2\alpha^2 \Phi_{yy} + \alpha^4 \Phi). \tag{9.16}$$

The properties of the mean flow, i.e., the tangential boundary-layer velocity profile, appear as $u(y)$ and its second derivative as $u_{yy} = d^2u/dy^2(y)$. The Reynolds number Re_δ on the right-hand side is defined locally with boundary-layer edge data and the boundary-layer thickness δ:[12]

$$Re_\delta = \frac{\rho_e u_e \delta}{\mu_e}. \tag{9.17}$$

Obviously stability or instability of a boundary layer depends locally on the mean-flow properties u, u_{yy}, and the Reynolds number Re_δ. A typical stability chart for flat-plate flow is sketched in Fig. 9.2. The boundary layer is temporally unstable in the hatched area ($c_i > 0$, see eq. (9.9)) for $0 < \alpha \leqq \alpha_{max}$ and $Re \geqq Re_{cr}$, Re_{cr} being the critical Reynolds number. For $Re \gg Re_{cr}$ we see that the domain of instability shrinks asymptotically to zero, the boundary layer becomes stable again.

For large Re_δ the right-hand side of eq.(9.16) can be neglected which leads to the Rayleigh equation:

$$(u - c)(\Phi_{yy} - \alpha^2 \Phi) - u_{yy}\Phi = 0. \tag{9.18}$$

Stability theory based on this equation is called "inviscid" stability theory. This is sometimes misunderstood. Although the viscous terms in the Orr-Sommerfeld equation are neglected, stability properties of viscous flow with adverse pressure gradient can properly be investigated with it.

---

[11] $()_{yy}$ et cetera stands for twofold differentiation with respect to y et cetera.

[12] The boundary-layer thickness δ is the 99 per cent thickness. In practical criteria usually the displacement thickness δ_1 or the momentum thickness δ_2 is employed.

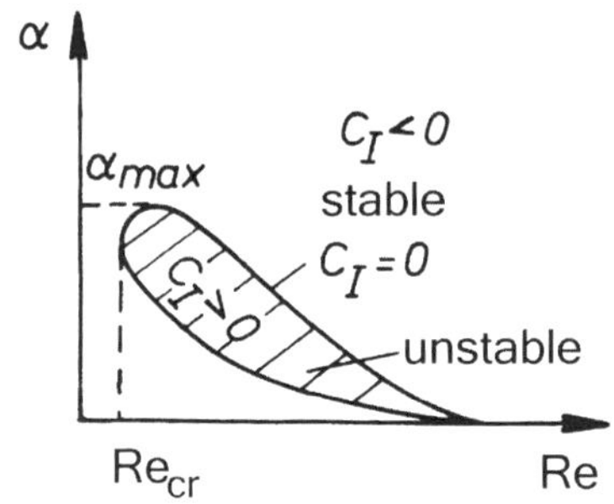

Fig. 9.2. Schematic of a temporal stability chart of a boundary layer at a flat plate ($c_I \equiv c_i$) [24].

9.2.3 Inviscid Stability Theory and the Point-of-Inflection Criterion

Inviscid stability theory gives insight into instability mechanisms with the point-of-inflection criterion, which follows from the Rayleigh equation. It says basically, [9, 12], that the presence of a point of inflection is a sufficient condition for the existence of amplified disturbances with a phase speed $0 \leqq c_r \leqq u_e$.[13] In other words, the considered boundary-layer profile $u(y)$ is unstable, if it has a point of inflection:[14]

$$\frac{d^2u}{dy^2} = 0 \tag{9.19}$$

lying in the boundary layer at y_{ip}:

$$0 < y_{ip} \leqq \delta. \tag{9.20}$$

The stability chart of a boundary layer with a point-of-inflection deviates in a typical way from that without a point of inflection. We show such a stability chart in Fig. 9.3.

For small Re the domain of instability has the same form as for a boundary layer without point of inflection, Fig. 9.2. This is the viscous instability part of the chart. For large Re its upper boundary reaches an asymptotic inviscid limit at finite wave number α. For large Re the boundary layer thus remains unstable.

When does a boundary-layer profile have a point of inflection? We remember Fig. 4.3 and Table 4.1 in Sub-Section 4.5. There the results of the discussion of the generalized wall-compatibility conditions, eqs. (4.57) and

[13] Because it is only a sufficient, not a necessary condition, a velocity profile without point of inflection can be unstable, see above the flat-plate case.

[14] The reader is warned that this is a highly simplified discussion. The objective is only to arrive at insights into the basic instability behavior, not to present detailed theory.

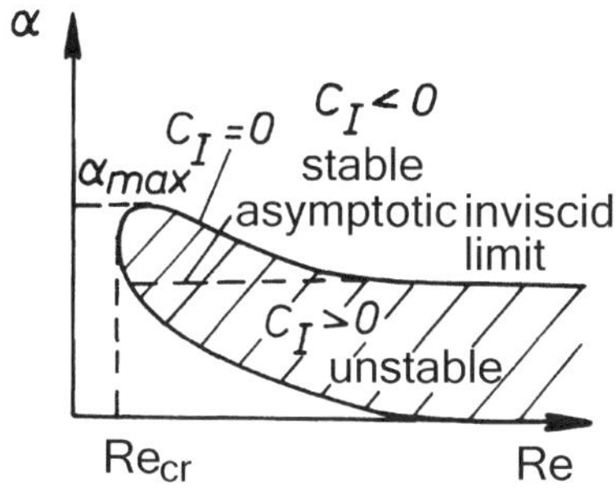

Fig. 9.3. Schematic of a temporal stability chart of a boundary-layer with inviscid instability ($c_I \equiv c_i$) [24].

(4.58), are given. We treat here only the two-dimensional case with the first of these equations, and recall that a point of inflection exists away from the surface at $y_{ip} > 0$, if the second derivative of $u(y)$ at the wall is positive: $d^2u/dy^2|_w > 0$. In a Blasius boundary layer, the point of inflection lies at $y = 0$.

With the help of Table 4.1 on page 94 we find:

- A boundary layer is destabilized by an adverse pressure gradient ($\partial p/\partial x > 0$), by heating,[15] i.e., a heat flux from the surface into the boundary layer ($\partial T_w/\partial y|_{gw} < 0$) and by blowing through the surface ($v_w > 0$).

The destabilization by an adverse pressure gradient is the classical interpretation of the point-of-inflection instability. Regarding turbulent boundary layers on bodies of finite length and thickness, it can be viewed in the following way: downstream of the location of the largest thickness of the body we have $\partial p/\partial x > 0$, hence a tendency of separation of the laminar boundary layer. Point-of-inflection instability signals the boundary layer to become turbulent, i.e., to begin the lateral transport of momentum (in general also of energy and mass, the latter in gas mixtures in chemical non-equilibrium [4]) by turbulence fluctuations towards the body surface. The ensuing time-averaged turbulent boundary-layer profile is fuller than the laminar one, Fig. 9.4, which reduces the tendency of separation.[16] Although the skin-friction drag goes up, total drag remains small, because the pressure drag remains small. The flat-plate boundary layer is a special case, where this does not apply.

If an adverse pressure gradient is too strong, the ordinary transition sequence will not happen. Instead the (unstable) boundary layer separates

[15] In this case the result holds only for an air or gas boundary layer, Section 4.5, page 93.

[16] Due to the larger amount of tangential momentum flux close to the surface, the turbulent boundary layer can negotiate a larger adverse pressure gradient than the laminar one.

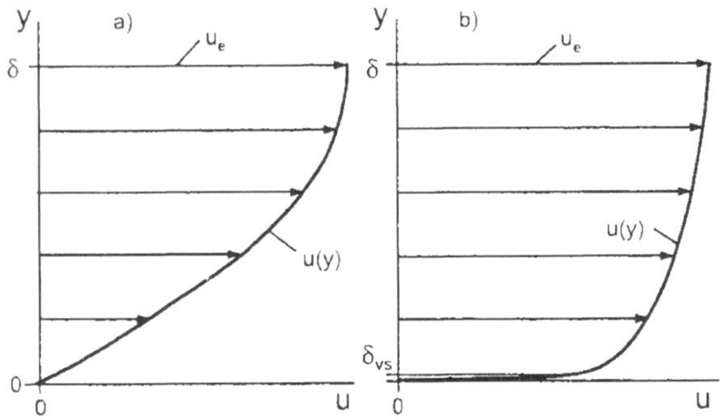

Fig. 9.4. Schematic of two-dimensional tangential velocity profiles and boundary-layer thicknesses δ of a) laminar and b) turbulent flow, as well as the thickness of the laminar sub-layer δ_{vs} of turbulent flow in b) [4] (the two boundary layers of course have not equal thicknesses).

and forms a usually very small and flat separation bubble. At the end of the bubble the flow re-attaches then turbulent (separation-bubble transition, see, e.g., [12] and [25] with further references).

- An air boundary layer is stabilized by a favorable pressure gradient ($\partial p/\partial x < 0$), by cooling, i.e., a heat flux from the boundary layer into the surface ($\partial T_w/\partial y|_{gw} > 0$) and by suction through the surface $v_w < 0$.

These results basically also hold for high speed/hypersonic flow, where a generalized point of inflection appears [4].

9.2.4 The Thermal State of the Surface, Compressible Flow

It is sometimes overlooked that the wall-temperature ratio T_w/T_∞ is also a similarity parameter, Section 3.4. Thermal surface effects in attached viscous flow were demonstrated in Sub-Section 2.3.3. Regarding laminar-turbulent transition we note that the thermal state of the surface—both the wall temperature and the temperature gradient in the gas normal to the wall—is an important parameter regarding stability or instability not only in compressible boundary layers, but in any boundary layer.

This becomes evident from the point-of-inflection criterion. The compatibility condition, eq. (4.57), which shows the influence of the flow parameters on the point of inflection, includes the wall-normal temperature gradient in the gas at the wall. Of course, this gradient interacts with the other parameters. The viscosity term $1/\mu$ in front of the bracket of that equation acts as a modifier, reflecting the influence of the wall temperature. Because both the temperature gradient and the wall temperature can play a role, the result is that the stability—also of incompressible boundary layers—can be affected by the thermal state of the surface.

If laminar-turbulent transition is important in the design of a flight vehicle, one must be aware of this fact. It concerns wind-tunnel measurements—with cryogenic tunnels playing an extra role—, possible boundary-layer control devices, and, in particular, viscous and thermo-chemical thermal surface effects in supersonic and hypersonic flight.

The original formulation of stability theory for compressible flow of L. Lees and C.C. Lin, [26], with the generalized point of inflection, led to the result that sufficient cooling can stabilize the boundary layer in the whole Reynolds and Mach number regime of flight [27].

This was an interesting finding. It was thought at that time that it could help to reduce the thermal load and drag problems of hypersonic flight vehicles, if such vehicles would fly with cryogenic fuel. An appropriate layout of the airframe surface as heat exchanger would combine both cooling of the surface and stabilization of the attached laminar viscous flow past it. This would be possible even for flow portions with adverse stream-wise pressure gradient, because, as we can see from eq. (4.57), the influence of the pressure-gradient term can be compensated by sufficiently strong cooling. If then the flow past the flight vehicle would not become turbulent, the heating and drag increments due to the occurrence of transition could be avoided.

Unfortunately this conclusion is not true. It was shown almost two decades later by L.M. Mack, [20, 28], that at supersonic and hypersonic Mach numbers higher modes (the so-called "Mack modes") appear, which cannot be stabilized by cooling, on the contrary, they are amplified by it. The first of these higher modes, the "second mode", if the low-speed mode is called first mode, in general is of largest importance at high boundary-layer edge Mach numbers, because it is most amplified. For an adiabatic, flat-plate boundary layer higher modes appear already at edge Mach numbers larger than $M_e \approx 2.2$. We do not discuss this topic further. The interested reader is referred to [4], where it is discussed in detail with many references given.

9.3 Real Flight-Vehicle Effects

The infinitely thin flat plate is the canonical configuration of boundary-layer theory and also of stability and transition research. Basic concepts and fundamental results are gained with and for the boundary-layer flow past it. However, real configurations have a finite volume, are inclined fully and/or in parts against the free-stream, hence pressure gradients are present. The attached viscous flow in general is influenced by a number of effects, which are not present in boundary layers on the flat plate.

We discuss in the following sub-sections some particular transition phenomena which are present in the flow past configurations of typical transonic transport airplanes. Such and other phenomena, which are present at supersonic and hypersonic flight vehicles, are treated in [4].

On the considered flight-vehicle configurations large flow portions exist, which are only weakly three-dimensional. Appreciable three-dimensionality of the viscous flow is found at leading edges of swept wings and stabilization surfaces, at fuselages during flight with angle of attack and at fuselage portions in level flight, see the examples in Chapter 10.

In the following sub-sections we discuss briefly the influence of the most important real flight-vehicle stability and transition effects. Other possible real-vehicle effects like noise of the propulsion system transmitted through the airframe and dynamic aeroelastic surface deformations (vibrations, panel flutter) are difficult to assess quantitatively. To comment on them is not possible in the frame of this book.

9.3.1 Attachment-Line Instability

Primary attachment lines exist at the leading edges of swept wings or stabilization surfaces as well as at the windward sides of fuselages at angle of attack. At large angles of attack—due to separation—secondary and tertiary attachment lines can be present at the leeward side of a fuselage. The canonical attachment line situation in aerodynamics corresponds to an attachment line along the leading edge of a swept wing with constant symmetric profile at zero angle of attack in the span-wise direction, or at the windward symmetry line of a circular cylinder at angle of attack or yaw.

At such attachment lines both inviscid and boundary-layer flow diverge symmetrically with respect to the upper and the lower side of the wing, and to the left and the right hand side of the cylinder at angle of attack, respectively. The infinitely extended attachment line is a useful approximation of reality, which can be helpful for basic considerations and for estimations of local flow properties. We have discussed flow properties of such cases in Section 7.4. We have noted that non-zero flow velocity and hence a boundary layer exists in the direction of the attachment line[17]. This boundary layer can be laminar, laminar unstable, transitional or turbulent. On the infinitely extended attachment line only one of these flow states can exist.

The simplest presentation of an infinite swept attachment-line flow is the swept Hiemenz boundary-layer flow, which is an exact solution of the incompressible continuity equation and the Navier-Stokes equations [9].

This attachment-line flow can be unstable. The (linear) stability model for this flow is the Görtler-Hämmerlin model, which in its extended form gives insight into the stability behavior of attachment-line flow, see, e.g., [29]. Attachment-line flow is the "initial condition" for the, however only initially, highly three-dimensional boundary-layer flow away from the attachment line to the upper and the lower side of the wing or cylinder (see above). The

[17] At a non-swept, infinitely long cylinder an attachment line exists, where the flow comes fully to rest. In the two-dimensional picture this attachment line is just the forward (primary) stagnation point.

correspondingly observed cross-flow instability, see below, has been connected by F.P. Bertolotti to the instability of the swept Hiemenz flow [30].

9.3.2 Leading-Edge Contamination

Consider transition in the boundary layer on a flat plate or on a non-swept wing of finite span. Instability will set in at a certain distance from the leading edge and downstream of it the flow will become fully turbulent by regular transition. If locally at the leading edge a disturbance[18] is present, the boundary layer can become turbulent just behind this disturbance. In that case a "turbulent wedge" appears in the otherwise laminar flow regime, with the typical half angle of approximately 7°, which downstream merges with the turbulent flow, Fig. 9.5 a). Only a small part of the laminar flow regime is affected.[19]

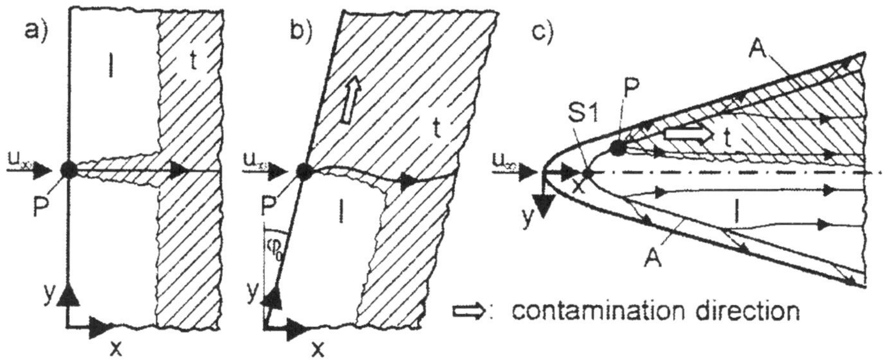

Fig. 9.5. Attachment-line contamination (schematic) [4]: transition forced by a surface disturbance P on a) a flat plate or unswept wing, b) the leading edge of a swept wing, and c) the primary attachment line at the lower side of a flat blunt-nosed delta wing or fuselage configuration (symbols: l: laminar flow, t: turbulent flow, S_1: forward stagnation point, A: primary attachment line). Note that in this figure the coordinates x, y, z are the usual aerodynamic body coordinates.

At the leading edge of a swept wing the situation can be very different, Fig. 9.5 b). A turbulent wedge can spread out in span-wise direction, "contaminating" the originally laminar flow regime between the disturbance location and the wing tip. On a real airplane with swept wings it is the turbulent boundary layer of the fuselage which contaminates the otherwise laminar flow at the leading edge.

[18] This can be a dent on the plate's surface, or an insect cadaver at the leading edge of the wing.

[19] Remember in this context the global characteristic properties of attached viscous flow discussed in Section 4.4.

The phenomenon of leading-edge contamination was discovered independently by W. Pfenninger [31] and M. Gaster [32], see the survey of the field in Section 1.3.

The low-speed flow criterion, of, e.g., N.A. Cumpsty and M.R. Head [33]:

$$Re_{CH} = \frac{0.4 \sin\varphi_0\, u_\infty}{\sqrt{\nu(\partial u_e/\partial x^*)|_{LE}}}, \tag{9.21}$$

illustrates well the physical background. Here $\sin\varphi_0\, u_\infty$ is the component of the external inviscid flow along the leading edge in the span-wise (wing-tip) direction, $\partial u_e/\partial x^*|_{LE}$ the gradient of the external inviscid flow in direction normal to the leading edge (x^*) at the leading edge, and ν the kinematic viscosity. Experimental data show that $Re_{CH} \gtrapprox 100 \pm 20$ is the critical value, and that for $Re_{CH} \gtrapprox 240$ "leading-edge contamination", as it was termed originally, fully happens.

Another criterion, [10], says that along the leading edge the flow remains laminar, if[20]

$$Re_{\delta_{2_{LE}}} = \frac{\sin\varphi_0\, u_\infty\, \delta_{2_{LE}}}{\nu} < 100. \tag{9.22}$$

Here δ_2 is the momentum thickness of the boundary layer along the attachment line, with w being the velocity component on the attachment line in span direction and y the coordinate normal to the surface:

$$\delta_{2_{LE}} = \int_0^\delta \frac{w}{w_e}\left(1 - \frac{w}{w_e}\right) dy|_{LE}. \tag{9.23}$$

For $100 \lessapprox Re_{\delta_{2_{LE}}} \lessapprox 150$, transition is likely, for $Re_{\delta_{2_{LE}}} \gtrapprox 150$, the flow along the leading will be turbulent.

We obtain from the criterion eq. (9.21) the following qualitative picture: the larger the external inviscid flow component in the span-wise (wing-tip) direction, and the smaller the acceleration of the flow normal to the leading edge, the larger the tendency of leading-edge, or more in general, attachment-line contamination. Otherwise only a turbulent wedge would show up from the location of the disturbance in the chord-wise direction, similar to that shown in Fig. 9.5 a), however skewed.

"Contamination" can happen on general attachment lines, for instance, on those at the lower side of a flat blunt-nosed delta wing or fuselage configuration, Fig. 9.5 c). If, for instance, the thermal protection system of a winged re-entry flight vehicle has a misaligned tile lying on the attachment line, turbulence can spread prematurely over a large portion of the lower side of the flight vehicle. This argument was brought forward by D.I.A. Poll [34] in order to explain transition phenomena observed on the Space Shuttle Orbiter during re-entry, see also the discussion in [35].

[20] See also Problem 9.6.

Poll [36] made an extensive study of attachment-line contamination at swept leading edges for both incompressible and compressible flows in the 1970s. Today, still all prediction capabilities concerning attachment-line contamination rely on empirical data.

Leading-edge contamination is an important topic of laminar flow control at swept wings and stabilizers [37]. Insect cadavers, surface distortions due to elements of the high-lift system or the anti-icing system—all to be considered in the context of permissible surface properties, Sub-Section 2.3.3—are potentially causing the contamination. Also the boundary layer of the fuselage which enters the leading edge at a swept wing's root, is a matter of concern, if a wing-root fairing is employed, Section 10.3.

To avoid this effect, a so-called turbulence diverter can be employed. Such a device was proposed in 1965 by M. Gaster [38]. The "Gaster bump" near the root of the leading edge induces a local forward stagnation point, from which a new laminar boundary layer along the leading edge is supposed to evolve. That the Gaster bump works was shown, for instance, in flight tests with the vertical stabilizer of an Airbus A320 [39].

Interesting in this regard is the situation at a forward swept wing. An airplane configuration with such a wing—discussed for instance in [40]—has three primary attachment points: the fuselage nose point and the two wing-tip leading-edge points. From a leading-edge tip attachment point a laminar flow evolves along the leading-edge without being polluted by the fuselage flow. Then permissible surface properties, see above, should be achieved along the leading edge in order to prevent the occurrence of leading-edge contamination.

But it is a question, whether a reduction of the span-wise flow component—now in wing-root direction—is possible, and an increase of the acceleration of the flow normal to the leading edge due to a reduction of the leading-edge radius. Both are flow properties affecting leading-edge contamination, as implied by the criterion eq. (9.21). See in this regard the discussion of Fig. 10.21. Of course, also the phenomenon of cross-flow instability plays a role.

9.3.3 Cross-Flow Instability

Three-dimensional boundary-layer flow is characterized by skewed boundary-layer profiles which can be decomposed into a main-flow profile and into a cross-flow profile, Fig. 2.1 b). With increasing cross flow, i.e. increasing three-dimensionality, the so-called cross-flow instability becomes a major instability and transition mechanism. This observation dates back to the early fifties, when transition phenomena on swept wings became research and application topics. It was found that the transition location with increasing sweep angle of the wing moves forward to the leading edge. The transition location then lies upstream of the location, which is found at zero sweep angle, and which is governed by Tollmien-Schlichting instability. Steady vortex patterns—initially visualized as striations on the surface—were observed in the boundary layer, with the vortex axes lying approximately parallel to the streamlines of the external inviscid flow, see, e.g., [36].

P.R. Owen and D.G. Randall, [41], proposed a criterion based on the properties of the cross-flow profile. The cross-flow Reynolds number χ reads:

$$\chi = \frac{\rho v_{n_{max}} \delta_q}{\mu}. \tag{9.24}$$

Here $v_{n_{max}}$ is the maximum cross-flow velocity in the local cross-flow profile, Fig. 2.3, and δ_q a somewhat vaguely defined boundary-layer thickness found from that profile:[21]

$$\delta_q = \int_0^\delta \frac{v_n}{v_{n_{max}}} \, dy. \tag{9.25}$$

If $\chi \gtrapprox 175$, transition due to cross-flow instability happens. If the main-flow profile becomes unstable first, transition will happen, in the frame of this ansatz, due to the Tollmien-Schlichting instability as in two-dimensional flow.

A criterion given in [10] states that transition due to cross-flow instability occurs, when the Reynolds number Re_{cf} reaches the critical value 150:[22]

$$Re_{\delta_{cf}} = \frac{v_{t_e}}{\nu} \delta_{cf} = 150, \tag{9.26}$$

where, Fig. 2.3,

$$\delta_{cf} = \int_0^\delta -\frac{v_n}{v_{t_e}} \, dy. \tag{9.27}$$

The phenomenon of cross-flow instability was studied so far mainly for low speed flow. Experimental and theoretical/numerical studies have elucidated many details of the phenomenon. Local and especially non-local stability theory has shown that the disturbance wave vector lies indeed approximately normal to the external inviscid streamlines [12]. The disturbance flow exhibits counter-rotating vortex pairs. Their superposition with the mean flow results in the experimentally observed co-rotating vortices. The critical cross-flow disturbances have wave lengths approximately 2 to 4 times of the boundary-layer thickness, compared to the critical Tollmien-Schlichting waves which have wave lengths approximately 5 to 10 times of the boundary-layer thickness. This is the reason why non-parallel effects and surface curvature effects must be regarded, which require non-local stability methods, which are better suited than local methods.

The subsequent transition to turbulent flow can be due to a mixture of cross-flow and Tollmien-Schlichting instability, see in this regard also Section 10.5. Typically cross-flow instability plays a role, if the local flow angle

[21] We take here y instead of z in that figure for the coordinate normal to the surface.

[22] See also Problem 9.6.

is larger than 30°. Tollmien-Schlichting instability comes into play earliest in the region of an adverse pressure gradient. However also fully cross-flow dominated transition can occur [42, 43].

9.3.4 Görtler Instability

The Görtler instability is a centrifugal instability which appears in flows over concave surfaces, but also in other concave flow situations, for instance in the stagnation region of a cylinder. It is an instability which can affect the transition process in coaction with other instabilities. In supersonic and hypersonic flows Görtler vortices can lead to high thermal loads in striation form, for instance at deflected trim or control surfaces, also on inlet ramps, see, e.g., [4]. However, striation heating can also be observed at other parts of a flight vehicle configuration [44].

Consider the boundary-layer flow past curved surfaces in Fig. 9.6. Although we have boundary layers with no-slip condition at the surface, we assume that we can describe the two flow cases with the lowest-order approximation:

$$\rho \frac{U^2}{|R|} = |\frac{dp}{dy}|. \tag{9.28}$$

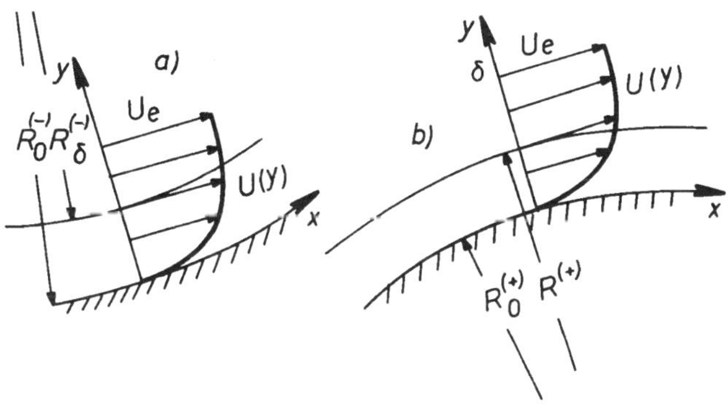

Fig. 9.6. Boundary-layer flow past curved surfaces [24]: a) concave surface, and b) convex surface.

With assumed constant pressure gradient $|dp/dy|$ and constant density, the term U^2 at a location inside the boundary layer $0 \leq y \leq \delta$ must become larger, if we move from R to $R + \Delta R$. This is the case on the convex surface, Fig. 9.6 b). It is not the case on the concave surface, Fig. 9.6 a). As a consequence in this concave case flow particles at R with velocity U attempt to exchange their location with the flow particles at the location $R + \Delta R$ where the velocity U - ΔU is present.

As a further consequence a vortical movement inside the boundary layer can be triggered which leads to stationary, counter-rotating pairs of vortices, the Görtler vortices, with axes parallel to the mean flow direction. They were first described by Görtler [45] in the frame of the laminar-turbulent transition problem (influence of surface curvature on flow instability). Results of early experimental investigations are found in [46] and [47].

Görtler vortices can appear in almost all concave flow situations, also if the body contour is convex, Fig. 9.7.

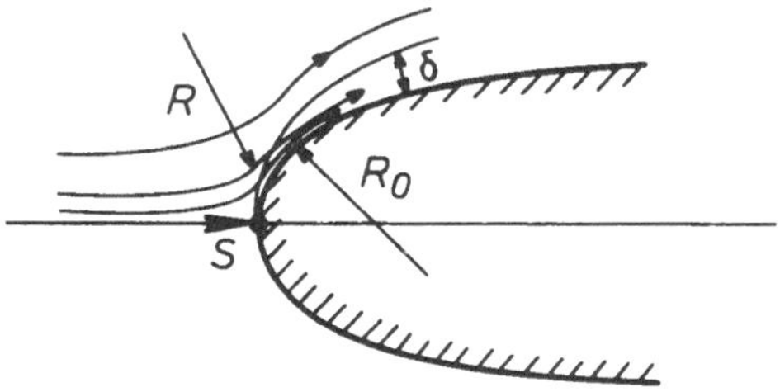

Fig. 9.7. Concave flow situation near a stagnation point lying on a convex surface portion [24].

Görtler vortices can appear in laminar, transitional and turbulent flow. With regard to laminar-turbulent transition, it seems to be an open question whether Görtler instability acts as an operation modifier on the linear amplification process in the sense of Morkovin, [13], or whether it can lead directly to transition (streak breakdown?). For overviews, see, e.g., [47, 48].

The Görtler instability can be treated in the frame of linear stability theory [9, 12]. The Görtler parameter G_ℓ reads:

$$G_\ell = \frac{\rho_e u_e \ell}{\mu_e} \left(\frac{\ell}{R} \right)^{0.5} , \tag{9.29}$$

where ℓ is a characteristic length, for instance the displacement thickness δ_1 of the mean flow. Based on this thickness, laminar-turbulent transition in a two-dimensional boundary layer at a concave surface can approximately be correlated with a critical value $G_{\delta_1} = 38$ [47].

9.3.5 Relaminarization

Relaminarization is a reverse transition process: a turbulent flow becomes laminar again. R. Narasimha [49] distinguishes three principal types of relaminarization: 1) Reynolds number relaminarization, due to a drop of the local (boundary-layer edge based) Reynolds number, 2) Richardson relaminarization, if the flow has to work against buoyancy or curvature forces, and 3) acceleration relaminarization, if the boundary-flow is strongly accelerated [50].

The latter type is of interest for us. Consider a swept wing with leading-edge contamination. The turbulent flow is accelerated from the attachment line away to the lower (pressure) side and the upper (suction) side of the wing.

For acceleration relaminarization in two-dimensional flow a criterion is, [51], see also [52]:

$$K_{crit} = \frac{\nu}{u_e^2}\frac{du_e}{dx} \gtrapprox 2 \cdot 10^{-6}. \tag{9.30}$$

To obtain effectively laminarized flow, $K > 5 \cdot 10^{-6}$ is necessary.

We have a situation, where the flow acceleration has two effects. With increasing acceleration of the flow away from the attachment line, leading-edge contamination becomes less likely, eq. (9.21). At the same time relaminarization would become effective, eq. (9.30). In [53] it is shown for two leading-edge sweep angles ($\varphi_{LE} = 30°$, $60°$) and the free-stream Reynolds number in the range of $Re_\infty = 3 \cdot 10^6$ to $72 \cdot 10^6$, with free-stream Mach numbers $M_\infty = 0.696$ to 1.298, that relaminarization will occur only in a few cases. This result must be seen with reservations, because the above criterion was found in two-dimensional flow, whereas here strong three-dimensionality exists.

The situation is different if the wing is in a high-lift situation. At the high angles of attack—$\alpha = 12°$ to $14°$, for instance, at take-off—the attachment line lies below the leading edge,[23] and the flow accelerates strongly around the convex leading edge and relaminarization is likely to occur. This is still a topic of research, see, e.g., [54].

9.4 Receptivity Issues

Laminar-turbulent transition is connected to disturbances like free-stream turbulence, surface roughness, noise, etc. which enter the laminar boundary layer. If the boundary layer is unstable, these disturbances excite eigenmodes which are at the begin of a sequence of events which finally lead to the turbulent state of the boundary layer, see, e.g., [18]. The kind of entry of the disturbances into the boundary layer is called boundary-layer receptivity.

We give an overview of the main receptivity issues, in particular surface roughness and free-stream turbulence. We treat these two items separately, although in reality they are simultaneously active receptivity mechanisms.

Very important is the fact that, for instance, surface roughness also influences strongly skin friction and the thermal state of the surface if the flow is turbulent, Section 9.6.

[23] High-lift devices usually employ a slat which complicates the flow situation at the leading edge.

9.4.1 Surface Irregularities and Transition

Mechanical surface properties like roughness, steps et cetera can be employed to trigger laminar-turbulent transition and also to enhance transport processes (e.g., cooling, heating), in particular if the flow is turbulent. Mechanical surface properties, however, often turn out to exhibit disadvantageous properties. The surface properties in general should be "sub-critical" in size in order not to influence the flow, i.e, not to lead to—premature—transition, to skin-friction and heat transfer increments. The reason is that the consequences can be unwanted increments of skin-friction drag of a flight vehicle and unwanted changes of the thermal state of the vehicle's surface.

We consider surface irregularities as a sub-set of surface properties, Section 2.3.3. In the context of laminar-turbulent transition "permissible" surface irregularities are sub-critical surface roughness, waviness, steps, gaps et cetera, which are also important in view of fully turbulent flow, Section 9.6.

If a given flight vehicle is transition sensitive, the magnitudes of all permissible values of surface irregularities should be well known, because structural surface tolerances should be as large as possible in order to minimize manufacturing costs.

Surface irregularities in general are not of much concern in fluid mechanics and aerodynamics, because flow past hydraulically smooth surfaces usually is at the center of attention. Surface irregularities are kind of a nuisance which comes with practical applications. Nevertheless, empirical knowledge is available, especially concerning single and distributed surface roughness in two-dimensional flow, see, e.g., [9, 10].

Surface roughness can be characterized by the ratio k/δ_1, where k is the height of the roughness and δ_1 the displacement thickness of the boundary layer at the location of the roughness x_k. The height of the roughness at which it becomes effective—with given δ_1—is the critical roughness height k_{cr}, with the Reynolds number at the location of the roughness, $Re_k = u_e(x_k)k/\nu$, playing a major role. For $k < k_{cr}$ the roughness is sub-critical and does not influence transition: the surface can be considered as hydraulically smooth. This does not necessarily rule out that the roughness influences the instability behavior of the boundary layer, and thus regular transition. For $k > k_{cr}$ the roughness triggers turbulence and we have forced transition. The question then is whether turbulence appears directly at the roughness (effective height of the roughness) or at a certain, finite, distance behind it (critical height of the roughness).

Since a boundary layer is thin at the front part of a flight vehicle, and becomes thicker in down-stream direction, a given surface irregularity may be critical at the front of the vehicle, and sub-critical further downstream.

Actually the effectiveness of surface irregularities to influence or to force transition depends on several flow parameters, including the Reynolds number and the thermal state of the surface, and on geometrical parameters, like configuration and spacing of the irregularities. Important is the observation that with increasing boundary-layer edge Mach number, the height of a roughness must increase drastically in order to be effective. For $M_e \gtrapprox 5$ to 8 the limit of effectiveness seems to be reached [9, 35], in the sense that it becomes extremely difficult, or even impossible, to force transition by means of surface roughness.

Another aspect is that of turbulence tripping in ground-simulation facilities, if the attainable Reynolds number is too small. Boundary-layer tripping in the lower speed regimes is already a problem.[24] In high Mach-number flows boundary-layer tripping might require roughness heights of the order of the boundary-layer thickness in order to trigger turbulence. In such a situation the character of the whole flow field will be changed (over-tripping). If moreover the Reynolds number is not large enough to sustain turbulent flow, the boundary layer will relaminarize.

For further details and also surface roughness/tripping effectiveness criteria, also in view of attachment-line contamination, see, e.g., [35, 56].

We have noted the necessity to define permissible surface properties, if a given flight vehicle is transition sensitive. Permissible surface properties in the sense that clear-cut criteria for sub-critical behavior in the real-flight situation are defined, are scarce. Such criteria generally are only available regarding distributed roughness effects on transition.

Permissible surface properties for low-speed turbulent boundary layers are given for instance in [9]. Data for supersonic and hypersonic turbulent boundary layers are not known. As a rule the height of a surface irregularity must be smaller than the viscous sub-layer thickness in order to have no effect on the wall shear stress and the heat flux in the gas at the wall.

9.4.2 Free-Stream Fluctuations and Transition

Under environment we understand either the atmospheric flight environment of a vehicle or the environment which the sub-scale model of the flight vehicle has in a ground-simulation facility. The question is how the respective environment influences instability and transition phenomena on the flight vehicle or on its sub-scale model [12, 57]. Ideally there should be no differences between the flight environment and the ground-facility environment. That we have to distinguish between these two environments points already to the fact that these environments have different characteristics and different

[24] Major issues are the location of the boundary-layer tripping device (roughness elements, et cetera) on the wind tunnel model, the effectiveness of the device, and the avoidance of over-tripping (by, for instance, too large roughness height), which would falsify the properties (displacement thickness, wall shear stress, heat flux) of the ensuing turbulent boundary layer, see, e.g., [55].

influences on transition. Both in view of scientific and practical, i.e. vehicle design issues, these different influences pose large problems.[25]

The atmosphere, through which a vehicle flies, poses a disturbance environment. Information about the environment appears to be available for the troposphere (up to altitudes of approximately 10 km), but not so much for the stratosphere (above 10 km altitude). Morkovin suggests, [13], see also [14], as a work hypothesis that distribution, intensities and scales of disturbances can be assumed to be similar in the troposphere and the stratosphere. Flight measurements in the upper troposphere (11 km altitude) have shown strong anisotropic air motions with very low dissipation and weak vertical velocity fluctuations [58]. How much the flight speed of the vehicle plays a role for triggering transition is not known. This will be partly a matter of the receptivity properties of the boundary layer.

Much is known of the disturbance environment in ground-simulation facilities, see, e.g., [13, 14, 35]. In supersonic tunnels the sound field radiated from the turbulent boundary layers of the tunnel wall poses a major problem. The quest to create in supersonic and hypersonic ground-simulation facilities a disturbance environment similar to that of free flight (whatever that is) has led to the concept of the "quiet" tunnel, see, e.g., [14] and also [4].

The disturbance environment of a flight vehicle or of its sub-scale wind-tunnel model is very important, because it provides for regular laminar-turbulent transition:

1. The "initial" conditions in flight (atmospheric fluctuations) and in the ground-simulation facility (free-stream turbulence).
2. The "boundary" conditions in flight (surface conditions, engine noise) and in the ground-simulation facility (tunnel-wall noise, support vibration, model surface conditions).

In the aerodynamic practice velocity fluctuations u', v', w', which are also called free-stream turbulence, are the entities of interest. The classical measure is the "level of free-stream turbulence":

$$Tu = \sqrt{\frac{\overline{u'^2} + \overline{v'^2} + \overline{w'^2}}{3u_\infty^2}}. \tag{9.31}$$

If $\overline{u'^2} = \overline{v'^2} = \overline{w'^2}$, this is called isotropic free-stream turbulence. At low speed, the level of free-stream disturbances governs strongly the transition process. The free-stream turbulence of wind tunnels even for industrial measurements should be smaller than $Tu = 0.001$.

[25] As noted above surface irregularities, like surface roughness, and environment aspects are often combined under one heading. We have treated surface irregularities in the preceding sub-section.

A rational and rigorous approach to identify types of disturbances is the consideration of the characteristic values of the system of equations of compressible stability theory, see, e.g., [22]. There the following types of disturbances are distinguished:

- Temperature fluctuations, T', also called entropy fluctuations.
- Vorticity fluctuations, ω_x', ω_y', ω_z'.
- Pressure fluctuations, p', or acoustic disturbances (noise). These are of large importance in supersonic/hypersonic wind tunnels for $M \gtrsim 3$, but also in transonic wind tunnels with slotted or perforated walls.

The environment (free-stream) disturbance properties are of large importance in particular for non-local non-linear instability methods, which are the basis of non-empirical transition prediction methods, see the following Section 9.5.2. These methods need a receptivity model. Actually all types of disturbance-transport equations (non-linear/non-local theories) need initial values in the form of free-stream disturbances. These are also needed for the direct numerical simulation (DNS) of stability and transition problems. In e^N-methods, see below, the N-factor can be chosen depending on, for instance, the degree of free-stream turbulence.

The state of the art regarding boundary-layer receptivity to free-stream disturbances is discussed in [59]. A comprehensive discussion of the problems of receptivity models, also in view of the influence of flight speed and flow-field deformation in the vicinity of the airframe is still missing.

We note in this context that for the computational simulation of turbulent flows by means of transport-equation turbulence models, for instance of $k-\varepsilon$ or $k-\omega$ type, initial values of the turbulent energy k, the dissipation ε or the dissipation per unit turbulent energy ω as free-stream values are needed, too, see, e.g., [60]. A typical value used in many computational methods for the turbulent energy is $k_\infty \approx (0.005\ u_\infty)^2$, whereas ω or ε should be "sufficiently small" [61, 62]. Large eddy simulation (LES) of turbulent flow also needs free-stream initial values. The question is whether in non-empirical transition prediction methods for the free-flight situation, apart from surface vibrations and engine noise (relevance of both?), this kind of "white noise" approach is a viable approach. For the ground-facility situation of course the environment, which the facility and the model pose, must be determined and incorporated in a prediction method [14].

9.5 Prediction of Stability/Instability and Transition

In this section we wish to acquaint the reader with the possibilities to actually predict stability/instability and transition. No review is intended, but a general overview is given with a few references to prediction theories and methods. Transition-prediction theories and methods based on experimental data are treated under the headings "semi-empirical" and "empirical" transition prediction in Sub-Section 9.5.2. Recent developments in non-local and non-linear instability and transition prediction are touched, too.

9.5.1 Stability/Instability Theory and Methods

Theory and methods presented here are in any case methods for incompressible and compressible, mainly two-dimensional but also three-dimensional flow. The thermal boundary conditions are usually only the constant surface temperature or the adiabatic-wall condition, although for most of the methods it should be no problem to implement non-constant boundary conditions.

Linear and Local Theory and Methods. The classical stability theory is a linear and local theory. It describes only the linear growth of disturbances (stage 2—see the footnote on page 205—in branch IIa, Fig. 9.1). Neither the receptivity stage is covered,[26] nor the saturation stage and the last two stages of transition. Extensions to include non-parallel effects are possible and have been made. The same is true for curvature effects. However, the suitability of such measures appears to be questionable, see, e.g., [63].

Linear and local theory is, despite the fact that it covers only stage 2, the basis for the semi-empirical e^N transition prediction methods, which are discussed in Sub-Section 9.5.2.

Linear and local stability methods for two-dimensional and three-dimensional incompressible and compressible flows are, for instance, COSAL (M.R. Malik, 1982 [64]), COSTA (U. Ehrenstein and U. Dallmann, 1989 [65]), CASTET (F. Laburthe, 1992 [66]), LST3D (M.R. Malik, 1997 [67]), COAST (G. Schrauf, 1992 [68], 1998 [69]), LILO (G. Schrauf, 2004 [70]).

Non-Local Linear and Non-Linear Theory and Methods. Non-local theory takes into account the wall-normal and the downstream changes of the mean flow as well as the changes of the amplitudes of the disturbance flow and the wave numbers. Non-local and linear theory also describes only stage 2 in branch IIa, Fig. 9.1. However, non-parallelism and curvature are consistently taken into account which makes it a better basis for e^N methods than local linear theory.

Non-linear non-local theory on the other hand describes all five stages, especially also stage 1, the disturbance reception stage, the latter however not in all respects. Hence, in contrast to linear theory, form and magnitude of the initial disturbances must be specified, i.e., a receptivity model must be employed, Sub-Section 9.4.

Non-local methods are (downstream) space-marching methods for two- or three-dimensional flow that solve a system of disturbance equations, which

[26] Note that the result of linear stability theory is the relative growth of (unstable) disturbances of unspecified small magnitude, eq. (9.11) or (9.12), only.

have space-wise parabolic character. Hence such methods are also called "parabolized stability equations (PSE)" methods. We do not discuss here the parabolization and solution strategies and refer the reader instead to the review article of Th. Herbert [71] and to the individual references given in the following.

Non-local linear stability methods are, for instance, xPSE (F.P. Bertolotti, linear and non-linear (the latter incompressible only), 1991 [72]), PSE method (linear and non-linear) of C.-L. Chang et al., 1991 [73], NOLOS (M. Simen, 1993 [74]).

Non-local non-linear stability methods are, for instance, NOLOT/PSE (M. Simen et al., 1994 [75], see also S. Hein [76]), CoPSE (M.S. Mughal and P. Hall, 1996 [77]), PSE3D (M.R. Malik, 1997 [67]), NELLY (H. Salinas, 1998 [78]), LASTRAC (C.-L. Chang, 2004 [79]).

9.5.2 Transition Models and Criteria

The knowledge of instability and transition phenomena today is rather good. Many stability/instability methods for three-dimensional attached viscous flows are available. The accurate and reliable prediction of the shape, the extent and the location of the transition zone,[27] i.e., the transition sub-branch IIb, Fig. 9.1, for the flight vehicles in the background of our considerations is possible, however, not in any case.

We distinguish three classes of means for transition prediction, namely

- empirical,
- semi-empirical,
- non-empirical

criteria and models. Of these the first two rely fully or partly on experimental data.

A common feature of empirical and semi-empirical criteria is that they are applicable only for the configurational class and in the parameter space for which the respective underlying data base was obtained. This holds in particular for regular but also for forced transition. A data base can have been obtained in a wind tunnel or in flight tests. In both cases limitations regarding accuracy and reliability of the data must be observed. This includes the relevant flow and vehicle surface properties.

We give a short overview over the three classes of criteria and models for transition prediction. Extended discussions of criteria, models and applications in particular for three-dimensional boundary layers are given, for instance, by T. Cebeci and the second author of this book [10]. D. Arnal et al.

[27] The transition zone in reality can be an arbitrarily shaped surface with rather small downstream extent wrapped around the configuration, see, for instance, Section 10.5.

give an outstanding survey of prediction methods for subsonic and transonic flows [18].

Empirical Transition Criteria. Empirical transition prediction is based on criteria derived from experimental data which are obtained in ground-simulation facilities but also in flight-test campaigns.

The empirical criteria usually are local criteria, i.e., they employ local integral boundary-layer and boundary-layer edge-flow properties. This means that the phenomena discussed in Sections 9.2 to 9.4 are only implicitly taken into account. Some of them are explicitly regarded via an employed boundary-layer integral quantity. Data from ground-simulation facilities are possibly falsified by tunnel effects. The lower and the upper transition location often are not explicitly given, i.e., length and shape of the transition sub-branch IIb are not specified. This holds also for the overshoot at the end of this branch.

Due to the nature of empirical criteria, predictions with them must be made in full awareness of the properties of the underlying data bases. If error bars are given, uncertainties of the derived aerodynamic data base and hence design margins of the flight vehicle can be established. In any case parametric guesses can and should be made.

Empirical criteria are discussed in some detail in, for instance, [24]. One of the earliest criterion for two-dimensional flow is that of R. Michel [80]. He correlated the transition location $x_{tr,u}$, Fig. 9.1, measured at airfoils, by plotting the transition Reynolds number based on the momentum-loss thickness δ_2 versus the Reynolds number based on the boundary-layer running length. Because the airfoils had similar surface pressure distributions, a good correlation was obtained.

P.S. Granville correlated flat-plate data with the level of free-stream turbulence Tu by taking into account the location of primary instability x_{cr}, Fig. 9.1 [81]. Criteria for flows with pressure gradient were published by, for instance, L.F. Crabtree [82], D.J. Hall and J.C. Gibbings [83], and J. Dunham [84]. The experimental data were correlated with the local pressure-gradient parameter

$$\lambda_2 = -\frac{(\delta_2)^2}{\nu}\frac{1}{u_e}\frac{dp}{dx}|_{x_{tr,u}}, \qquad (9.32)$$

where ν is the kinematic viscosity and u_e the velocity of the external inviscid flow. Granville, [81], made a more refined correlation with the mean value of the pressure gradient parameter in the transition regime $x_{tr,u}$ - x_{cr}. Dunham as well as Hall and Gibbings introduced in their criteria the level of free-stream turbulence as additional parameter.

Criteria for three-dimensional flows are discussed above in Sub-Sections 9.3.2 (leading-edge contamination) and 9.3.3 (cross-flow instability). We note further the swept-wing criterion for cross-flow instability by D. Arnal et al. [85], who correlate the cross-flow displacement thickness with the stream-wise (t) shape factor $H_{tt} = \delta_{1t}/\delta_{2t}$ for different leading-edge sweep angles. Criteria which take into account effects of three-dimensionality are discussed also in [86].

Semi-empirical Transition Prediction. Semi-empirical transition prediction methods go back to J.L. van Ingen [87], and A.M.O. Smith and N. Gamberoni [88]. They observed independently in the frame of local linear stability theory that, for a given boundary-layer mean flow, the envelope of the most amplified disturbances correlates observed transition locations.

For airfoils it turned out that on average the value, see Sub-Section 9.2.2,

$$\ln \frac{A}{A_0} = \int_{t_0}^{t} \omega_i dt = 9, \tag{9.33}$$

best correlates the measured data, hence the name e^9 criterion. Unfortunately later the "universal constant" 9 turned out to be a "universal variable" N. Already in the data of Smith and Gamberoni the scatter was up to 20 per cent. Now we speak of the e^N criterion, which can be a reliable and accurate transition prediction tool for well defined two-dimensional low-speed flow classes with good experimental data bases, see, e.g., [89].

Basically the e^N method describes only sub-branch IIa, Fig. 9.1. However, compared to the empirical criteria, it takes into account locally—via the shapes of the velocity profiles in that branch—the pressure gradient and also the thermal state of the surface. The disturbance environment can be regarded to some degree. The free-stream turbulence Tu of a wind tunnel, for instance, can be taken into account by introducing $N_{tr} = N_{tr}(Tu)$ [90]. Surface roughness can be treated, too.

The classical solution approach is a local one. This makes it possible to apply data-base methods, i.e., tabulated values found with stability computations for self-similar tangential velocity profiles. With such methods the computation effort can be drastically reduced. Examples for incompressible flow are by H.-W. Stock et al. [91] and for compressible flow by D. Arnal et al. [92].

The e^N method can be used by employing the parabolized stability equations (PSE) approach [18]. The result is a linear, non-local method. With this method surface curvature and non-parallel effects can be taken into account.

For three-dimensional flows the situation becomes complex. Several strategies for the determination of the N-factor or -factors have been proposed [18]. The $N_{TS} - N_{CF}$ method, for instance, works with different N-factors for the Tollmien-Schlichting modes (N_{TS}) and the cross-flow modes (N_{CF}). The different receptivity mechanism—free-stream turbulence and surface roughness—can be taken into account. The interaction between Tollmien-Schlichting and cross-flow waves cannot be described by local linear theory, but can be modelled by reduction schemes [93].

The $N_{TS} - N_{CF}$ method, coupled with boundary-layer or Navier-Stokes codes, today can be applied to flows past realistic flight vehicle configurations, see, e.g., [94]. The example of a flow past an inclined ellipsoid is discussed below in Section 10.5.

Non-empirical Transition Prediction. It appears that non-local non-linear instability (PSE) methods have the potential to become non-empirical transition prediction methods for practical purposes [18]. They are not yet mature enough and still need too large computation power.

The present state of development of non-local non-linear methods appears to permit the prediction of the location of stage 5, i.e., the begin of sub-branch IIb, Fig. 9.8. However, two different combinations of disturbance modes (receptivity problem) lead to small, but significant differences between the solutions (location and initial shape of sub-branch IIb).

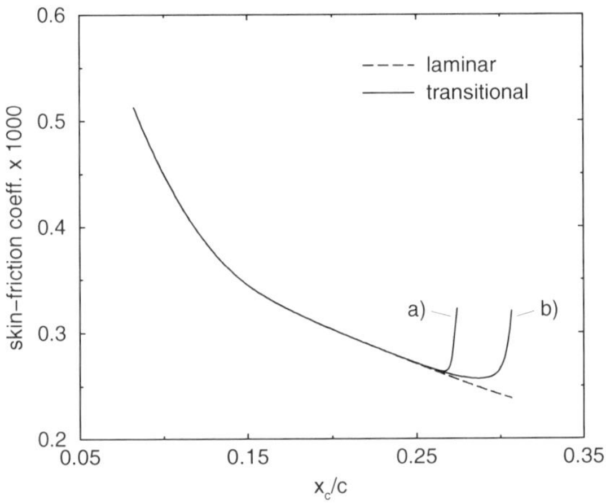

Fig. 9.8. Result of a non-local non-linear method [76]: rise of the skin-friction coefficient in stage 5, i.e., at the begin of sub-branch IIb. Swept wing, $\varphi_{LE} = 21.75°$, $M_\infty = 0.5$, $Re_\infty = 27{\cdot}10^6$. Curves a) and b): two different disturbance mode combinations.

The result of Fig. 9.8 is for a low-speed case, similar results for the high-speed flows of interest are available to a certain extent. It can be expected, in view of the references given in Sub-Section 9.5.1, that at least results similar to those shown can be obtained, after additional research has been conducted, especially also with regard to the receptivity problem.

Very encouraging is in this context that the problem of surface irregularities (transition triggering, permissible properties) seems to become fully amenable for non-empirical prediction methods, see [95], at least for transonic flow past swept wings[28]. That would allow to take into account the influence

[28] An alternative approach with a large potential also for optimization purposes, e.g., to influence the instability and transition behavior of the flow by passive or active means, is to use adjoint equation systems [96].

of weak surface irregularities on regular transition, but also to model transition forced, for example, by strong surface irregularities.

9.6 Turbulence Phenomena and Models

Our main emphasis lies on three-dimensional attached viscous flow. This, however, includes phenomena like shock-wave/boundary-layer interaction at transonic flight, as well as flow-off separation at trailing edges. If the flow past a configuration has become turbulent or is considered to be fully turbulent, turbulence models have to be applied in boundary-layer (BL) or Navier-Stokes methods. The latter are the so-called (statistical) Reynolds-Averaged Navier-Stokes (RANS) methods. The basic modelling approaches for RANS and BL methods were discussed in Sections 3.5 and 4.3, respectively.

We can not discuss here statistical turbulence models for three-dimensional flow. We only note that they are incorporated as zero-, one-, and two-equation models in the many BL and RANS codes in use today. For turbulence data and details of modelling see, e.g., [10, 60], [97]–[101].[29]

Generally it can be said that for turbulent attached viscous flow RANS codes with statistical turbulence models can be used. This also holds, if mild separation is present. Of course, regarding the attached flow domain it is a question how far the locality principle holds. In any case the results and recommendations of assessments and best practice guides (for both see below) must be heeded.

For flow cases with massive separation, hybrid methods, which combine RANS and large-eddy simulation (LES), appear to have the best prospects. The developments in this field can be followed in, e.g., [102] and [103]. A detailed discussion of the present view of the simulation capabilities of hybrid approaches is given in [104].[30]

Of interest for the practitioner is the assessment of turbulence models. Here we point to the proceedings of several European projects, where turbulence models and their applications to two- and three-dimensional, subsonic, transonic and supersonic flow fields past aeronautical shapes are reported [105]–[107]. Regarding hybrid approaches we mention also the newest project "Advanced Turbulence Simulation for Aerodynamic Application Challenges (ATAAC)" [108].

Dedicated in particular to the assessment of drag determination of whole airplane configurations with turbulent flow at transonic flight Mach numbers is the AIAA CFD Drag Prediction Workshop series [109]. In Section 10.3 we

[29] The matter of non-isentropic turbulence in three-dimensional flows was touched in Section 1.3, page 14.

[30] Considered are combinations of RANS and so-called scale resolving simulation (SRS) models, like scale-adaptive simulation (SAS), detached eddy simulation (DES), wall-modelled large-eddy simulation (WMLES) and so on.

discuss for one case the skin-friction line topology on the so-called Common Research Model of this workshop series.

Best practice guides are available, for instance that of ERCOFTAC (http://www.ercoftac.org) and also that of the ATAAC project.

When looking at design problems, one has to depart from the consideration of turbulent flow past hydraulically smooth surfaces. Much is known about the influence of surface roughness on turbulent flow. This knowledge concerns mainly two-dimensional flow but is applicable to weakly three-dimensional flow, too. Surface roughness enhances turbulent transport of momentum, energy (and mass in the case of multi-component flow) toward the wall. Resulting are strong increments of skin-friction and thermal surface effects [9, 10]. This phenomenon is important in flight vehicle design. Accordingly extensions of turbulence models to include the effect of surface roughness were made, see, e.g., [110, 111].

Another phenomenon are density fluctuations in compressible boundary-layer flow. These fluctuations can occur even if stream-wise pressure changes are small, and also in low-speed flows, if large temperature gradients normal to the surface are present.

In turbulence modelling density fluctuations in a turbulent boundary layer can be neglected, if they are small compared to the mean-flow density: $\rho' \ll \rho_{mean}$. Morkovin's hypothesis [112] states that this holds for boundary-layer edge-flow Mach numbers $M_e \lessapprox 5$ in attached viscous flow. Hence we do not consider further this issue.

Since long it is known that the Prandtl number is not constant in attached high speed turbulent flows [113]. Measured turbulent Prandtl numbers in attached low supersonic flow are in the range $0.8 \lessapprox Pr_{turb} \lessapprox 1$. In turbulence models usually a mean constant Prandtl number $Pr_{turb} \approx 0.9$ is employed. It is advisable to check with parametric variations whether the solution for a given flow reacts sensitively to the choice of the (constant) Prandtl number.

9.7 Boundary-Layer Control

The need to employ boundary-layer control in order to reduce skin-friction drag—as discussed briefly in Section 1.3, page 14—will become urgent again.[31] Boundary-layer control on the one hand regards laminar flow control, mainly on wings, empennage surfaces and engine nacelles. Turbulent flow control on the other hand regards mainly the turbulent flow on the fuselage.

Laminar flow control, i.e. moving the transition location on the wing downstream in chord direction, can be achieved by an appropriate shaping of the wing section: natural laminar flow (NLF) control. If in addition suction through the wing surface must be employed, we have hybrid laminar flow (HLF) control. As we have seen in Section 9.3, the different transition

[31] Flow control regarding separation, noise and mixing phenomena is another important topic: active flow control (AFC), see. e.g., [114].

mechanisms pose very large challenges to achieve laminar flow in the complex environment of a swept wing's leading edge.

The possibility of laminar flow control on swept wings or stabilization surfaces in principle has been proven. Turbulent flow control by, for instance riblets or microfabricated electro-mechanical systems (MEMS) has been shown to be very complex too in view of the large surfaces to be covered and the general flight environment.

The in-depth discussion of these topics is out of the scope of this book. We refer the reader to, for instance, [115, 116]. We note however, that many of boundary-layer control devices work well in two-dimensional flow. On flight vehicle configurations in any case a profound knowledge of three-dimensional attached viscous flow is necessary in order to make boundary-layer control a viable and economic option.

9.8 Problems

Problem 9.1. Consider a two-dimensional boundary-layer flow. a) Plot a typical laminar-turbulent transition pattern in terms of the wall shear stress. b) What is the characteristic thickness δ_{char} in the laminar and in the turbulent branch of the boundary layer? c) Give the proportionalities to the running length x from the boundary-layer origin for the thickness of a laminar and a turbulent boundary layer, and of the viscous sub-layer. d) Plot the typical pattern of the characteristic thicknesses.

Problem 9.2. Plot the typical stability charts for viscous and inviscid instability in terms of the wave number α as function of the Reynolds number. Name the governing equation of classical stability theory, and that of inviscid stability theory. What is the basic assumption regarding the boundary-layer mean flow? What is the consequence of this assumption.

Problem 9.3. We consider a two-dimensional incompressible flow on a body such that the external inviscid velocity u_e is given by

$$u_e = k\,x^m \text{ with } m = 1/6.$$

The transition criterion proposed by R. Michel, Sub-Section 9.5.2, is

$$Re_{\delta_{2_{tr}}} = 1,535\,Re_{x_{tr}}^{0,444}.$$

The index 'tr' refers to the transition location which is understood to be the upper location 'tr, u', Fig. 9.1. The two Reynolds numbers are based on the velocity u_e and the kinematic viscosity coefficient ν.

1. Give the values of the Reynolds numbers $Re_{\delta_{2tr}}$ and $Re_{x_{tr}}$ at the transition location. Compare these values to those obtained for the flat plate.

   Hint: From the Falkner-Skan self-similarity solutions, [10], the evolution of the Reynolds number Re_{δ_2} for $u_e = kx^m$ is given by

   $$Re_{\delta_2} = C\sqrt{Re_x},$$

   where C is a function of m. With $m = 1/6$ we get $C = 0.509$. For the flat plate flow, we have $m = 0$ and get $C = 0.664$.
2. What is your observation?
3. a) What does one have to keep in mind when using an empirical transition criterion? b) What is a particularity of Michel's criterion?

Problem 9.4. Sketch a profile $u(y)$, its first $u_y(y)$ and second $u_{yy}(y)$ derivative each for accelerated and decelerated two-dimensional boundary layer flow. What is the wall condition which allows to make the assertion about the profiles? Does it regard only the boundary-layer equation?

Problem 9.5. The shape of the tangential flow profile $u(y)$ governs the inviscid stability behavior. What affects the shape and in particular the function $u_{yy}(y)$ and how is the stability behavior affected? Consider two-dimensional flow.

Problem 9.6. Consider the attachment-line flow of an infinite swept wing with sweep angle φ_0, Fig. 9.9, see also Section 8.3. The velocity component v_∞ is the free-stream velocity, the velocity component normal to the leading edge is $u_\infty = v_\infty \cos\varphi_0$, the component parallel to it $w_\infty = v_\infty \sin\varphi_0$. We assume incompressible flow.

The external inviscid flow component on the wing's surface in x-direction in the vicinity of the leading edge is assume to be $u_e = k\,x$, with $k = 2\,u_\infty/R_N$ (R_N is the radius of the inscribed cylinder of the leading edge), whereas the component in z-direction is $w_e = w_\infty = \text{const.}$, see also Problem 8.1 of Chapter 8.

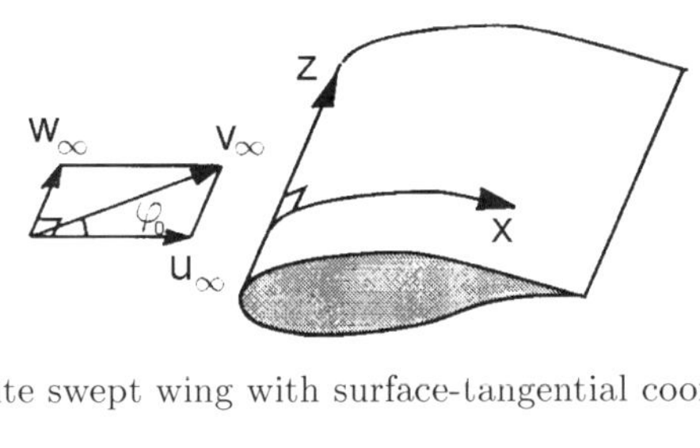

Fig. 9.9. Infinite swept wing with surface-tangential coordinates x and z.

The objective of this problem is to show that the boundary layer can be turbulent along the leading edge of a swept wing and to determine the conditions under which this can happen.

The main-flow and the cross-flow direction in the vicinity of the leading edge are indicated in Fig. 9.10. The angle ψ is related to the velocity components by

$$\sin\psi = \frac{w_e}{v_{t_e}}, \quad \cos\psi = \frac{u_e}{v_{t_e}}.$$

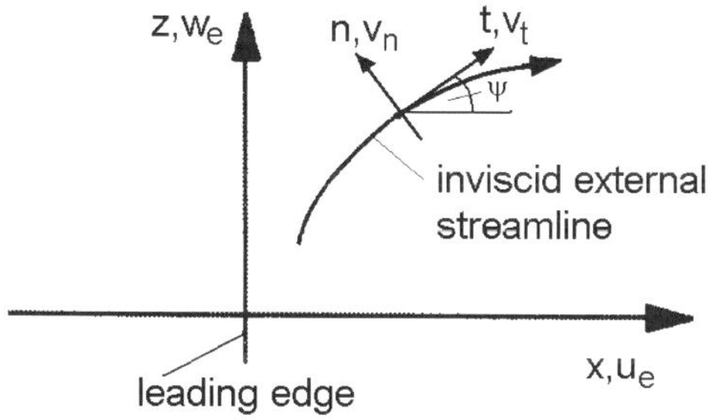

Fig. 9.10. Main-flow (t) and cross-flow (n) direction as well the main-flow (v_t) and the cross-flow (v_n) velocity components, see also Fig. 2.3.

The similarity solution, [10], for the velocity components in the external streamline-oriented coordinate system (the surface-normal coordinate is the y-coordinate, ν is the kinematic viscosity)

$$\frac{u}{u_e} = f'(\eta), \quad \frac{w}{w_e} = g'(\eta) \quad \text{with} \quad \eta = y\sqrt{\frac{k}{\nu}}$$

yields for the functions $f'(\eta)$ and $g'(\eta)$:

$$\int_0^\infty (1 - f')\,\mathrm{d}\eta = 0.6479,$$

$$\int_0^\infty (1 - g')\,\mathrm{d}\eta = 1.0265,$$

$$\int_0^\infty g'(1 - g')\,\mathrm{d}\eta = 0.4044.$$

A possible mechanism of laminar-turbulent transition in the vicinity of the leading edge of a swept wing is due to the *instability of the cross-flow velocity profile* $v_n(y)$, Sub-Section 9.3.3.

Transition occurs when the Reynolds number

$$Re_{\delta_{cf}} = \frac{v_{t_e}}{\nu}\delta_{cf}$$

reaches the critical value $Re_{\delta_{cf}} = 150$.

The thickness δ_{cf} is defined by

$$\delta_{cf} = \int_0^\delta -\frac{v_n}{v_{t_e}}\, dy.$$

Another possible transition mechanism is the *leading-edge contamination*, Section 9.3.2. Along the leading edge, the boundary layer remains laminar if

$$Re_{\delta_{2_{LE}}} = \frac{\sin\varphi_0\, u_\infty\, \delta_{2_{LE}}}{\nu} < 100.$$

The momentum thickness of the boundary layer along the attachment line $\delta_{2_{LE}}$ is defined by:

$$\delta_{2_{LE}} = \int_0^\delta \frac{v_t}{v_{t_e}}\left(1 - \frac{v_t}{v_{t_e}}\right) dy|_{LE} = \int_0^\delta \frac{w}{w_e}\left(1 - \frac{w_e}{w_e}\right) dy|_{LE}.$$

1. Give the expression of $Re_{\delta_{cf}}$ as function of $v_\infty R_N/\nu$, φ_0, and x/R_N, and also the expression of $Re_{\delta_{2_{LE}}}$ as function of $v_\infty R_N/\nu$ and φ_0.
2. Deduce that if the boundary layer remains laminar from the criterion eq. (9.22) (i.e. no leading-edge contamination), the boundary layer remains necessarily laminar from the criterion eq. (9.26) (i.e. no transition due to cross-flow instability).
3. Let c be the chord length of the wing and choose $R_N/c = 0.03$. Draw the curve $Re_{\delta_{2_{LE}}}$ as function of φ_0 for $Re_c = v_\infty c/\nu = 15 \cdot 10^6$; $Re_c = 25 \cdot 10^6$; $Re_c = v_\infty c/\nu = 35 \cdot 10^6$.
4. Determine the maximum sweep angle of the wing for which the boundary layer stays laminar along the leading edge for the three values of the Reynolds number.

Problem 9.7. Which parameters besides the flow parameters can influence the laminar-turbulent transition in a ground-simulation facility?

Problem 9.8. The empirical criterion $(u_k k/\nu)^{0.5} = f(d/k)$ by A.E. von Doenhoff and A.L. Braslow, [117], can be used to predict the triggering of transition by isolated three-dimensional roughnesses.

In the criterion, k is the height of roughness and d is its transverse dimension. The velocity u_k is the value of the velocity in the laminar boundary layer at the distance of the wall $y = k$.

We study the flow of air past a flat plate for the velocity $u_e = 100$ m s^{-1}. The flow can be considered as incompressible.

We want to trigger transition by means of small spheres placed at the distance $x = 75$ mm from the leading edge. These spheres are considered as three-dimensional isolated roughnesses with $k = d$. We seek to calculate the minimum dimension of spheres which trigger transition at $x = 75$ mm.

The velocity profile in the laminar boundary layer, for $y \leqq \delta$, is approximately represented by

$$\frac{u}{u_e} = \sin\left(\frac{\pi}{2}\frac{y}{\delta}\right),$$

where δ is the boundary layer thickness.

1. Calculate the displacement thickness δ_1 at the location $x = 75$ mm. The kinematic viscosity of air in the studied conditions is $\nu = 1.5 \cdot 10^{-5}$ m^2 s^{-1}. We note, Appendix B.3.1, that the displacement thickness of a flat plate boundary layer is given by

$$\delta_1 = 1,721 \frac{x}{\sqrt{Re_x}}.$$

2. By means of the above relation of u/u_e calculate the ratio δ_1/δ. Deduce the value of the boundary layer thickness δ at the point $x = 75$ mm.
3. By means of the transition criterion, determine the minimum height of roughness necessary to trigger transition at the location $x = 75$ mm.

References

1. Hirschel, E.H., Stock, H.-W., Cousteix, J.: Current Turbulence Modelling in Aircraft Design. In: Rodi, W., Martelli, F. (eds.) Proc. 2nd International Symposium on Engineering Turbulence Modelling and Measurements 2, Florence, Italy, May 31-June 2, pp. 665–690. Elsevier Science Publishers B.V., Amsterdam (1993)
2. Hirschel, E.H.: Present and Future Aerodynamic Process Technologies at Dasa Military Aircraft. Viewgraphs presented at the ERCOFTAC Industrial Technology Topic Meeting, Florence, Italy, October 26. Dasa-MT63-AERO-MT-1018, Ottobrunn, Germany (1999)
3. Shea, J.F.: Report of the Defense Science Board Task Force on the National Aerospace Plane (NASP). Office of the Under Secretary of Defense for Acquisition, Washington, D. C. (1988)
4. Hirschel, E.H.: Basics of Aerothermodynamics, AIAA, Reston, Va. Progress in Astronautics and Aeronautics, vol. 204. Springer, New York (2004)
5. Van der Bliek, J.A.: ETW, a European Resource for the World of Aeronautics. The History of ETW in the Context of European Aeronautical Research and Development Cooperation. ETW, Köln-Porz (1996)
6. Green, J., Quest, J.: A Short History of the European Transonic Wind Tunnel ETW. Progress in Aerospace Sciences 47, 319–368 (2011)
7. Polhamus, E.C., Kilgore, R.A., Adcock, J.B., Ray, E.J.: The Langley Cryogenic High Reynolds Number Wind-Tunnel Program. Astronautics and Aeronautics 12(10) (1974)
8. Paryz, R.W.: Upgrades at the NASA Langley Research Center National Transonic Facility. AIAA-Paper 2012-0102 (2012)
9. Schlichting, H., Gersten, K.: Boundary Layer Theory, 8th edn. Springer, Heidelberg (2000)
10. Cebeci, T., Cousteix, J.: Modeling and Computation of Boundary-Layer Flows, 2nd edn. Horizons Publ., Springer, Long Beach, Heidelberg (2005)

11. N.N.: Advances in Laminar-Turbulent Transition Modelling. NATO Research and Technology Organisation (RTO), RTO-EN-AVT-151 (2008) ISBN 978-92-837-0900-6
12. Schmid, P.J., Henningson, D.S.: Stability and Transition in Shear Flows. Springer, Heidelberg (2001)
13. Morkovin, M.V.: Critical Evaluation of Transition from Laminar to Turbulent Shear Layers with Emphasis on Hypersonically Travelling Bodies. Air Force Flight Dynamics Laboratory, Wright-Patterson Air Force Base, AFFDL-TR-68-149 (1969)
14. Reshotko, E.: Boundary-Layer Stability and Transition. Annual Review of Fluid Mechanics 8, 311–349 (1976)
15. Schubauer, G.B., Skramstadt, H.K.: Laminar Boundary Layer Oscillations and Transition on a Flat Plate. NACA Rep. 909 (1943)
16. Kendall, J.M.: Supersonic Boundary-Layer Stability Experiments. In: McCauley, W.D. (ed.) Proceedings of Boundary Layer Transition Study Group Meeting, vol. II, Air Force Report No. BSD-TR-67-213 (1967)
17. Mack, L.M.: Boundary-Layer Stability Theory. JPL 900-277 (1969)
18. Arnal, D., Casalis, G., Houdeville, R.: Practical Transition Prediction Methods: Subsonic and Transonic Flows. In: Advances in Laminar Turbulent Transition Modeling. Von Kármán Inst. Lecture Series, Rhode-St-Genèse, Belgium, pp. 7-1–7-33 (2008)
19. Menter, F.R., Langtry, R.B., Likki, S.R., Suzen, Y.B., Hung, P.G., Völker, S.: A Correlation-Based Transition Model Using Local Variables. Part I–Model Formulation. ASME Paper No. GT2004–53452 (2004)
20. Mack, L.M.: Stability of the Compressible Laminar Boundary Layer According to a Direct Numerical Solution. AGARDograph 97(Pt. I), 329–362 (1965)
21. McDonald, H., Fish, R.W.: Practical Calculations of Transitional Boundary Layers. Int. Journal of Heat and Mass Transfer 16, 25–53 (1973)
22. Mack, L.M.: Boundary-Layer Linear Stability Theory. AGARD R-709, 3-1–3-81 (1984)
23. Arnal, D.: Laminar Turbulent Transition. In: Murthy, T.K.S. (ed.) Computational Methods in Hypersonic Aerodynamics, pp. 233–264. Computational Mechanics Publications and Kluwer Academic Publishers (1991)
24. Hirschel, E.H.: Entstehung der Turbulenz in Grenzschichten - Eine Einführung. DFVLR IB 252 - 79 A 08 (1979)
25. Lang, M., Marxen, O., Rist, U., Wagner, S.: A Combined Numerical and Experimental Investigation of Transition in a Laminar Separation Bubble. In: Wagner, S., Kloker, M., Rist, U. (eds.) Recent Results in Laminar-Turbulent Transition. NNFM, vol. 86, pp. 149–164. Springer, Heidelberg (2004)
26. Lees, L., Lin, C.C.: Investigation of the Stability of the Boundary Layer in a Compressible Fluid. NACA TN 1115 (1946)
27. Lees, L.: The Stability of the Laminar Boundary Layer in a Compressible Fluid. NACA TN 876 (1947)
28. Mack, L.M.: Early History of Compressible Linear Stability Theory. In: Fasel, H.F., Saric, W.S. (eds.) Proc. IUTAM Symposium on Laminar-Turbulent Transition, Sedona, AZ, USA, pp. 9–34. Springer, Heidelberg (1999, 2000)
29. Theofilis, V., Fedorov, A.V., Obrist, D., Dallmann, U.C.: The Extended Görtler-Hämmerlin Model for Linear Instability of Three-Dimensional Incompressible Swept Attachment-Line Boundary-Layer Flow. J. Fluid Mechanics 487, 271–313 (2003)

30. Bertolotti, F.P.: On the Connection between Cross-Flow Vortices and Attachment-Line Instability. In: Fasel, H.F., Saric, W.S. (eds.) Proc. IUTAM Symposium on Laminar-Turbulent Transition, Sedona, AZ, USA, 1999, pp. 625–630. Springer, Heidelberg (2000)
31. Pfenninger, W.: Flow Phenomena at the Leading Edge of Swept Wings. Recent Developments in Boundary Layer Research - Part IV. AGARDograph 97 (1965)
32. Gaster, M.: On the Flow along Swept Leading Edges. The Aeronautical Quarterly XVIII, 165–184 (1967)
33. Cumpsty, N.A., Head, M.R.: The Calculation of Three-Dimensional Turbulent Boundary Layers. Part II: Attachment Line Flow on an Infinite Swept Wing. The Aeronautical Quarterly XVIII(Pt. 2), 99–113 (1967)
34. Poll, D.I.A.: Boundary Layer Transition on the Windward Face of Space Shuttle During Re-Entry. AIAA-Paper 85-0899 (1985)
35. Arnal, D.: Laminar-Turbulent Transition Problems in Supersonic and Hypersonic Flows. AGARD R-761, 8-1–8-45 (1988)
36. Poll, D.I.A.: Some Aspects of the Flow near a Swept Attachment Line with Particular Reference to Boundary Layer Transition. Doctoral thesis, Cranfield, U. K., CoA Report 7805/L (1978)
37. Thiede, P. (ed.): Aerodynamic Drag Reduction Technologies. Proc. of the CEAS/DragNet European Drag Reduction Conference, Potsdam, Germany, June 19-21, 2000. NNFM, vol. 76. Springer, Heidelberg (2001)
38. Gaster, M.: A Simple Device for Preventing Turbulent Contamination on Swept Leading Edges. J. Royal Aeronautical Soc. 69, 788 (1965)
39. Henke, R.: Airbus A 320 HLF Fin Flight Test. In: Thiede, P. (ed.) Aerodynamic Drag Reduction Technologies. Proc. CEAS/DragNet European Drag Reduction Conference, Potsdam, Germany, June 19-21. NNFM, vol. 76, pp. 31–38. Springer, Heidelberg (2001)
40. Seitz, A., Kruse, M., Wunderlich, T., Bold, J., Heinrich, L.: The DLR Project LamAiR: Design of a NLF Forward Swept Wing for Short and Medium Range Transport Application. AIAA-Paper 2011-3526 (2011)
41. Owen, P.R., Randall, D.G.: Boundary Layer Transition on a Swept Back Wing. R.A.E. TM 277 (1952), R.A.E. TM 330 (1953)
42. Bippes, H.: Basic Experiments on Transition in Three-Dimensional Boundary Layers Dominated by Crossflow Instability. Progress in Aerospace Sciences 35(3-4), 363–412 (1999)
43. Saric, W.S., Reed, H.L., White, E.B.: Stability and Transition of Three-Dimensional Boundary Layers. Annual Review of Fluid Mechnics 35, 413–440 (2003)
44. Kipp, H.W., Helms, V.T.: Some Observations on the Occurance of Striation Heating. AIAA-Paper 85-0324 (1985)
45. Görtler, H.: Über den Einfluß der Wandkrümmung auf die Entstehung der Turbulenz. ZAMM 20, 138–147 (1940)
46. Liepmann, H.W.: Investigation of Boundary-Layer Transition on Concave Walls. NACA ACR 4J28 (1945)
47. Tani, I., Aihara, Y.: Görtler Vortices and Boundary-Layer Transition. ZAMP 20, 609–618 (1969)
48. Saric, W.S.: Görtler Vortices. Annual Review of Fluid Mechnics 26, 379–409 (1994)

49. Narasimha, R.: The Three Archetypes of Relaminarisation. In: Proc. 6th Canadian Conf. of Applied Mechanics, vol. 2, pp. 503–518 (1977)
50. Narasimha, R., Sreenivasan, K.R.: Relaminarisation in Highly Accelarated Turbulent Boundary Layers. J. Fluid Mechanics 61, 417–447 (1973)
51. Launder, B.E., Jones, W.P.: On the Prediction of Laminarisation. ARC CP 1036 (1969)
52. White, F.M.: Viscous Fluid Flow, 2nd edn. McGraw-Hill, New York (1991)
53. Hirschel, E.H.: The Influence of the Free-Stream Reynolds Number on Transition in the Boundary Layer on an Infinite Swept Wing. AGARD R-602, 1-1–1-11 (1973)
54. Mukund, R., Viswanath, P.R., Crouch, J.D.: Relaminarization and Retransition of Accelerated Turbulent Boundary Layers on a Convex Surface. In: Fasel, H.F., Saric, W.S. (eds.) Proc. IUTAM Symposium on Laminar-Turbulent Transition, Sedona, AZ, USA, 1999, pp. 243–248. Springer, Heidelberg (2000)
55. N. N.: Boundary Layer Simulation and Control in Wind Tunnels. AGARD-AR-224 (1988)
56. Poll, D.I.A.: Laminar-Turbulent Transition. AGARD-AR-319 I, 3-1–3-20 (1996)
57. Reshotko, E.: Environment and Receptivity. AGARD R-709, 4-1–4-11 (1984)
58. Schumann, U., Konopka, P., Baumann, R., Busen, R., Gerz, T., Schlager, H., Schulte, P., Volkert, H.: Estimate of Diffusion Parameters of Aircraft Exhaust Plumes near the Tropopause from Nitric Oxide and Turbulence Measurements. J. of Geophysical Research 100(D7), 14,147–14,162 (1995)
59. Saric, W.S., Reed, H.L., Kerschen, E.J.: Boundary-Layer Receptivity to Free-stream Disturbances. Annual Review of Fluid Mechnics 34, 291–319 (2002)
60. Wilcox, D.C.: Turbulence Modelling for CFD. DCW Industries, La Cañada, CAL., USA (1998)
61. Menter, F.R.: Influence of Freestream Values on $k - \omega$ Turbulence Model Predictions. AIAA J. 33(12), 1657–1659 (1995)
62. Celic, A.: Performance of Modern Eddy-Viscosity Turbulence Models. Doctoral Thesis, Universität Stuttgart, Germany (2004)
63. Schrauf, G.: Industrial View on Transition Prediction. In: Wagner, S., Kloker, M., Rist, U. (eds.) Recent Results in Laminar-Turbulent Transition. NNFM, vol. 86, pp. 111–122. Springer, Heidelberg (2004)
64. Malik, M.R.: COSAL—A Black Box Compressible Stability Analysis Code for Transition Prediction in Three-Dimensional Boundary Layers. NASA CR 165925 (1982)
65. Ehrenstein, U., Dallmann, U.: Ein Verfahren zur linearen Stabilitätsanalyse von dreidimensionalen, kompressiblen Grenzschichten. DFVLR IB 221-88 A 20 (1988)
66. Laburthe, F.: Problème de stabilité linéaire et prévision de la transition dans des configurations tridimensionelles, incompressibles et compressibles (Problems of Linear Stability and Prediction of Transition in Three-Dimensional, Incompressible and Compressible Flows). Doctoral Thesis, ENSAE, Toulouse, France (1992)
67. Malik, M.R.: Boundary-Layer Transition Prediction Toolkit. AIAA-Paper 97-1904 (1997)
68. Schrauf, G.: Curvature Effects for Three-Dimensional, Compressible Boundary Layer Stability. Zeitschrift Für Flugwissenschaften und Weltraumforschung (ZFW) 16, 119–127 (1992)

69. Schrauf, G.: COAST3—A Compressible Stability Code. User's Guide and Tutorial. Deutsche Airbus, TR EF-040/98 (1998)
70. Schrauf, G.: LILO 2.1. User's Guide and Tutorial. GSSC Technical Report 6 (2004)
71. Herbert, T.: Parabolized Stability Equations. Annual Review of Fluid Mechanics, Palo Alto 29, 245–283 (1997)
72. Bertolotti, F.P.: Linear and Nonlinear Stability of Boundary Layers with Streamwise Varying Properties. Doctoral Thesis, Ohio State University, USA (1991)
73. Chang, C.-L., Malik, M.R., Erlebacher, G., Hussaini, M.Y.: Compressible Stability of Growing Boundary Layers Using Parabolized Stability Equations. AIAA-Paper 91-1636 (1991)
74. Simen, M.: Lokale Und nichtlokale Instabilität hypersonischer Grenzschichtströmungen (Local and Non-Local Instability of Hypersonic Boundary-Layer Flows). Doctoral Thesis, Universität Stuttgart, Germany (1993)
75. Simen, M., Bertolotti, F.P., Hein, S., Hanifi, A., Henningson, D.S., Dallmann, U.: Nonlocal and Nonlinear Stability Theory. In: Wagner, S., Periaux, J., Hirschel, E.H. (eds.) Computational Fluid Dynamics 1994, pp. 169–179. John Wiley and Sons, Chichester (1994)
76. Hein, S.: Nonlinear, Nonlocal Transition Analysis. Doctoral Thesis. Universität Stuttgart, Germany (2004)
77. Mughal, M.S., Hall, P.: Parabolized Stability Equations and Transition Prediction for Compressible Swept-Wing Flows. Imperial College for Science, Technology and Medicine, final report on DTI contract ASF/2583U (1996)
78. Salinas, H.: Stabilité linéaire et faiblement non linéaire d'une couche limite laminaire compressible tridimensionelle par l'approche PSE (Linear and Weakly Non-Linear Stability of a Laminar, Compressible Three-Dimensional Boundary Layer with the PSE Approach). Doctoral Thesis, ENSAE, Toulouse, France (1998)
79. Chang, C.-L.: Langley stability and transition analysis code (LASTRAC), version 1.2, user manual. NASA TM-2004-213233 (2004)
80. Michel, R.: Etude de la transition sur les profiles d'aile—Établissement d'un critère de détermination du point de transition et calcul de la trainée de profil en incompressible. ONERA Rapport 1/1578 A (1951)
81. Granville, P.S.: The Calculation of the Viscous Drag of Bodies of Revolution. David Taylor Model Basin Report 849 (1953)
82. Crabtree, L.F.: Prediction of Transition in the Boundary Layer on an Airfoil. J. Royal Aeronautical Soc. 62, 525–527 (1958)
83. Hall, D.J., Gibbings, J.C.: Influence of Free-Stream Turbulence and Pressure Gradient Upon Boundary Layer Transition. J. Mechanical Eng. Science 14, 134–146 (1972)
84. Dunham, J.: Predictions of Boundary Layer Transition on Turbomachines. AGARD PEP Ad-hoc Study, Paris (1972)
85. Arnal, D., Habiballah, M., Coustols, E.: Laminar Stability Theory and Transition Criteria in Two- and Three-Dimensional Flow. Rech. Aérospatiale 2 (1984)
86. Poll, D.I.A., Tran, P., Arnal, D.: Capabilities and Limitations of Available Transition Prediction Tools. Aerospatiale TX/AP no. 114 779 (1994)
87. Van Ingen, J.L.: A Suggested Semi-Empirical Method for the Calculation of the Boundary-Layer Transition Region. Reports UTH71 and UTH74, Delft, The Netherlands (1956)

88. Smith, A.M.O., Gamberoni, N.: Transition, Pressure Gradient and Stability Theory. Douglas Report No. ES 26388 (1956)
89. Arnal, D.: Boundary-Layer Transition: Predictions Based on Linear Theory. AGARD-R-793, 2-1–2-63 (1994)
90. Mack, L.M.: Transition Prediction and Linear Stability Theory. AGARD CP-224, 1-1–1-22 (1977)
91. Stock, H.-W., Haase, W.: Some Aspects of Linear Stability Calculations in Industrial Applications. In: Henkes, R.A.W.M., van Ingen, J.L. (eds.) Transitional Boundary Layers in Aeronautics, pp. 225–238. North-Holland Press, Amsterdam (1996)
92. Arnal, D.: Transition Prediction in Transonic Flow. In: Zierep, J., Oertel, H. (eds.) Symposium Transonicum III, pp. 253–262. Springer, Heidelberg (1988)
93. Stock, H.-W.: e^N Transition Prediction in Three-Dimensional Boundary Layers on Inclined Prolate Spheroids. AIAA J. 44, 108–118 (2006)
94. Krimmelbein, N., Radespiel, R.: Transition Prediction for Three-Dimensional Flows Using Parallel Computation. Computers & Fluids 38, 121–136 (2009)
95. Bertolotti, F.P.: The Equivalent Forcing Model for Receptivity Analysis with Application to the Construction of a High-Performance Skin-Perforation Pattern for Laminar Flow Control. In: Wagner, S., Kloker, M., Rist, U. (eds.) Recent Results in Laminar-Turbulent Transition. NNFM, vol. 86, pp. 25–36. Springer, Heidelberg (2004)
96. Hill, D.C.: Adjoint Systems and their Role in the Receptivity Problem for Boundary Layers. J. Fluid Mechanics 292, 183–204 (1995)
97. Dussauge, J.-P., Fernholz, H.H., Smith, R.W., Finley, P.J., Smits, A.J., Spina, E.F.: Turbulent Boundary Layers in Subsonic and Supersonic Flow. AGARDograph 335 (1996)
98. Aupoix, B.: Introduction to Turbulence Modelling for Turbulent Flows. In: Benocci, C., van Beek, J.P.A.J. (eds.) Introduction to Turbulence Modeling, VKI, Rhode Saint, Genèse, Belgium. VKI Lecture Series 2002-02 (2002)
99. Pope, S.B.: Turbulent Flows. Cambridge University Press, Cambridge (2000)
100. Piquet, J.: Turbulent Flows: Models and Physics. Springer, Heidelberg (2001) revised second printing
101. Durbin, P.A., Pettersson Reif, B.A.: Statistical Theory and Modeling for Turbulent Flow, 2nd edn. John Wiley, Hoboken (2011)
102. Peng, S.-H., Doerffer, P., Haase, W. (eds.): Progress in Hybrid RANS-LES Modelling. NNFM, vol. 111. Springer, Heidelberg (2010)
103. Fu, S., Haase, W., Peng, S.H., Schwamborn, D. (eds.): Progress in Hybrid RANS-LES Modelling. Papers contributed to the 4th Symp. on Hybrid RANS-LES Methods. NNFM, vol. 117. Springer, Heidelberg (2012)
104. Menter, F.R., Schütze, J., Gritskevich, M.: Global vs. Zonal approaches in hybrid RANS-LES turbulence modelling. In: Fu, S., Haase, W., Peng, S.-H., Schwamborn, D. (eds.) Progress in Hybrid RANS-LES Modelling. NNFM, vol. 117, pp. 15–28. Springer, Heidelberg (2012)
105. Haase, W., Chaput, E., Elsholz, E., Leschziner, M.A., Müller, U.R. (eds.): ECARP—European Computational Aerodynamics Research Project. Validation of CFD Codes and Assessment of Turbulence Models. NNFM, vol. 58. Vieweg, Braunschweig Wiesbaden (1997)
106. Dervieux, A., Braza, M., Dussauge, J.-P. (eds.): Computation and Comparison of Efficient Turbulence Models for Aeronautics—European Research Project ETMA. NNFM, vol. 65. Vieweg, Braunschweig Wiesbaden (1998)

107. Haase, W., Aupoix, B., Bunge, U., Schwamborn, D. (eds.): Schwamborn, D. FLOMANIA—A European Initiative on Flow Physics Modelling. Results of the European Union funded project, 2002–2004. NNFM, vol. 94. Springer, Heidelberg (2006)
108. Schwamborn, D., Strelets, M.: ATAAC—An EU-Project Dedicated to Hybrid RANS/LES Methods. In: Fu, S., Haase, W., Peng, S.-H., Schwamborn, D. (eds.) Progress in Hybrid RANS-LES Modelling. NNFM, vol. 117, pp. 59–75. Springer, Heidelberg (2012)
109. Vassberg, J.C., Tinoco, E.N., Mani, M., Rider, B., Zickuhr, T., Levy, D.W., Brodersen, O.P., Eisfeld, B., Crippa, S., Wahls, R.A., Morrison, J.H., Mavriplis, D.J., Murayama, M.: Summary of the Fourth AIAA CFD Drag Prediction Workshop. AIAA-Paper 2010-4547 (2010)
110. Aupoix, B., Spalart, P.R.: Extensions of the Spalart-Allmaras Model to Account for Wall Roughness. Int. J. of Heat and Fluid Flow 24, 454–462 (2003)
111. Knopp, T., Eisfeld, B., Calvo, J.B.: A New Extension for k-ω Turbulence Models to Account for Wall Roughness. Int. J. of Heat and Fluid Flow 30, 54–65 (2009)
112. Morkovin, M.V.: Effects of Compressibility on Turbulent Flows. Colloque International CNRS No. 108, Mécanique de la Turbulence, Editions CNRS (1961)
113. Meier, H.U., Rotta, J.C.: Temperature Distributions in Supersonic Turbulent Boundary Layers. AIAA J. 9, 2149–2156 (1971)
114. King, R. (ed.): Active Flow Control II. Papers contributed to the Conference "Active Flow Control II". NNFM, vol. 108. Springer, Heidelberg (2010)
115. Bushnell, D.M., Hefner, J.M. (eds.): Viscous Drag Reduction in Boundary Layers. Progress in Astronautics and Aeronautics, vol. 123. AIAA, Reston (1990)
116. Stanewsky, E., Délery, J., Fulker, J., de Matteis, P. (eds.): Drag Reduction by Shock and Boundary Layer Control. Results of the Project EUROSHOCK II, Supported by the European Union, 1996–1999. NNFM, vol. 80. Springer, Heidelberg (2002)
117. von Doenhoff, A.E., Braslow, A.L.: The Effect of Distributed Surface Roughness on Laminar Flow. In: Lachmann, V. (ed.) Boundary Layer Control, Its Principles and Application, vol. 2, pp. 657–681. Pergamon Press, Oxford (1961)

10

Illustrating Examples

This chapter is devoted to the illustration of several of the phenomena and flow properties which we have studied in some of the preceding chapters. It is in the character of attached three-dimensional viscous flow that predominantly skin-friction line patterns are used for such illustrations. We present five different examples.

The first example concerns the locality principle. It is demonstrated with the flow past a helicopter fuselage, Sub-Section 10.1. The inviscid flow field was determined with a panel method without a modelling of the separation region at the aft of the fuselage. On that basis a three-dimensional boundary-layer computation was performed. The computed skin-friction line pattern is compared with an oil-flow picture. The good agreement of the two patterns demonstrates well the validity of the locality principle which was put forward in Section 1.2. Regarding boundary-layer computations at finite-span wings a short consideration in Sub-Section 10.1 shows how the locality principle can come to its limits.

Example two is a consideration of mainly inviscid flow patterns upstream of and at trailing edges of lifting wings with large aspect ratio, Section 10.2. We identify important properties of such flows, and we discuss in particular a compatibility condition between the circulation distribution and the inviscid flow field pattern. This enables us to judge the quality of computed viscous flow data and to understand the different flow and skin-friction line patterns observed at back and forward swept lifting wings.

We discuss the third example in Section 10.3, the computed distributions of the wall pressure and of the skin-friction coefficient, as well as skin-friction line patterns at selected locations of an airplane configuration. This is in view of aspects of flow topology which were presented in Chapter 7. The airplane configuration is the NASA/Boeing Common Research Model (CRM) which is a test configuration of the AIAA CFD Drag Prediction Workshop series.

The fourth example illustrates the occurrence of extrema of the thermal state of the surface due to the relative extrema of the characteristic boundary-layer thickness across attachment and separation lines. We study the case of three-dimensional viscous flow past a generic re-entry configuration, the Blunt Delta Wing (BDW), Section 10.4. The external surfaces of such vehicles are radiation cooled and we observe hot-spot and cold-spot phenomena at

E.H. Hirschel, J. Cousteix, and W. Kordulla, *Three-Dimensional Attached Viscous Flow*,
DOI: 10.1007/978-3-642-41378-0_10,

the different attachment and separation lines present in the flow past the configuration.

Laminar-turbulent transition may play a large role with regard to aerodynamic properties of and thermal and mechanical loads on a given flight vehicle. If a transition sensitivity exists, the knowledge and the determination of the location of the transition region is important. In the fifth example, Section 10.5, it is shown that in three-dimensional attached viscous flow the different transition mechanisms may be of widely different relative importance and that the transition region can have a very complex shape.

We provide for each case—if available—besides the free-stream Mach and Reynolds number as the third defining parameter the free-stream temperature T_∞. If the wall temperature is not specified, we assume $T_w = T_\infty$ or an adiabatic wall.

10.1 The Locality Principle: Flow Past a Helicopter Fuselage and Past Finite-Span Wings

The Helicopter Case

We study computed and measured skin-friction line patterns at the surface of a generic helicopter fuselage, Fig. 10.1 [1, 2].

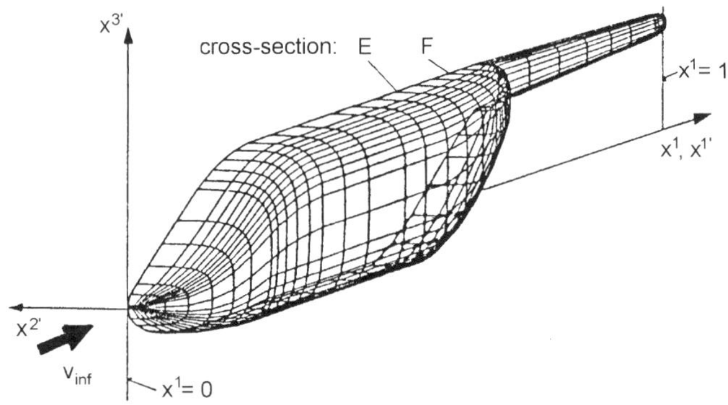

Fig. 10.1. Schematic of a helicopter fuselage with cross-sections E ($x^{1'} = 0.799$ m, $x^1 = 0.49$) and F ($x^{1'} = 0.962$ m, $x^1 = 0.59$), left half [2].

The flow parameters are given in Table 10.1. They are those of the wind-tunnel investigations reported in [3]. The authors of that report also calculated the external inviscid flow field by means of a panel method. The separation region, presumably present at the aft of the fuselage, was not modelled in any way.

The three-dimensional boundary layer over the fuselage was computed with the integral method of Cousteix-Aupoix in the MBB version [4]. The

Table 10.1. Parameters of the generic helicopter fuselage computation case. L is the length of the fuselage including the tail boom, see Fig. 10.1.

M_∞	Re_L	L [m]	T_∞ [K]	T_w [K]	α [°]	β [°]	boundary-layer
0.184	$6.558 \cdot 10^6$	1.63	300	300	-5	0	fully turbulent

experiment had shown laminar-turbulent transition to appear close to the nose and along the front columns, Fig. 10.1, there in the form of bubble-type transition, see the lower part of Fig. 10.4.

To ease the boundary-layer computation, fully turbulent flow was assumed with approximate initial data prescribed at $x^1 = 0.03$. In this way the problem of the metric singularity at $x^1 = 0$ of the employed cross-section coordinate system was avoided. This approach made use of the observation that initial data for boundary-layer computations placed in regions with sufficiently large favorable pressure gradient, soon loose their influence on the solution.[1]

In Fig. 10.2 we show in the cross-sections E and F the contours of the boundary-layer thickness δ and the—three-dimensional—displacement thickness δ_1. Contours at locations upstream of E and F can be found in [1].

The helicopter was considered to fly in forward motion, therefore the angle of attack is negative, Table 10.1. Hence the upper symmetry line is a weak attachment line. However, a relative minimum of δ and δ_1 is not discernible there.

The boundary-layer accumulates at the side and below the fuselage. The bulging of both δ and δ_1 at location a) seen in Fig. 10.2 a) points to imminent separation. This is corroborated by the convergence of the skin-friction lines at location a) in Fig. 10.3.

The convergence in the vicinity of location a) is accompanied by divergence in the vicinity of locations b) and c) and consequently by an indentation of the contours of δ and δ_1, Fig. 10.2. Note the negative displacement thickness at location c).

At cross-section F the flow has already separated at the lower rear side of the fuselage and no data are available for that region.

The δ and δ_1 contours and the skin-friction line pattern indicate the presence of a separation line as shown in the oil-flow picture in the lower part of Fig. 10.4. The location of the separation line, as indicated by the computation (location a) in the upper part of the figure), appears to lie somewhat downstream of that one seen in the oil-flow picture. This might be due to the fact that neither the local nor the global interaction of the separated with the inviscid flow was taken into account.

[1] If approximate initial data are placed in regions with zero or even adverse pressure gradient their influence does not vanish in down-stream direction.

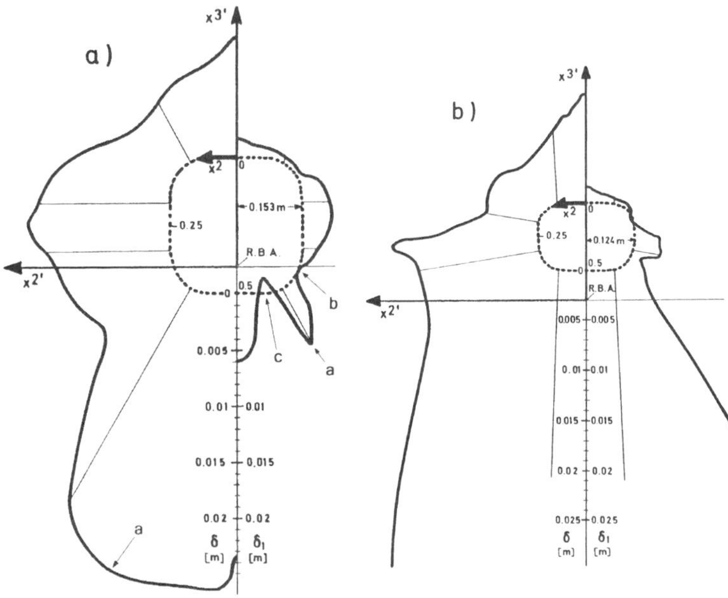

Fig. 10.2. Boundary-layer thickness $\delta(x^2)$ and displacement thickness $\delta_1(x^2)$ at a) cross-section E and b) cross-section F (R.B.A. ≙ reference body axis) [2].

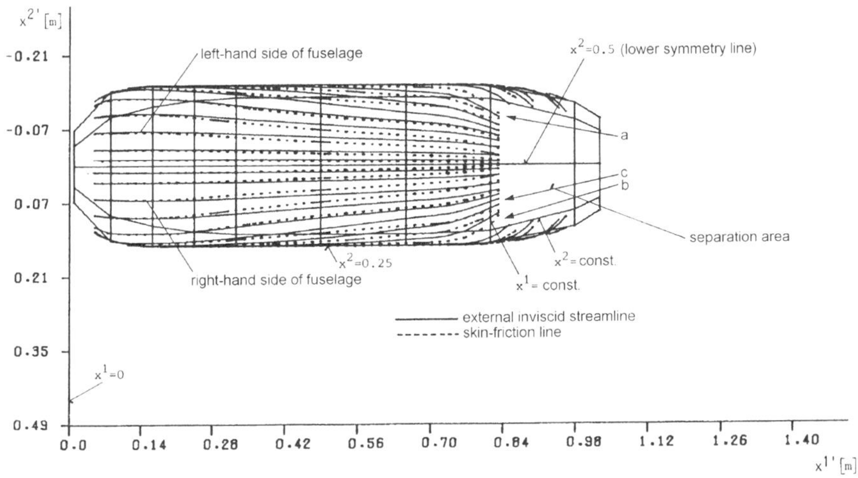

Fig. 10.3. External inviscid streamlines and skin-friction lines at the lower side of the helicopter fuselage [2].

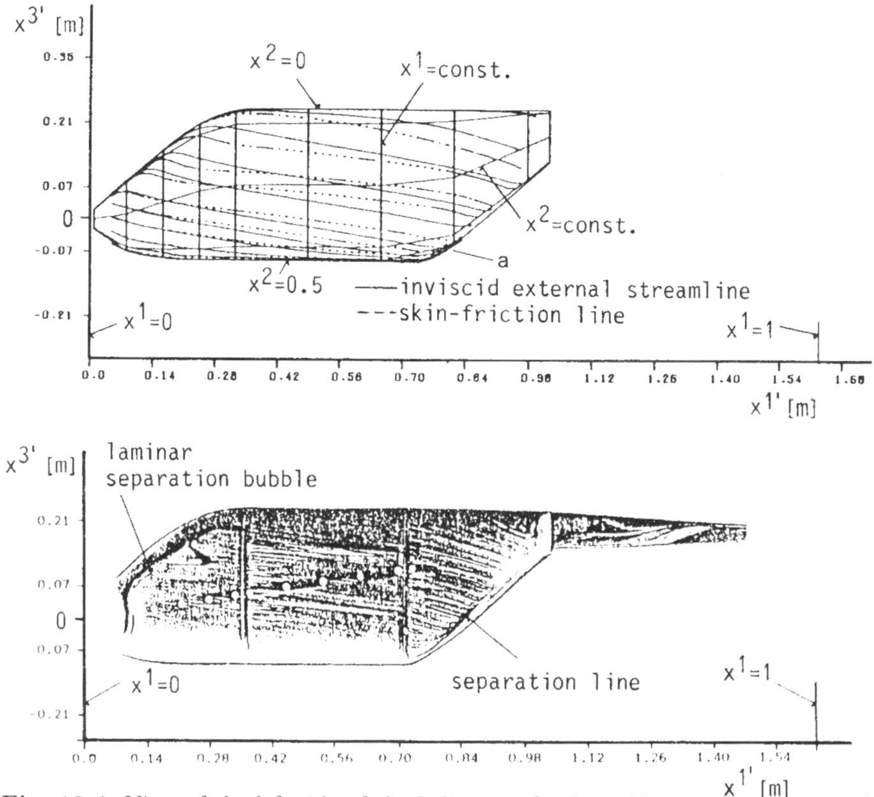

Fig. 10.4. View of the left side of the helicopter fuselage. Upper part: pattern of computed external inviscid streamlines and skin-friction lines [1, 2]. Lower part: oil-flow picture [3].

Nevertheless, the computed skin-friction line pattern upstream of that location compares well with the oil-flow pattern, Fig. 10.4, up to the separation location. The separation bubble at the front column, which in the experiment leads to laminar-turbulent transition, is not present in the calculated skin-friction field. The agreement between the computed and the measured skin-friction field is strikingly good and corroborates fully the locality principle which was put forward in Section 1.2.

The separation region covers the whole lower rear side of the fuselage as is evident from Fig. 10.5 which is by courtesy of Th. Schwarz, DLR-Institute of Aerodynamics and Flow Technology. The wind-tunnel model with its sting is shown from behind. The oil-flow picture reveals that a pair of separating vortices is present which forms what in [6] is called an owl face.

The upstream influence, the global interaction, Section 1.2, is small to the extent that the boundary-layer solution based on the panel method solution for the inviscid flow shows very good agreement with the experiment, at least in terms of the skin-friction line patterns. At the separation line we have

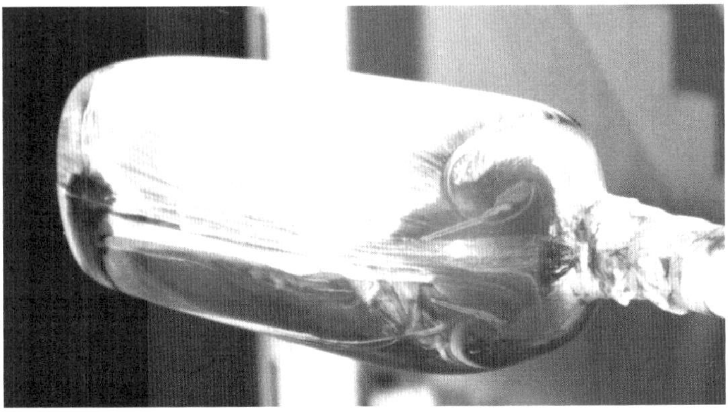

Fig. 10.5. Oil-flow picture of the lower back side of the helicopter fuselage [5].

locally strong interaction. This of course does not permit to find the exact location of that line in the frame of the present boundary-layer approach.

Results like this, also results for cases, see, e.g., [7], where it was possible to compare measured and computed flow parameters, further confirm the locality principle.

The Finite-Span Wing Case

In order to widen the consideration on the locality principle, we look briefly also at the case of a finite-span lifting wing. Theory permits to determine the induced angle of attack α_i which is due to the vortex sheet leaving—by flow-off separation—the wing's trailing edge [8]. If the span-wise circulation distribution is elliptical, the induced angle of attack for incompressible flow around unswept wings is found to be

$$tan\,\alpha_i = \frac{D_i}{L}, \tag{10.1}$$

where D_i is the induced drag and L the lift. The induced angle of attack reduces the geometrical angle of attack to the effective one.

We consider now a swept wing at a flight Mach number $M_\infty \approx 0.8$. The lift-to-drag ratio may be $L/D \approx 25$, the lift coefficient $C_L \approx 0.5$ and the slope of the lift curve $d\,C_L/d\alpha \approx 0.1/$(degree angle of attack)—compare, for instance, with Sub-Section 6.4.2.

We assume that the induced drag of the wing is approximately half of its total drag and that we can employ eq. (10.1) to make a guess of the induced angle of attack. The result is $\alpha_i \approx 1.1°$. This induced angle of attack reduces the lift coefficient by $\triangle\,C_L \approx 0.11$.

If we now use for a boundary-layer study on the considered wing at, let us say a mid-span location, an infinite-swept wing approach, we can either take the pressure distribution of the two-dimensional airfoil at that location, or the pressure distribution of the whole wing which, when properly computed, includes the induced angle of attack due to the vortex layer. In the latter case, the global interaction is adequately modelled.

In our case the lift difference is $\triangle C_L \approx 0.11$ which at $C_L \approx 0.5$ means a probably barely tolerable change of the pressure field at the mid-span location, so that the two-dimensional airfoil pressure distribution could be used. The global interaction effect on the pressure distribution appears to be small enough so that it can be neglected—which in this case would need to be confirmed. This is also an aspect of the locality principle. The larger the aspect ratio, the smaller is the lift difference. With a small-aspect ratio wing in any case one has to check whether the influence of this difference on the pressure distribution can be tolerated/neglected.

Prerequisite is that the flow field is structurally stable. Consider the wing as operating close to but below the critical Mach number. A small rise of the angle of attack (or of the flight Mach number) may change the flow past the suction side of the wing from pure subsonic flow to a flow with an embedded supersonic pocket which might be terminated with a shock wave. This then would mean a structural change of the flow field. In such a case the above consideration is no longer valid.

10.2 Flow Patterns Upstream of and at Trailing Edges of Lifting Wings with Large Aspect Ratio

In this section we discuss mainly inviscid flow patterns that typically occur at lifting wings with large aspect ratio. In Sub-Section 7.4.2 we investigated properties of attachment-line flow, here we concentrate on the flow near the wing's trailing edge.

We assume flight at sub-critical Mach numbers and at angles of attack, where the viscous flow is fully attached and leaves the wing by flow-off separation at the trailing edge. The attached viscous flow in such cases is mainly governed by the external inviscid flow field past the wing. Such flow can be prescribed by, for instance, linearized potential flow theory [8].

We follow basically the presentation of the topic as given in [9]. The classical theory of lifting wings tells us that a vortex sheet leaves the trailing edge of a lifting wing of finite span, see, e.g., [8]. This sheet carries kinematically active and inactive vorticity [10]. Locally the strength of the kinematically active vorticity is proportional to the change of circulation in the span-wise direction.

In [11] it is shown that the inviscid flow field at the trailing edge of a lifting wing has to be compatible with the circulation distribution in a particular way. The way of looking rigorously at the properties of the inviscid flow field only, as for instance in [12], however was abandoned in favor of a more realistic, but still idealized picture.

Consider the flow as it locally leaves the trailing edge of a back-swept finite-span lifting wing, Fig. 10.6. We disregard strong interaction phenomena and a possible finite, if small, thickness of the trailing edge. The flow over the lower side generally has a direction toward the wing tip, that over the upper side away from the wing tip.[2] We assume that the velocity vectors in the figure lie in the skeletal plane of the vortex sheet which leaves the trailing edge.

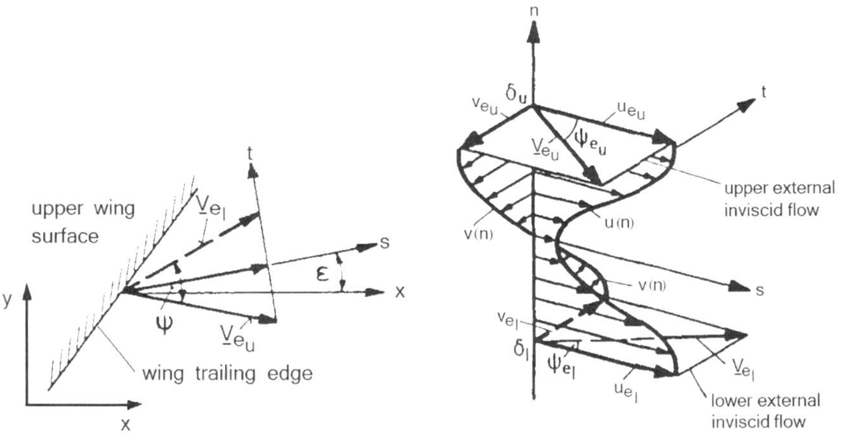

Fig. 10.6. Idealized situation at the trailing edge of a back-swept finite-span wing [9]. Left part: Detail of the inviscid velocity components at the trailing edge of a lifting wing. The coordinate y points in span direction, the coordinate x in free-stream and in chord direction. Right part: wake profiles just downstream of the trailing edge.

The static pressure, like the total pressure, can be assumed to be the same at the upper and the lower side of the trailing edge. Then the magnitudes of the external inviscid velocity vectors at the upper side ($\underline{V}_{e_u}$) and at the lower side ($\underline{V}_{e_l}$) of the trailing edge are the same: $|\underline{V}_{e_u}| = |\underline{V}_{e_l}|$. We decompose the two vectors in such a way that they have the components $u_{e_u} = u_{e_l}$ in s-direction (bi-sector direction) and $v_{e_u} = -v_{e_l}$ in t-direction (normal to the bi-sector direction), left part of Fig. 10.6. The s-direction obviously is the direction of a vortex line.

The angle ε between the s-direction and the chord direction, the vortex-line angle, is small but not necessarily zero. Its sign is governed by the sweep of the trailing edge [12]: positive (in wing-tip direction) at back-swept edges, negative (in wing-root direction) at forward swept edges.

The different deflections of the vortex lines are due to the different general properties of the pressure fields. At a back-swept wing, the flow at the wing's

[2] This is a general trend of the flow over bodies and wings at positive angle of attack, see, e.g., Fig. 7.2. However, on a body with flat lower side, this trend is not observed, Fig. 7.6.

leading edge is already directed toward the wing tip. At a forward-swept wing, the flow is directed toward the wing root.

The magnitude of ε at given lift depends on the thickness of the wing. In potential-flow models of lifting wings usually $\varepsilon = 0$ is assumed, see, e.g., [8].

We define the angle between the two external inviscid velocity vectors $\underline{V}_{e_u}$ and $\underline{V}_{e_l}$ as the (local) trailing-edge flow (TEF) shear angle $\psi = |\psi_{e_u}| + |\psi_{e_l}|$, with

$$tan\psi_{e_u} = \frac{v_{e_u}}{u_{e_u}}, \; tan\psi_{e_l} = \frac{v_{e_l}}{u_{e_l}}, \tag{10.2}$$

where the u_e are the components of the two velocity vectors in s-direction and the v_e those in n-direction. Note that $\psi_{e_u} = -\psi_{e_l}$.

The flow profile of the wake just downstream of the trailing edge, right part of Fig. 10.6, can be decomposed into that in s-direction and that in t-direction [10]. We obtain the profile $u(n)$ in s-direction and the profile profile $v(n)$ in t-direction.

The vorticity vector $\underline{\omega}$ reads with only the boundary-layer terms kept:

$$\underline{\omega} = [\omega_s, \, \omega_t, \, \omega_n]^T = [-\frac{\partial v}{\partial n}, \, \frac{\partial u}{\partial n}, \, 0]^T. \tag{10.3}$$

The vorticity content, [9, 10], of the profile $u(n)$ is

$$\Omega_u = \int_{\delta_l}^{\delta_u} \omega_t(n) \, dn = u_{e_u} - u_{e_l} = 0, \tag{10.4}$$

i.e, this profile carries kinematically inactive vorticity like the wake of a lifting or non-lifting airfoil in steady flow.

On the other hand the vorticity content of the profile $v(n)$ is

$$\Omega_v = \int_{\delta_l}^{\delta_u} \omega_s(n) \, dn = v_{e_u} - v_{e_l} = 2v_{e_u} \neq 0, \tag{10.5}$$

i.e, this profile carries kinematically active vorticity.

If the vortex-line angle ε is small, the connection of the TEF shear angle $\psi(y)$ to the span-wise circulation distribution $\Gamma(y)$ is given via its gradient in y-direction in form of the compatibility condition [9, 10]:

$$\frac{d\Gamma}{dy}(y) = 2|V_{e_u}(y)| \, tan\psi_u(y) = \Omega_v(y). \tag{10.6}$$

$\Omega_v(y)$ is locally the kinematically active vorticity content of the wake. We note immediately that at the trailing edge of an infinite swept wing, Section 8.3, no kinematically active vorticity leaves the wing surface: $d\Gamma/dy(y) = 0$, since the flow properties do not change in span-wise direction. Because $|V_{e_u}(y)|$ is finite there, this means that the TEF shear angle ψ is zero. In the vicinity of the trailing edge of an infinite swept wing the flow pattern hence is not representative of the pattern of a finite-span wing!

Toward the wing tip the circulation $\Gamma(y)$ decreases with $d\Gamma/dy(y)$ increasing. Eq. (10.6) implies that then either $|V_{e_u}(y)|$ must increase or the TEF shear angle ψ. In reality in most cases the TEF shear angle increases toward the wing tip. This is a basic property of flow past a lifting wing and is already established in the ideal inviscid flow past the flat ellipsoid at angle of attack, Fig. 7.2 on page 137.

We consider now the flow fields over two wings, Fig. 10.7, the Kolbe wing, [13] and a forward swept wing which was studied in [14].

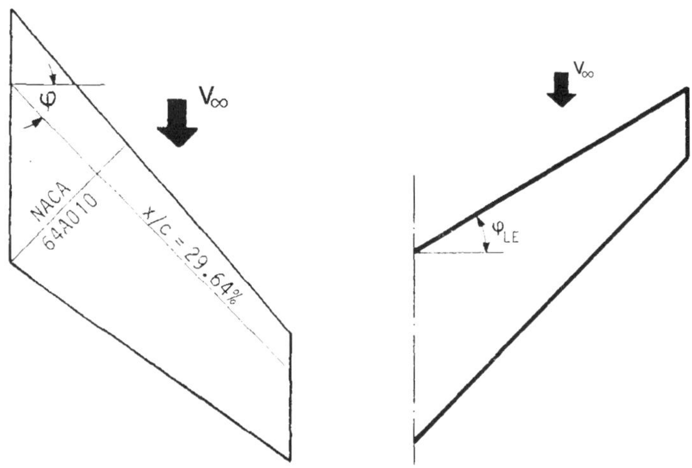

Fig. 10.7. View of two wing planforms [9]. Left part: Kolbe wing [13]. Right part: forward swept wing [14].

In the following figures we will find references to panel method 1 and to other panel methods (2, 3, etc.). Panel method 1 is the HISSS- (Higher-Order Subsonic-Supersonic Singularity-) method developed by L. Fornasier at the beginning of the 1980s [15]. This higher-order panel method has linear source distributions and quadratic doublet distributions in both chord- and span-wise directions. The other methods are lower-order methods. They have stepwise constant doublet distributions—or equivalent—in both chord- and span-wise directions. This leads to an erroneous determination of the v-component of the velocity vector near a wing's trailing edge. A similar problem is reported in [16], where the failure of low-order panel methods to compute the flow past very thin wings was attributed to an inadequate doublet distribution in the chord-wise direction.[3]

[3] Earlier boundary-layer studies with such erroneous inviscid flow fields even led authors to the conclusion that the boundary-layer flow over swept wings is predominantly two-dimensional.

Consider now Fig. 10.8. It shows the dimensionless span-wise circulation distribution Γ ($= \Gamma/(c_m u_\infty)$, with c_m being the mean chord), its derivative $d\Gamma/dy$, and the vorticity content Ω. The results of method 1 and method 2 do not differ much, also regarding lift and induced drag. The point-wise investigation of the velocity components at the trailing edge with eq. (10.6) yields values of Ω ($= \Omega/u_\infty$) which for the higher-order method 1 agree very well with $d\Gamma/dy$, whereas for the first-order method 2 they are completely wrong.

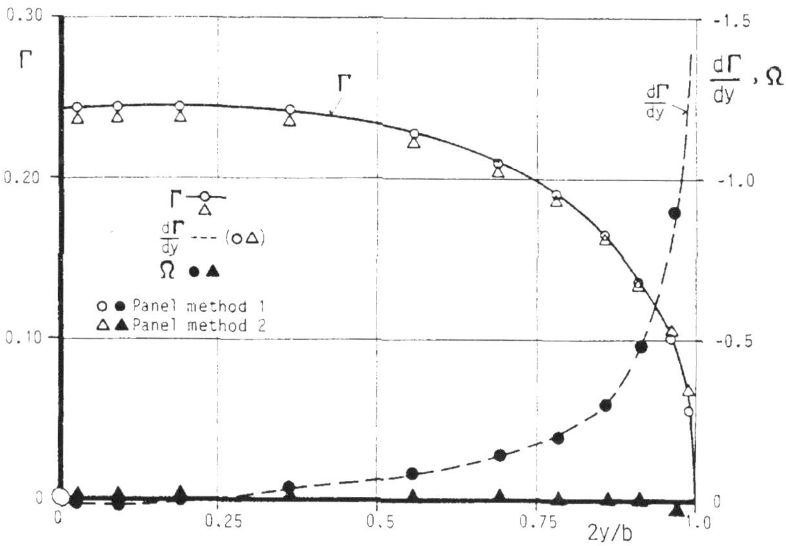

Fig. 10.8. Kolbe wing at $M_\infty = 0.25$, $\alpha = 8.2°$. Comparison of results of two panel methods [9]. Circulation distribution Γ, its derivative $d\Gamma/dy$, and the vorticity content Ω as functions of the half-span coordinate $2y/b$.

This result is reflected by the results for the TEF shear angle ψ in Fig. 10.9. Panel method 1 shows the expected rise of ψ_l ($= \psi/2$) toward the wing tip, whereas method 2 gives a nearly zero TEF shear angle. In contrast to that, the vortex-line angle ε is the same for both methods. The vortex line is deflected in wing-tip direction by $\varepsilon \approx 5°$. At the root and at the tip this angle is approximately zero.

The application of the higher-order panel method 1 and of other lower-order panel methods to a forward-swept wing, which was studied in [14], gives a similar result, Fig. 10.10. All methods agree rather well regarding the span-wise circulation distribution $\Gamma(y)$. However, only for method 1 the derivative $d\Gamma/dy(y)$ and the vorticity content $\Omega(y)$ are compatible.

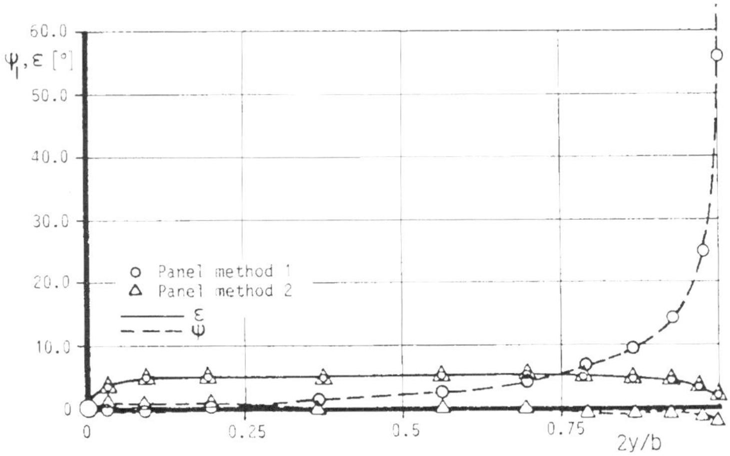

Fig. 10.9. Kolbe wing at $M_\infty = 0.25$, $\alpha = 8.2°$. Comparison of results of two panel methods [9]. TEF shear angle ψ_l and vortex-line angle ε as functions of the half-span coordinate $2y/b$.

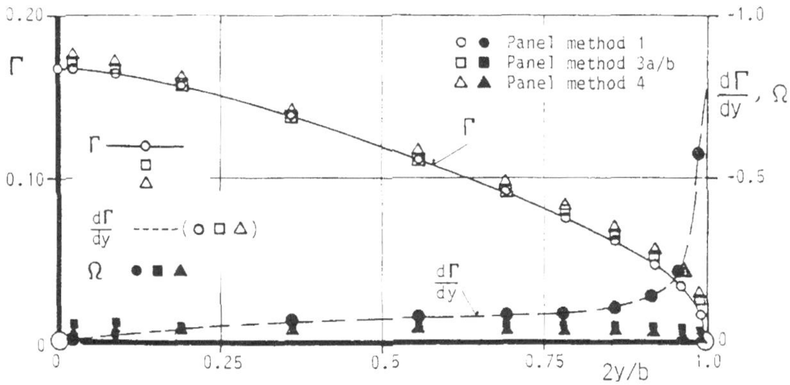

Fig. 10.10. Forward swept wing at $M_\infty = 0$, $\alpha = 4°$. Comparison of results of several panel methods [9]. Circulation distribution Γ, its derivative $d\Gamma/dy$, and the vorticity content Ω as functions of the half-span coordinate $2y/b$.

The TEF shear angle ψ_l $(= \psi/2)$ in Fig. 10.11 again shows the expected trend only for method 1. The vortex-line angle is negative with $\varepsilon \approx -7°$, i.e., the vortex line is deflected in wing-root direction. The results of the lower-order methods show much scatter, depending on where the kinematic flow condition is implemented, either on the skeleton plane of the wing, or on the true wing surface.

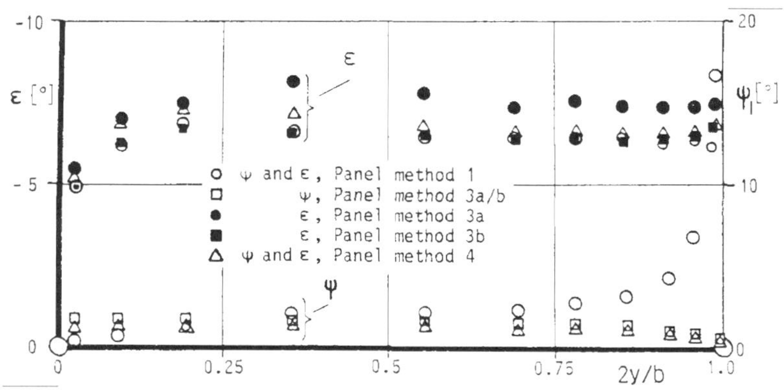

Fig. 10.11. Forward swept wing at $M_\infty = 0$, $\alpha = 4°$. Comparison of results of several panel methods [9]. TEF shear angle ψ_l and vortex-line angle ε as functions of the half-span coordinate $2y/b$.

Three-dimensional boundary-layer computations were performed with the external inviscid flow fields found with panel method 1 and 2 [9]. The boundary-layer method was the integral method of Cousteix-Aupoix in the MBB version [4].

The resulting streamlines of the external inviscid flow and the skin-friction lines of the turbulent boundary layer are shown in Fig. 10.12.

The inviscid streamlines found with panel method 2 approach the trailing edge nearly in chord direction. This holds in particular for the pressure side. The inviscid streamlines found with method 1 show directions according to the vortex-line angle ε and the TEF shear angle ψ_l in Fig. 10.9.

Note that only method 1 yields the points of inflection of the inviscid streamlines near approximately 75 per cent chord length at almost the whole pressure side, Fig. 10.12 a). The error in the inviscid flow field and therefore in the boundary-layer solution found with method 2 extends over approximately 50 per cent of both the suction and the pressure side.

Concluding this section we repeat and emphasize that the flow fields near the trailing edges of lifting wings have two particular properties which depend on the magnitude of the lift, respectively on the angle of attack.

- One property is the shear between the external inviscid flow vectors at the suction and the pressure side. The shear represents the kinematically active vorticity in the lifting wing's wake. It is—in terms of the TEF shear angle ψ of the inviscid flow part—connected via the compatibility condition, eq. (10.6), to the span-wise gradient of the circulation distribution and increases with that toward the wing tip.

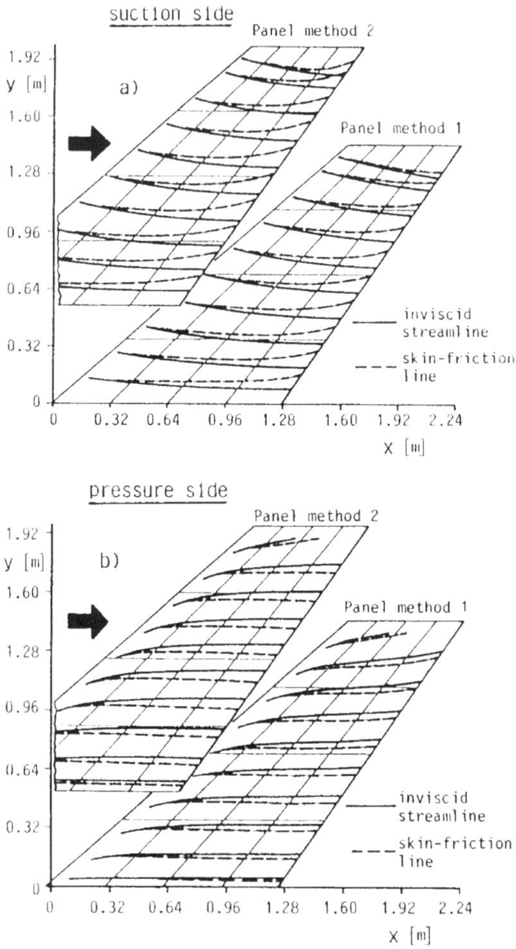

Fig. 10.12. Kolbe wing at $M_\infty = 0.25$, $Re_c = 18{\cdot}10^6$, $c = 1$ m, $\alpha = 8.2°$. Comparison of the results of two panel methods and a three-dimensional boundary-layer method [9]. Streamlines of the external inviscid flow and skin-friction lines of the turbulent boundary layer: a) upper (suction) side, b) lower (pressure) side of the wing.

The wake—symmetric to the symmetry plane of the wing—hence is a vortex layer which has the tendency to roll up behind the wing by self-induction. The result is the pair of discrete counter-rotating trailing vortices behind the wing. The axes coincide nearly with the free-stream direction.[4]

[4] We do not discuss here the effects which the vortex sheet and the trailing vortices have, the induced drag etc., but point to the literature, e. g., [8, 17]

In Prandtl's lifting-line wing theory the trailing vortices are located, fully developed, at the wing tips [18]. However, in reality the (initial) horizontal distance b' of the vortex axes is smaller than the wing span b. For a wing with elliptical circulation distribution the ratio is [8]

$$\frac{b'}{b} = \frac{\pi}{4}. \tag{10.7}$$

Only for wings with very large aspect ratios we find asymptotically $b'/b \to 1$. The roll-up process of the vortex sheet toward the pair of trailing vortices on the other hand is completed only at a few half-span distances downstream of the wing, depending on the magnitude of the lift.

– The other property is the general deflection of the flow vectors—in terms of the vortex-line angle ε of the wing's wake—at the rear part of the wing, either in wing-tip direction for back-swept wings, or in wing-root direction for forward-swept wings.

  The latter is the reason for the well observed unwelcome accumulation of boundary-layer material at the wing root and fuselage. This can lead to adverse separation phenomena and, with rear-mounted engines at the aft end of the fuselage, makes special measures necessary regarding the position of the engines. The general deflection of course is also a property of the vortex sheet which leaves the wing's trailing edge.

The pattern of the attached viscous flow—visualized in the form of the skin-friction line pattern—reflects the two particular properties, see for instance in Fig. 10.20 of the following section the patterns of the skin-friction lines at the upper and the lower side of the backward swept CRM wing. The respective patterns which are found at a forward swept wing are shown in Fig. 10.21.

We have discussed the two particular properties—the TEF shear angle ψ and the vortex-line angle ε—with the help of results of potential-flow and three-dimensional boundary-layer theory. This poses no restriction regarding the validity of the result. The two properties can be found already in any proper Euler solution and in all Navier-Stokes/RANS solutions as well as in any wind-tunnel or flight experiment regarding lifting surfaces.

10.3 Aspects of Skin-Friction Line Topology: Flow Past an Airplane Configuration

Introduction

In this section the distributions of the wall pressure and the skin-friction coefficients, as well as skin-friction line patterns at selected locations of an airplane configuration are discussed using the contents of Chapter 7—the topology chapter—as guideline.

The considered airplane configuration is the NASA/Boeing Common Research Model (CRM), [19], which serves as a test configuration of the AIAA CFD Drag Prediction Workshop series [20]–[22].

The CRM shape is a generic one, resembling a simplified transport airplane configuration, Fig. 10.13. The wing span is $b = 58.765$ m, the mean aerodynamic chord $C_{ref} = 7.00532$ m, the leading-edge sweep $\varphi_{LE} = 37.5°$. The wing is attached without wing-fuselage fairing.

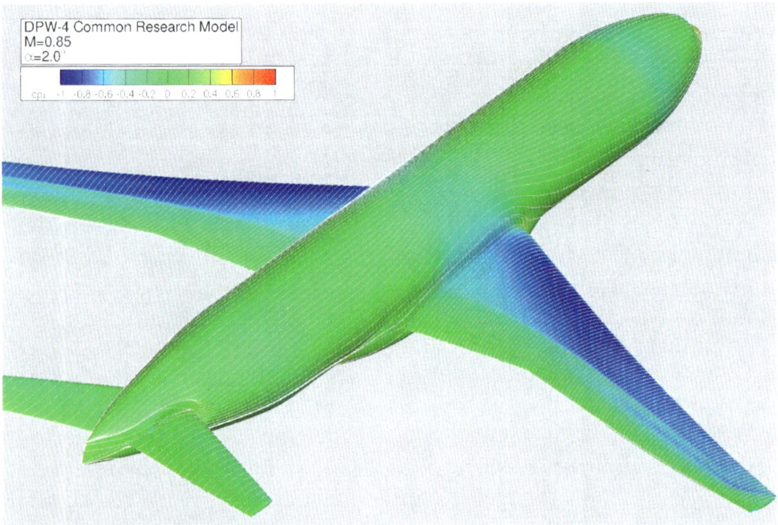

Fig. 10.13. Common Research Model (CRM) configuration [23]. Distribution of the surface pressure coefficient c_p.

The computation parameters, Table 10.2, correspond to that of "Case 1b" of the workshop, with the setting angle of the horizontal tail plane $\alpha_{HTP} = 0°$.

Table 10.2. Parameters of the CRM computation case ($L_{ref} = C_{ref}$) [23].

M_∞	$Re_{L_{ref}}$	L_{ref} [m]	T_∞ [K]	T_w [K]	φ_{LE} [°]	α [°]	β [°]	boundary-layer
0.85	$5 \cdot 10^6$	7.00532	310.93	310.93	37.5	2	0	fully turbulent

All results shown are for the angle of attack $\alpha = 2°$. Some results are available also for $\alpha = 4°$. These are not presented in detail, but are partly mentioned in the discussions that follow.

A few results from an investigation of a generic large transport airplane (LTA) configuration are discussed, too. They regard the flow at the wing root. The LTA wing has a leading-edge sweep $\varphi_{LE} = 35°$. The computation parameters are given in Table 10.3.

Table 10.3. Parameters of the LTA computation case [23].

M_∞	$Re_{L_{ref}}$	L_{ref} [m]	T_∞ [K]	T_w [K]	φ_{LE} [°]	α [°]	β [°]	boundary-layer
0.89	$100.9 \cdot 10^6$	12.3	228.2	260.73	35	2.1	0	fully turbulent

All figures in this section, except for Figs. 10.17, 10.21, and 10.22, are by courtesy of O.P. Brodersen, DLR-Institute of Aerodynamics and Flow Technology, member of the AIAA CFD Drag Prediction Workshop committee. He performed the present computations with DLR's TAU-code, which solves the RANS equations on hybrid grids, see, e.g., [24]. The flow was assumed to be fully turbulent, the one-equation turbulence model of P.R. Spalart and S.R. Allmaras (SA-model), [25], was employed. The size of the hybrid grid used for the CRM configuration was 13.3 million grid points, and for the generic LTA configuration 22.4 million points.

The CRM results show small separation phenomena at both the wing root (vicinity of the leading and the trailing edge) and the wing tip. These separation structures—singular points at the wing root, singular lines at the wing tip—appear to be topologically sound. Nevertheless, in view of the employed SA-model, they must be seen with some reservations. The SA-model, being basically a linear eddy-viscosity model—well suited for predominantly attached wing-type flows at high Reynolds numbers—seems to produce exaggerated separation zones, as reported, for instance, in [26].

The grid of the present computation was very fine, but not specifically adapted to the resolution of the rather small separation structures. Nevertheless, we discuss the computed separation phenomena without further analysis of potential inaccuracies in the prediction methodology.

The wall pressure coefficient is, as usual, defined by $c_p = (p - p_\infty)/q_\infty$, with $q_\infty = \rho_\infty v_\infty^2/2$ being the dynamic pressure. At the forward stagnation point—the primary attachment point—the pressure coefficient is $c_{p_{stag}} = 1.1939$ (eq. (7.29), perfect gas, $\gamma = 1.4$). The vacuum value, see, e.g., [27], is $c_{p_{vac}} = -2/(\gamma\, M_\infty^2) = -1.9773$. The skin-friction coefficient is defined by $c_f = |\tau_w|/q_\infty$.

We discuss in the following for the right side of the airplane configuration the computed distributions of the wall pressure and the skin-friction coefficient, together with the related skin-friction line patterns at the fuselage nose, the leading edge of the wing root, the wing leading edge, the wing trailing edge and the wing tip. All results are CRM results, only for the leading edge of the wing root some LTA results are presented for comparison and for the trailing edge a result from a forward-swept wing case.

Fuselage Nose

The CRM nose is a typical airplane nose, Fig. 10.14.[5] The primary attachment point lies for the small angle of attack of $\alpha = 2°$ more or less at the nose apex.

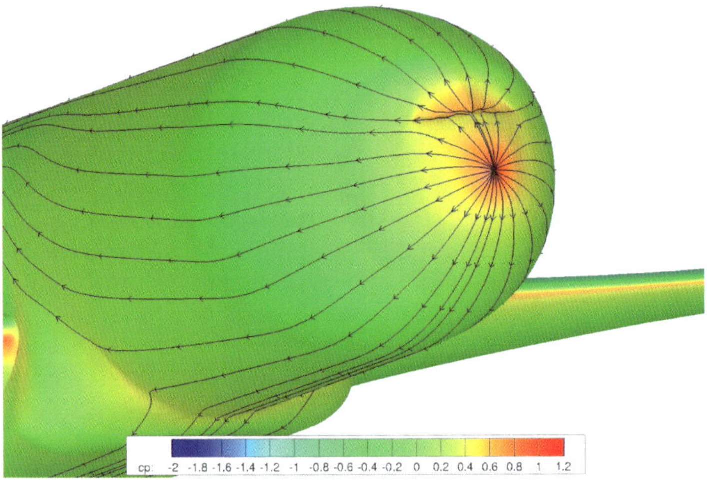

Fig. 10.14. CRM fuselage nose [23]: pattern of skin-friction lines and distribution of the surface pressure coefficient c_p.

A closer look at the nose region reveals the expected results, Fig. 10.15. The skin-friction lines show the typical nodal-point pattern, Fig. 7.3. The node is a general node, not a star node. The latter would be structurally unstable. The computation with $\alpha = 4°$ angle of attack (not shown) has indicated only minor changes in the pattern, i.e. a stable behavior of the node.

The upper part of Fig. 10.15 shows the absolute wall pressure maximum at the attachment point, $c_{p_{stag}} = 1.1939$. In the lower part of the figure the skin friction has there its absolute minimum, $c_f = 0$. Away from the attachment point we see a severe pressure drop—the flow is strongly accelerated—accompanied by a strong skin-friction increase.

Wing Root: Leading Edge

At the wing root, which has no wing-fuselage fairing, Fig. 10.16, a separation pattern is present, resembling that ahead of an obstacle on a flat surface.

We see in the pattern of the skin-friction lines a saddle at the side of the fuselage and a node at the wing's leading edge at a small distance from the

[5] The dent in the contour above the stagnation point in Fig. 10.14 indicates the cockpit location.

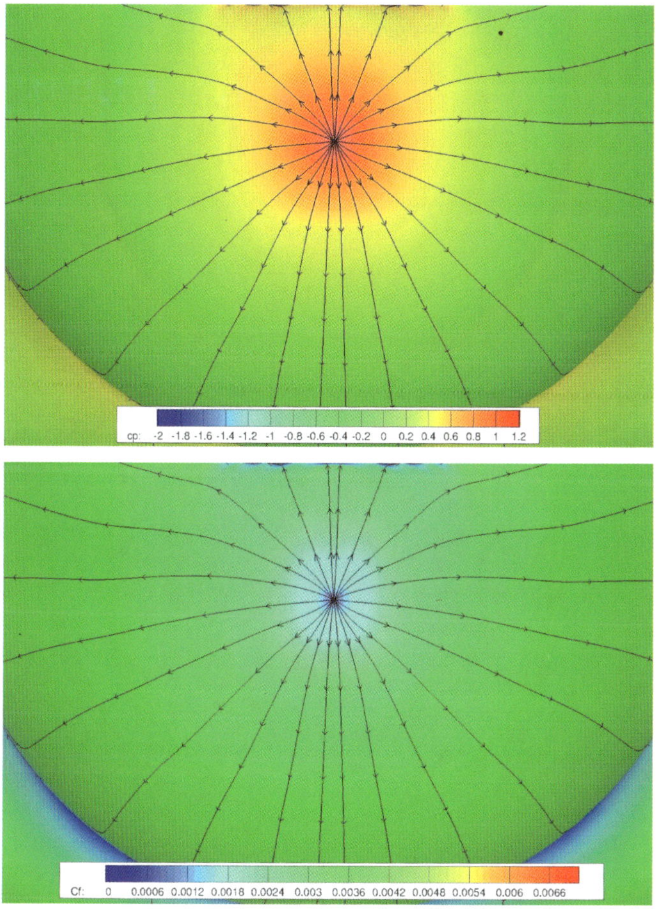

Fig. 10.15. CRM fuselage nose [23]. Upper part: pattern of skin-friction lines and distribution of the surface pressure coefficient c_p. Lower part: pattern of skin-friction lines and distribution of the skin-friction coefficient c_f.

root. The saddle is a separation point, and the node an attachment point, Fig. 10.17. The topology is structurally stable in the investigated angle-of-attack range. The solution for $\alpha = 4°$ (not shown) indicates only minor changes in the locations of the two singular points.[6] The attachment line along the wing's leading edge has its origin at the node.

[6] This is in contrast to the flow field at the suction side of the wing. There the increase from $\alpha = 2°$ to $\alpha = 4°$ results in a structural change of the flow field. For the smaller angle of attack we do not see a trace of a shock wave in the skin-friction pattern, Fig. 10.20. At the larger angle a change has occurred. Now the skin-friction line pattern shows a terminating shock wave extending from approximately one third span to full span at chord locations of 30 to 50 per cent.

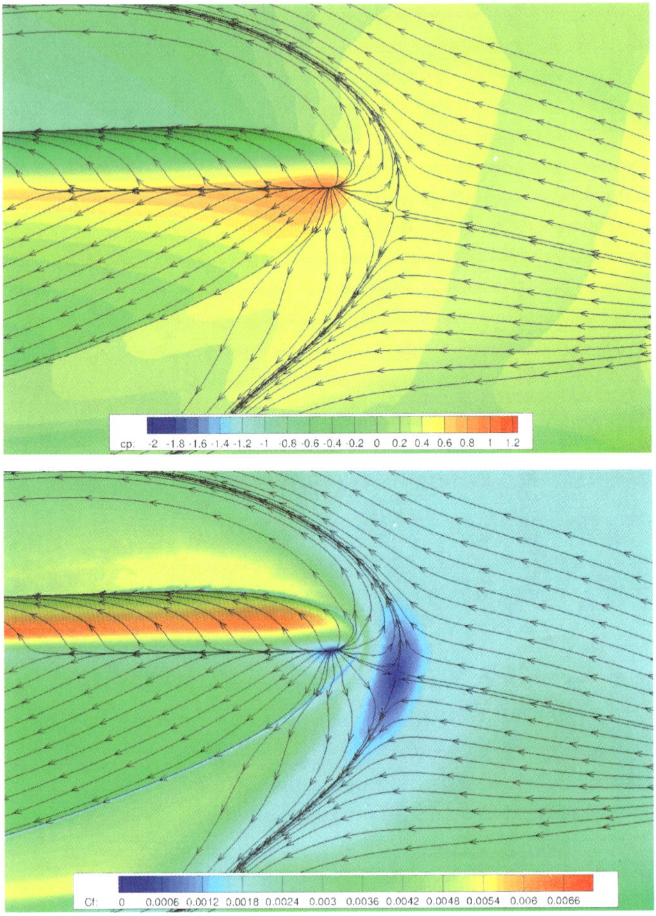

Fig. 10.16. CRM leading edge of the wing root [23]. Upper part: pattern of skin-friction lines and distribution of the surface pressure coefficient c_p. Lower part: pattern of skin-friction lines and distribution of the skin-friction coefficient c_f.

A count of the singular points on the vehicle's surface gives one node at the nose of the fuselage, one saddle ahead of each wing root and one node at each leading edge. If we assume in summary one node at the aft of the configuration, our count results in four nodes and two saddles. This fulfills the topological rule 1, page 144.

At the attachment point a relative pressure maximum is discernible, like on the leading edge the relative pressure maximum across the attachment line, upper part of Fig. 10.16. Due to the chosen color bar scale the p_{max}-line—in terms of c_p—is only weakly expressed. The same is true regarding the $|\tau_w|_{min}$-line—in terms of c_f—, lower part of Fig. 10.16. The skin-friction coefficient is zero at both the separation and the attachment point.

A schematic overview of the flow topology is given in Fig. 10.17. The separation point at the side of the fuselage ahead of the wing root's leading edge is the forward point of the separating fuselage boundary layer. The emanating free streamline is the forward portion of the separating sheet. The count of the singular points on the surface gives one center (node) ahead of each wing root, and four half-saddles (separation and attachment points). At the nose of the fuselage we have a half-saddle, too. We assume again summarily one half-saddle at the aft of the configuration. Our count then results in two nodes and four half-saddles. This fulfills the topological rule 2, page 144.

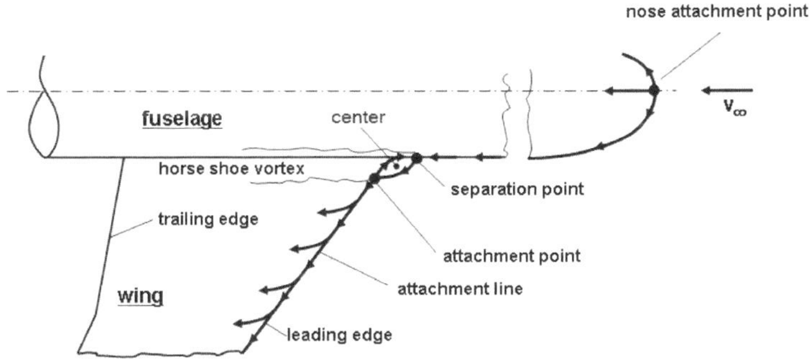

Fig. 10.17. CRM wing-root junction. View from above: schematic overview of the flow topology.

These considerations point to the existence of a wing-root vortex. The so-called horse-shoe vortices at wing roots carry away kinetic energy and hence lead to a drag increment. This could be called an induced drag of the second kind.

Wing-root horse-shoe vortices can lead to buffeting at large angle of attack. This observation instigated work by Th. von Kármán and co-workers which finally led to the introduction of smooth fairings (fillets) at wing roots [28]. In this way, horse-shoe vortices at the wing roots are eliminated. Wing-root fairings are the rule today for large transport airplanes. The generic LTA configuration has an elaborately shaped fairing. This can be seen from the computed data shown in Fig. 10.18.

The skin-friction lines exhibit a smooth transition of the flow from the fuselage side towards the leading edge. At the latter an open type attachment line is formed. The pressure field in the upper part of the figure exhibits a small relative maximum ahead of the leading edge, probably at the beginning of the fairing. The same is seen at the beginning of the leading edge below the developing attachment line. Both relative maxima correspond to relative minima of the skin friction, indicated by the darker blue spots in the lower part of the figure. Separation, however, does not occur.

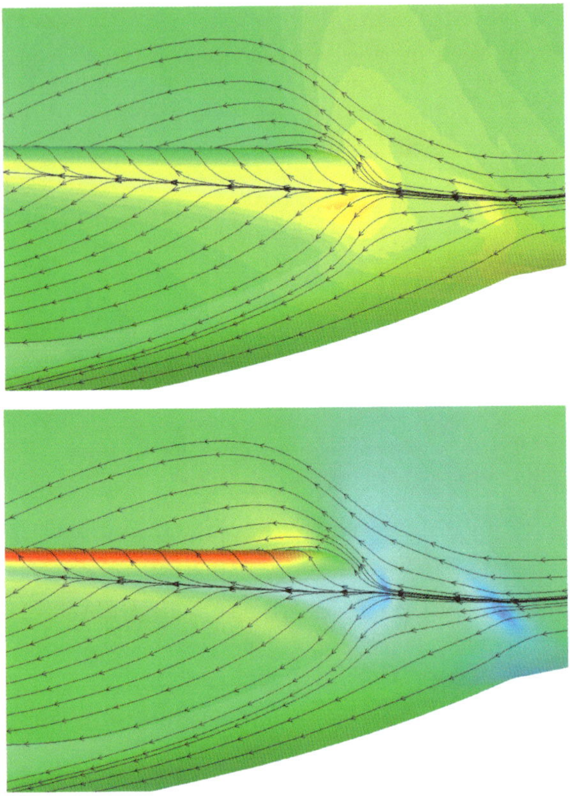

Fig. 10.18. LTA leading edge with wing-root fairing [23]. Upper part: pattern of skin-friction lines and distribution of the surface pressure coefficient c_p. Lower part: pattern of skin-friction lines and distribution of the skin-friction coefficient c_f.

At the trailing edge of the wing root—not shown—the CRM flow seems to have very small embedded separation phenomena. This was found at DLR also with other turbulence models (kw-SST, RSM) and with different grid resolutions. Brodersen reports that the computations of the workshop participants do not show unambiguous results in this regard, also not the experiments. Separation phenomena are not present in the LTA flow. There a smooth flow-off happens.

The computed data demonstrate clearly the value of the wing-root fairing, in particular at the wing root's leading edge. Of course one has to ask, whether the fairing has side effects which make a trade-off necessary. In our case, this is not obvious. However, when considering the topic of laminar flow control, an interesting question shows up. In Sub-Section 9.3.2 the matter of leading-edge contamination is treated. One of the problems in this regard is that the turbulent boundary-layer flow of the fuselage enters the wing's leading edge, if a wing-root fairing is present. This is evident from Fig. 10.18.

A means to avert this effect, is a turbulence diverter. Such a device was proposed in 1965 by M. Gaster [29]. The example of the CRM configuration without wing-root fairing suggests that with the attachment point at the leading edge an effect similar to that of the Gaster bump effect can be expected. We do not further discuss this rather academic question, but point instead to Sub-Section 9.3.2.

Wing Leading Edge

The attachment-line appears to lie on a geodesic, Fig. 10.19. How close the extrema lines, Sub-Section 7.4.2, are located to it, is not visible.

In the upper part of the figure the strong acceleration of the flow towards the suction side of the wing is clearly discernible. The pressure coefficient appears to be nearly constant in the span-wise direction, although a slight acceleration seems to happen in that direction. The skin-friction coefficient, lower part of the figure, rises from zero in the attachment point, Fig. 10.17, and becomes constant only further away in the wing-tip direction.

Wing Trailing Edge

In Section 10.2 the flow patterns at the upper and the lower side upstream and at the trailing edge of lifting large-aspect ratio wings were discussed. For the CRM wing the patterns of the skin-friction lines on both the upper and the lower side of the wing are shown in Fig. 10.20. At the trailing edge the patterns reflect well the behavior of the external inviscid flow in terms of the vortex-line angle ε and the TEF shear angle ψ.

At the outer part of the wing generally and in particular at the trailing edge the skin-friction lines show the typical outward tendency at the lower side and the inward tendency at the upper side. The lines indicate also clearly the positive vortex-line angle and that the local shear increases towards the wing tip.

This result is typical for a back-swept wing. At the forward-swept wing shown in Fig. 10.21 the computed skin-friction line patterns look quite different. The figure is by courtesy of Th. Kilian, DLR-Institute of Aerodynamics and Flow Technology. He performed the computation with DLR's TAU-code, see above, fully turbulent with the one-equation turbulence model of Spalart and Allmaras.

The wing geometry is derived from the DLR design of a natural laminar flow (NLF) forward-swept wing for a medium range transport airplane (DLR Project LamAiR [31, 32]). The with $\varphi_{LE} = -17°$ forward swept wing has a design Mach number $M_{\infty_{design}} = 0.78$ with an off-design capability of 0.80. The computation parameters, Table 10.4, are those of a wind-tunnel test.

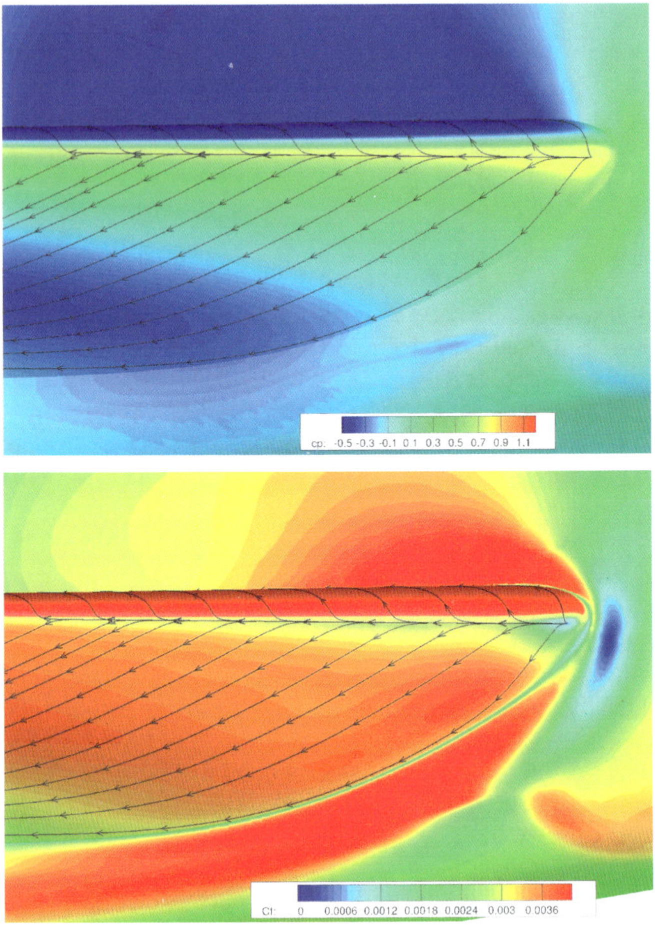

Fig. 10.19. CRM wing leading edge [23]. Upper part: pattern of skin-friction lines and distribution of the surface pressure coefficient c_p. Lower part: pattern of skin-friction lines and distribution of the skin-friction coefficient c_f.

Table 10.4. Parameters of the forward swept wing computation case (wind-tunnel situation) [30].

M_∞	$Re_{L_{ref}}$	L_{ref} [m]	T_∞ [K]	T_w [K]	φ_{LE} [°]	α [°]	β [°]	boundary-layer
0.26	$1.35 \cdot 10^6$	0.224	294	294	-17	4	0	fully turbulent

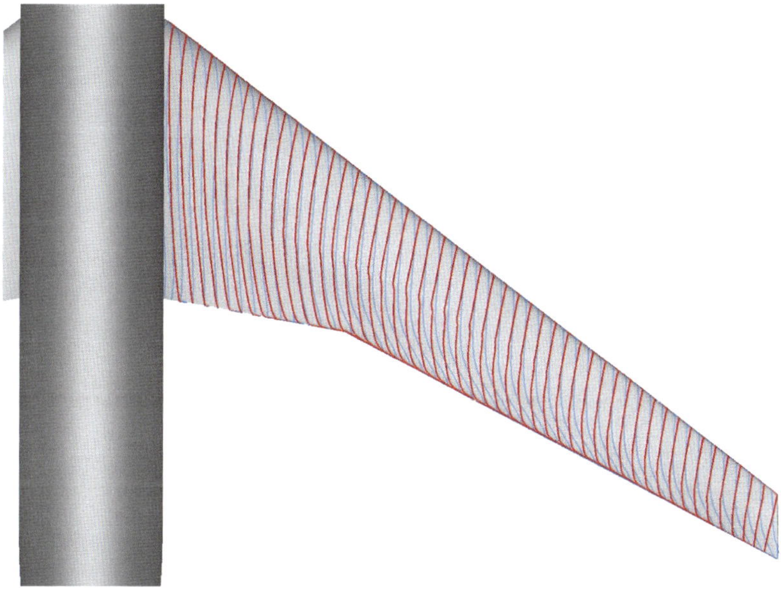

Fig. 10.20. CRM wing at angle of attack $\alpha = 2°$, view from above [23]. Skin-friction line pattern at the upper side (red lines) and at the lower side (blue lines). The free-stream direction is from the top to the bottom of the figure.

The wing's chord section has a pressure distribution such that (a) leading-edge contamination, Sub-Section 9.3.2, is avoided due to a large flow acceleration in direction normal to the leading edge, that (b) the cross-flow instability, Sub-Section 9.3.3, is inhibited and that (c) the growth of Tollmien-Schlichting waves, Sub-Section 9.2.2, is retarded as much as possible. That holds for both the suction and the pressure side of the wing. The pressure distributions are reflected in the computed skin friction line patterns up to about 66 per cent chord of the wing, Fig. 10.21: the lines bend immediately out of the attachment line and show almost over the whole wing span a nearly two-dimensional behavior.

Upstream of and at the trailing edge the skin-friction line patterns reflect the forward-swept wing behavior of the external inviscid flow in terms of the negative (!) vortex-line angle ε and the TEF shear angle ψ, which increases towards the wing tip, Section 10.2.

Wing Tip

The flow at the wing tip demands a particular discussion. At a lifting wing with a rounded wing tip the flow expands from the lower side around the tip to the upper side. If at the upper side the external inviscid streamlines are curved towards the wing tip, the boundary-layer streamlines are curved even stronger. This then leads to a convergence of the boundary-layer streams

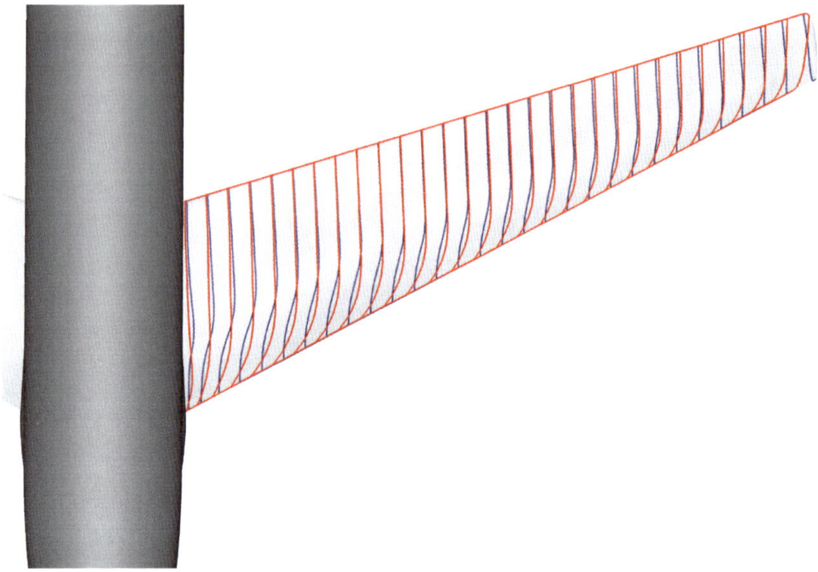

Fig. 10.21. Forward swept wing at angle of attack $\alpha = 4°$, view from above [30]. Skin-friction line pattern at the upper side (red lines) and at the lower side (blue lines). The free-stream direction is from the top to the bottom of the figure.

from below and above and to squeeze-off separation on the upper side near the wing tip, Fig. 10.22. (If the wing tip is sharp-edged, flow-off separation happens along the edge of the wing tip.) The resulting discrete tip vortex generally has no long life. It is merged downstream of the trailing edge into the vortex sheet respectively the trailing vortex.

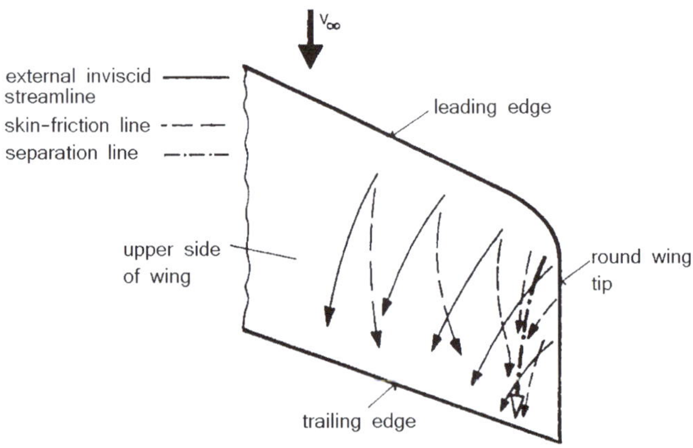

Fig. 10.22. Formation of the tip vortex at a rounded wing tip. Flow, schematically, at the upper side of the wing [33].

The CRM wing tip is well rounded, Fig. 10.23. The open-type squeeze-off separation line is formed at approximately one third chord length.

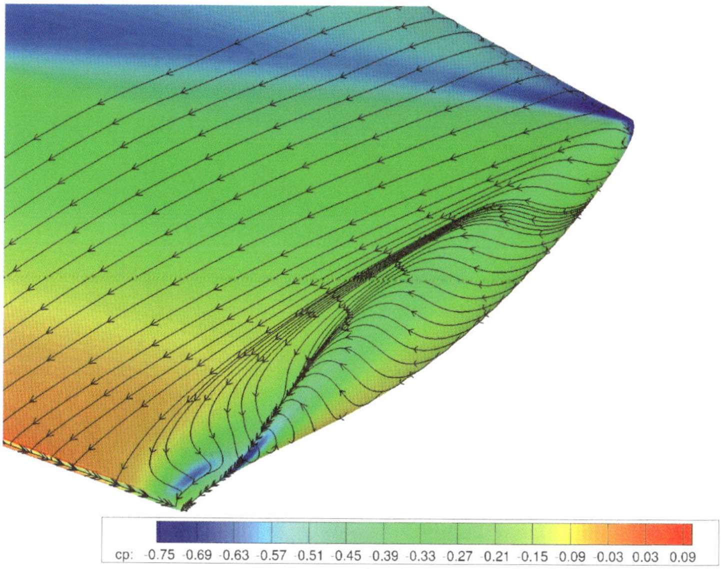

Fig. 10.23. CRM wing tip, view from above [23]. Pattern of skin-friction lines, distribution of the surface pressure coefficient c_p, primary and secondary open-type separation line.

The round wing tip, however, becomes sharp towards the wing's trailing edge. The sharp-edged part causes a locally strong transport of kinematically active vorticity into the tip vortex. This is indicated by the drop of the surface pressure (blue color) just ahead of the sharp tip portion. The flow field beneath the tip vortex becomes rearranged similar to what happens at the lee-side of a delta wing at angle of attack, Fig. 10.26 b).

This is indicated in Fig. 10.24 by the strong outward bending of the skin-friction lines near the trailing edge (there also the pressure is very low). The bending is terminated by a pressure rise and an open-type secondary separation line is formed. All is reflected too in the skin-friction field in the lower part of the figure.

The result is that a secondary vortex is formed just inboard of the wing tip. Topological considerations demand the presence of an attachment line between the two separation lines which is clearly indicated. The attachment line is of open type. The skin-friction extrema lines are not visible in the illustration.

In view of wing-tip devices (winglets), which often are applied in order to reduce the induced drag, we point to the necessity to distinguish between the tip vortices and the trailing vortices. In the literature often these are mixed

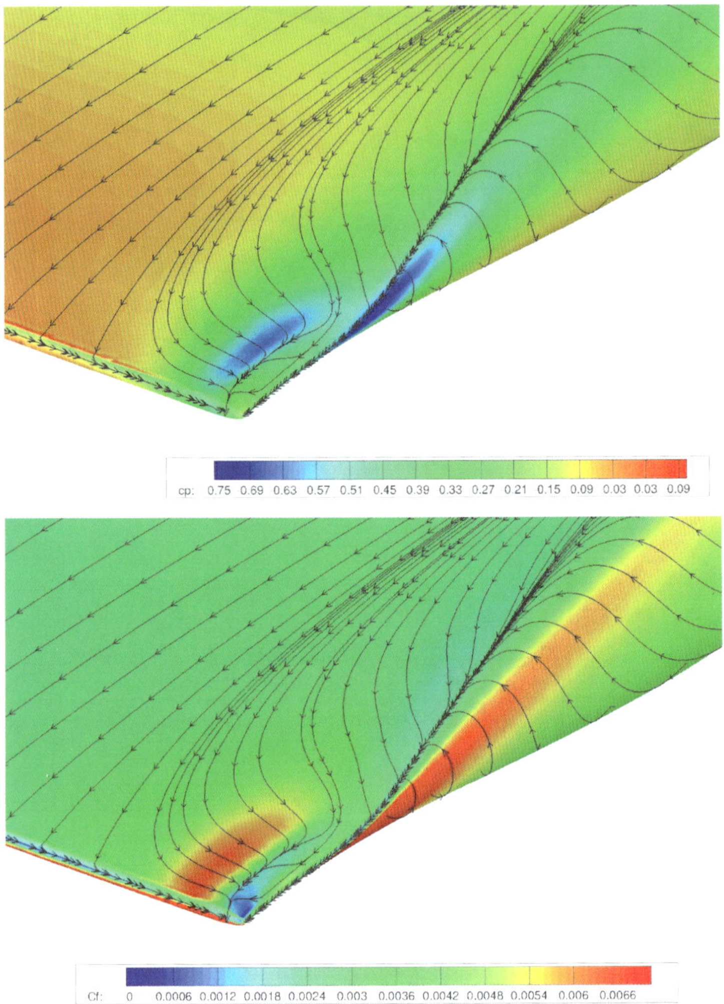

Fig. 10.24. CRM wing tip. Detail of Fig. 10.23 with the two separation lines and the attachment line between them [23]. Upper part: pattern of skin-friction lines and distribution of the surface pressure coefficient c_p. Lower part: pattern of skin-friction lines and distribution of the skin-friction coefficient c_f.

up and it is neglected that it is first a vortex sheet which is shed from the wing's trailing edge. From this sheet the trailing vortices are formed at a certain distance downstream of the trailing edge with a horizontal distance of their axes smaller than the wing span, see the discussion on page 257 ff. of the preceding section.

What is to be influenced by a wing-tip device firstly is the shed vortex sheet and, of course, also the tip vortex and that without the formation of additional vortices. The efficiency of a wing tip devices will depend on a detailed balance between vortex sheet, tip vortices and any other vortices which must be optimized for a specific flight condition of a given configuration.

If a wing design has a span restriction, a wing-tip device may help to reduce total drag. If there is no span restriction instead the wing span should be increased.

10.4 Extrema of the Thermal State of the Surface: Flow Past a Blunt Delta Wing

In Section 7.4 we have discussed that across attachment and separation lines relative extrema of the characteristic boundary-layer thicknesses and hence of the thermal state of a body surface occur. We present results regarding this phenomenon at a generic re-entry configuration, the Blunt Delta Wing (BDW). This configuration is a very strongly simplified re-entry vehicle configuration flying at moderate angle of attack. We follow closely parts of the discussion of this case which was given in [34].

At the beginning of the 1990s S. Riedelbauch [35] (see also [36]) studied the aerothermodynamic properties of hypersonic flow past the radiation-cooled surface of the BDW configuration [37]. The configuration is a simple slender delta wing with a blunt nose, Fig. 10.25. The lower side has a dihedral ($\gamma = 15°$, lower part of Fig. 10.25) and therefore is only approximately flat.

Navier-Stokes computations with perfect-gas assumption were performed with the parameters given in Table 10.5. The flow is laminar throughout, the vehicle surface is radiation cooled, [34], with a surface emissivity coefficient $\varepsilon = 0.85$.

Table 10.5. Computation parameters of the Blunt Delta Wing [35].

M_∞	H [km]	T_∞ [K]	Re^u_∞ [m^{-1}]	L [m]	φ_0 [°]	α [°]	ε	boundary layer
7.15	30	226.506	$2.69 \cdot 10^6$	14	70	15	0.85	laminar

We look at the topology of the computed skin-friction field, Fig. 10.26, in order to identify attachment and separation lines in the vicinity of which we expect extrema of the thermal state of the surface, Section 7.4.

At the lower (windward) side of the configuration in Fig. 10.26 a) we see the classical skin-friction line pattern present at the lower side of a delta wing. Because in our case this side is not fully flat, the flow exhibits a slight three-dimensionality between the two primary attachment lines. The latter are marked by strongly divergent skin-friction lines. The forward stagnation point, which is a node, Sub-Section 7.2.3, lies also on the lower side, at about

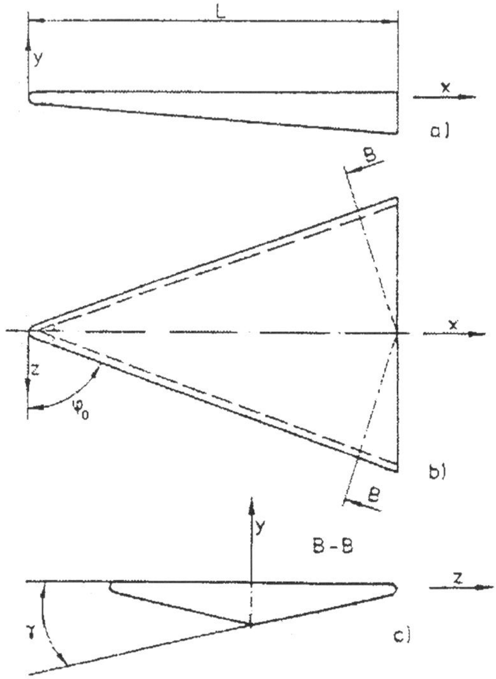

Fig. 10.25. Configuration of the BDW and the coordinate convention [35]: a) side view, b) view from above, c) cross-section B - B.

3 per cent of the body length. The primary attachment lines are almost from the beginning parallel to the leading edges, i.e. they do not show a conical pattern.

The situation is quite different at the upper (leeward) side of the wing, Fig. 10.26 b). Here we see on the left-hand side of the wing (from the leading edge towards the symmetry line) along the vertical line a succession of separation and attachment lines: the primary separation line S_1, the secondary attachment line A_2, a secondary separation line S_2, and a tertiary attachment line A_3. All is mirrored on the right-hand side of the wing. Again a conical pattern is not discernible, except for a small portion near to the nose. However, the secondary separation lines are almost parallel to the single tertiary attachment line along the upper symmetry line of the wing.

Both the primary and the secondary separation lines are of the type "open separation", i.e. the separation line does not begin in a singular point on the surface, Sub-Section 7.1.4. Fig. 10.27 shows this, as well as that all attachment lines are of open type, too.

With these surface patterns we construct qualitatively the structure of the leeward-side flow, Fig. 10.28. By marking the points, where the streamlines of the vortex-feeding layers and of the attaching stream surfaces penetrate

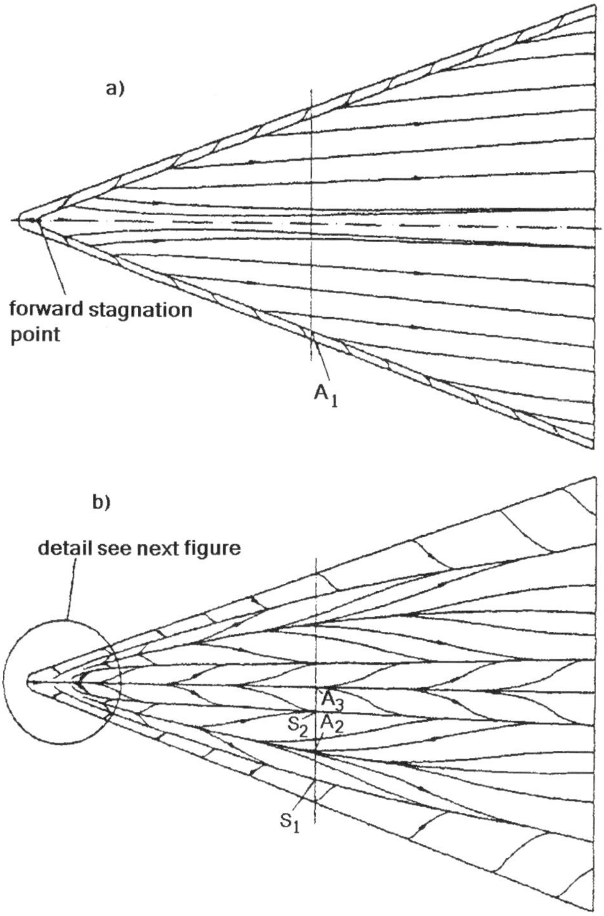

Fig. 10.26. Selected computed skin-friction lines at the surface of the BDW [35]: a) look at the lower side, b) look at the upper side of the configuration. The free-stream comes from the left.

a surface normal to the x-axis, one finds the Poincaré surface (Section 7.3), Fig. 10.29. The computed cross-flow shocks are indicated.

In Fig. 10.29 the attachment and separation lines are marked as "half-saddles (S')" (note that the primary attachment lines are "quarter-saddles (S'')", because the flow between them is (more or less) two-dimensional). The axes of the primary and the secondary vortices are marked as "foci (F)", which are counted as "nodes (N)". Finally a "saddle (S)" is indicated above the wing. This pattern obeys the topological rule 2', Section 7.3:

$$\left(4+\frac{1}{2}0\right)-\left(\sum 1+\frac{1}{2}7+\frac{1}{4}2\right)=-1, \tag{10.8}$$

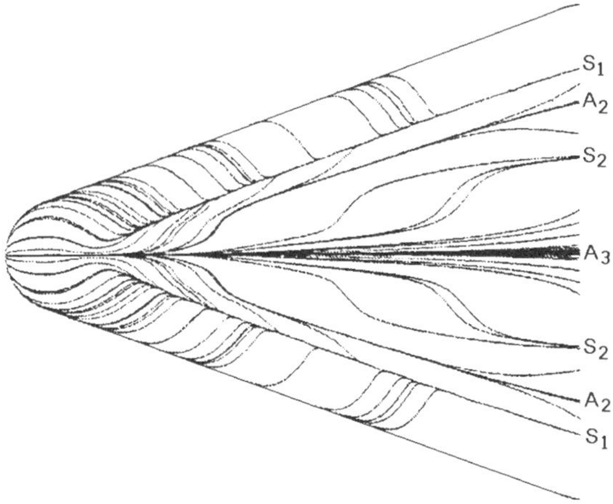

Fig. 10.27. Selected computed skin-friction lines at the upper side of the BDW near the nose (detail of Fig. 10.26 b)) [35].

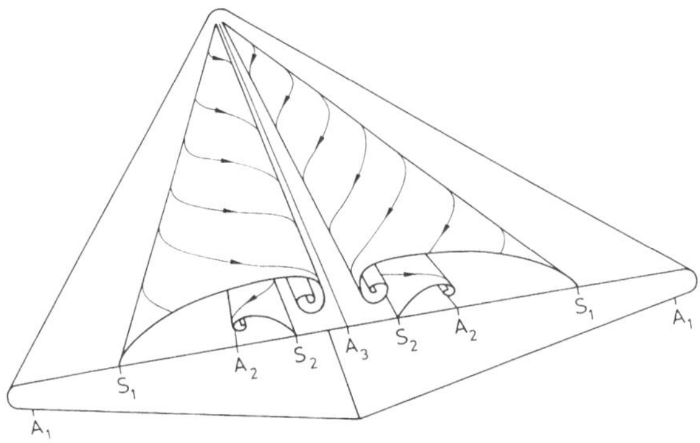

Fig. 10.28. Sketch of the leeward-side flow topology of the BDW [35]. A_1: primary attachment lines, A_2: secondary attachment lines, A_3: tertiary attachment line, S_1: primary separation lines, S_2: secondary separation lines.

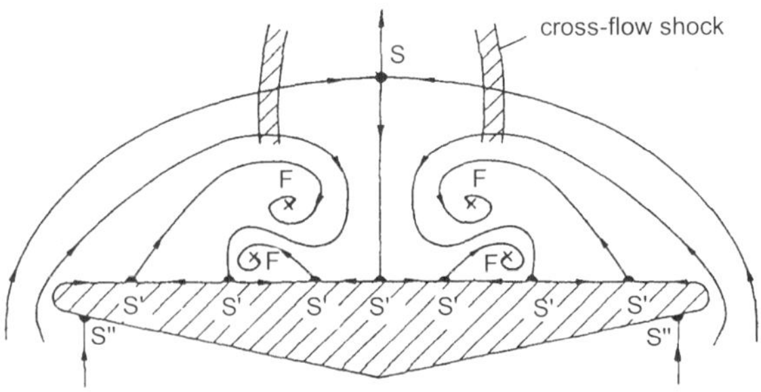

Fig. 10.29. Sketch of the topology of the BDW velocity field in the Poincaré surface at $x/L = 0.99$ [35].

and therefore is a valid topology.

Fig. 10.30 gives an overview of the results in terms of the thermal state of the surface on the lower and the upper surface. Unfortunately the color scales are not the same in the two parts a) and b) of the picture.

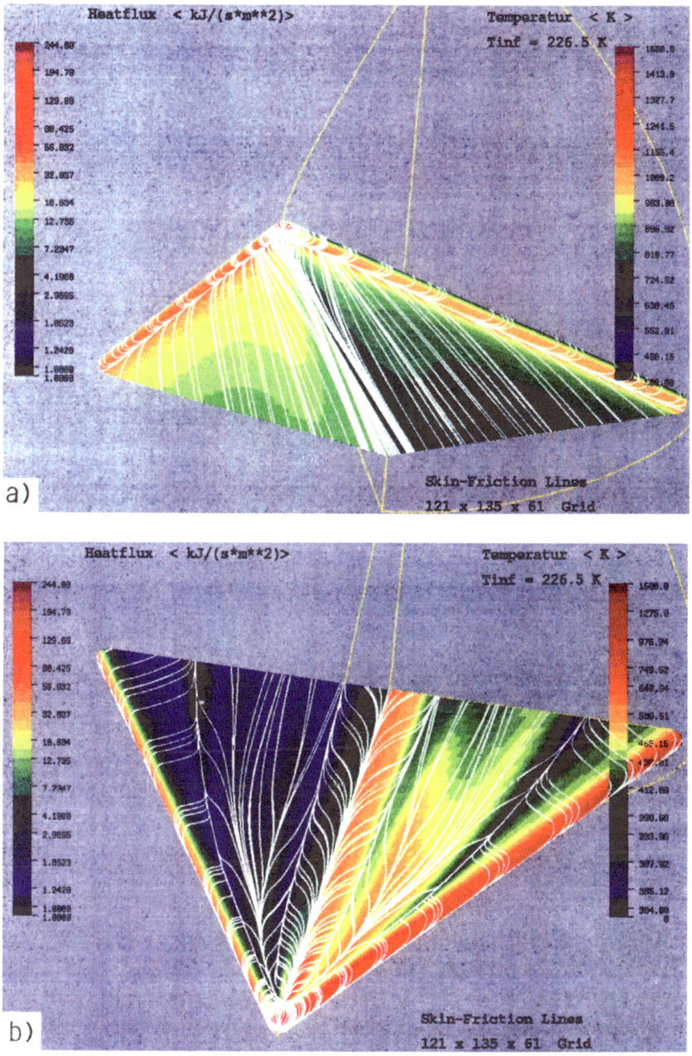

Fig. 10.30. Computed skin-friction lines, and distributions of the surface radiation heat flux q_{rad} (left) and the radiation-adiabatic surface temperature T_{ra} (right) at **a**) the lower (windward) side, and **b**) the upper (leeward) side of the BDW [35].

We disregard the radiation heat-flux distributions on the left-hand sides of the figure and concentrate instead on the distributions of the radiation-adiabatic temperatures on the right-hand sides. Part a) of Fig. 10.30 shows at the lower side of the wing the almost parallel flow between the primary attachment lines. Along the attachment lines heating ensues with a nearly constant temperature of approximately 1,100 K. This hot-spot situation is the consequence of the relative minimum of the characteristic boundary-layer thickness there, Section 7.4. Between the attachment lines the radiation adiabatic temperature reduces in the downstream direction as expected. On the larger portion of the lower side it lies around 800 K.

On the upper side, part b), along the round leading edge we see a nearly constant temperature of about 1,050 K. This high temperature is due to the small boundary-layer thickness, which is a result of the strong expansion of the flow around the leading edge. At the primary separation line the temperature drops fast and a real cold-spot situation develops. This cold-spot situation is the consequence of the relative maximum of the characteristic boundary-layer thickness there, Section 7.4.

The secondary attachment line seems to taper off at about 40 per cent body length. Possibly a tertiary vortex would develop, if the wing length would be increased (non-conical behavior). At the secondary separation line again a cold-spot situation develops, however weaker than that at the primary separation line. The tertiary attachment line shows the expected attachment-line heating with an almost constant temperature of approximately 650 K along the upper symmetry line.

These results demonstrate well the effect of the relative extrema of the characteristic boundary-layer thicknesses across attachment and separation lines. They result in this hypersonic flow case in relative extrema of the thermal state of the (radiation cooled) surface which pose serious hot-spot problems for the vehicle designer. The reader interested in hypersonic flow problems is referred to the extended discussion of this case in [34].

10.5 The Location of Laminar-Turbulent Transition: Flow Past an Ellipsoid at Angle of Attack

If the location of laminar-turbulent transition is of importance for the design of a given flight vehicle, see the discussion in the introduction to Chapter 9, we may have to deal with a very intricate situation. This is in stark contrast to two-dimensional flow problems, where the transition location geometrically is rather simply defined (although still predicted with some difficulty).

We consider as an example the flow past a body of low geometrical complexity, an ellipsoid at angle of attack. We have touched this flow case already in Sub-Section 7.4.3, results for it are given also in [38]. In a recent paper N. Krimmelbein and R. Radespiel presented results of a computation method for the prediction of transition lines on general three-dimensional configurations [39]. They treated also the flow past the 1:6 ellipsoid at 10° angle of attack.

The computations were made with the TAU-code, see, e.g., [24]. Laminar-turbulent transition is predicted with a special module which treats the boundary-layer stability in terms of Tollmien-Schlichting and cross-flow instability, see Sub-Sections 9.2.2 and 9.3.3. The transition location is determined with an e^N approach for three-dimensional flow. Once the flow is considered to have become fully turbulent, the standard Spalart-Allmaras turbulence model is employed.

The computation parameters from [39] are given in Table 10.6. The flow field was resolved in the direction normal to the wall with about 130 points, with 60 to 100 points in the laminar boundary layer. In the stream-wise direction about 300 points were chosen, so that the total number of grid points is approximately 2.8 million.

Table 10.6. Parameters of the ellipsoid computation case [39].

M_∞	Re_L	L [m]	T_∞ [K]	T_w [K]	α [°]	boundary-layer state
0.13	$6.5 \cdot 10^6$	2.4	273.15	273.15	10	according to transition module

The figures shown in this section are not from the cited publication. They are by courtesy of O.P. Brodersen and N. Krimmelbein, DLR-Institute of Aerodynamics and Flow Technology. The flow in all figures comes from the left-hand side. The surface pressure coefficient and the skin-friction coefficient are defined as in Section 10.3, page 261.

In Fig. 10.31 we see the general flow behavior in terms of the skin-friction lines. The stagnation point is at the left, somewhat below the nose of the ellipsoid. At the upper right we see an accumulation of the skin-friction lines and a separation pattern at the end of the ellipsoid. These phenomena will not be discussed in further detail here.

The surface pressure drops fast from the stagnation point value ($c_{p_{stag}} = 1.0042$) to values of about $c_p = -0.2$ at the upper side of the ellipsoid and then undergoes a recompression. The iso-pressure lines are forming oblique bands, extending roughly from the upper left to the lower right side.

In the lower part of the figure we see above the stagnation-point region the initial strong rise of the skin-friction coefficient of the laminar flow. This rise is due to the fast expansion around the nose to the upper side of the ellipsoid. The skin-friction then drops slowly until laminar-turbulent transition occurs.

The transition line is marked by the strong rise of the skin friction, visible in the change from the green to the brown/red color (note that the color bar scales are different in the different figures). The small yellow strip—yellow only in Fig. 10.31—between the laminar and the turbulent region is a kind of intermittency region, due to the diffusive properties of the turbulence model. The intermittency region as such is not modelled in the computation method.

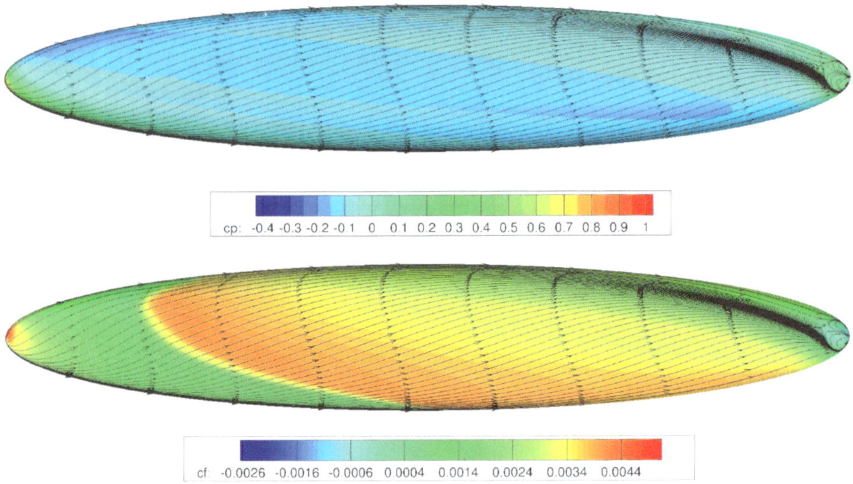

Fig. 10.31. View of the left side of the ellipsoid [23]. Upper part: pattern of skin-friction lines and distribution of the surface pressure coefficient c_p. Lower part: pattern of skin-friction lines and distribution of the skin-friction coefficient c_f.

The location of the transition line is not correlated with the locations of the surface-pressure bands. Across the transition line the skin friction rises to the high turbulent values which then drop only slowly in the stream-wise direction.[7]

Fig. 10.32 permits a closer look at the nose region. From the forward stagnation point, a node, the attachment line lying at the lower symmetry line emanates. The divergence of the skin-friction lines is well visible, less well visible are the relative pressure maximum and the relative skin-friction minimum across the attachment line. At the upper symmetry line the skin-friction lines converge only weakly. The fast drop of the surface pressure and the initially strong rise of the skin-friction coefficient along the upper symmetry line are well pronounced.

Fig. 10.33 gives the view of the lower side of the ellipsoid. The pressure drops in the flow direction fast from the high value at the stagnation point—note that here the color bar scale of the surface pressure is different from that in Fig. 10.31—to much lower values. The skin-friction line divergence indicates well the presence of the attachment line along the lower symmetry line of the ellipsoid. The relative pressure maximum at the attachment line is barely visible.

Also not visible with the given resolution is the relative minimum of the skin-friction coefficient at the attachment line (lower part of the figure). However, the attachment line is reflected by the small decrease of the skin-friction

[7] Regarding the different dependencies of laminar and turbulent flow on the boundary-layer running length see Appendix B.3.1.

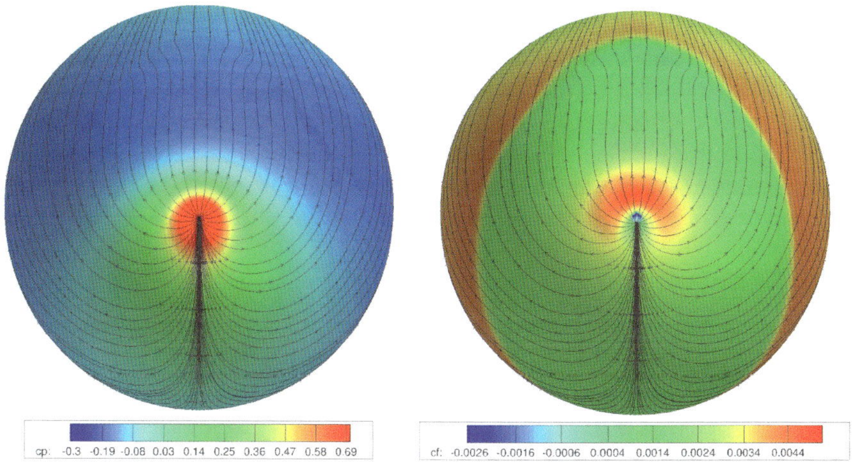

Fig. 10.32. Front view of the nose of the ellipsoid [23]. Left part: pattern of skin-friction lines and distribution of the surface pressure coefficient c_p. Right part: pattern of skin-friction lines and distribution of the skin-friction coefficient c_f.

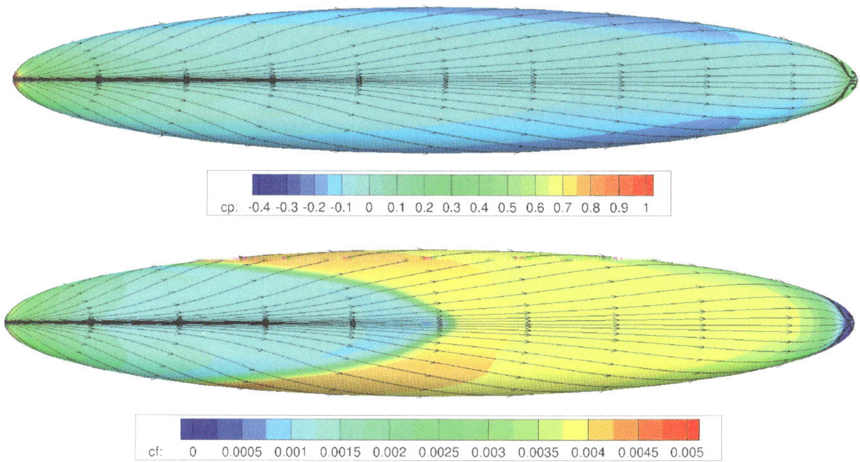

Fig. 10.33. View of the lower side of the ellipsoid [23]. Upper part: pattern of skin-friction lines and distribution of the surface pressure coefficient c_p. Lower part: pattern of skin-friction lines and distribution of the skin-friction coefficient c_f.

coefficient in the stream-wise direction. The transition line extends far downstream in the form of a narrowing tongue.

The view of the upper side gives a different picture, Fig. 10.34. The skin-friction lines first converge slightly towards the upper symmetry line. Approximately from mid-length on they diverge. The expansion around the nose is discernible in the surface-pressure distribution (upper part of the figure) as well as in the distribution of the skin-friction coefficient (lower part). Across the upper symmetry line relative extrema of the surface pressure and the skin-friction coefficient are indicated, however only weakly. The transition line extends not as far downstream as on the lower side, but forms also a narrow tongue.

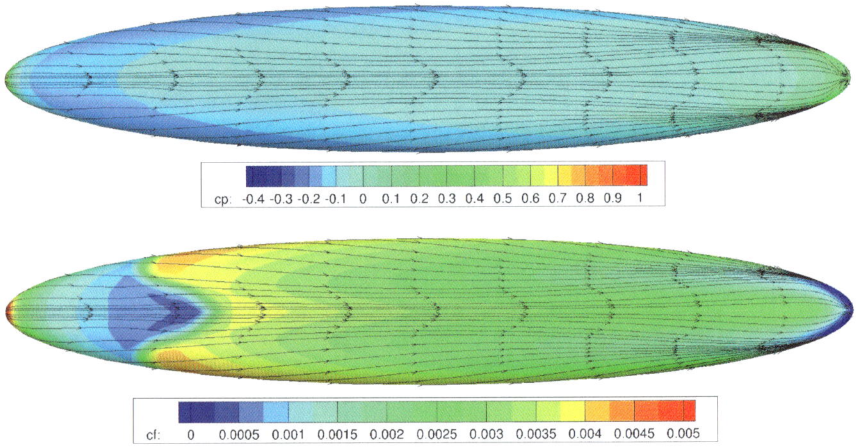

Fig. 10.34. View of the upper side of the ellipsoid [23]. Upper part: pattern of skin-friction lines and distribution of the surface pressure coefficient c_p. Lower part: pattern of skin-friction lines and distribution of the skin-friction coefficient c_f.

This case demonstrates how complex the form of a laminar-turbulent transition zone can be. Equally complex is the transition mechanism. Both Tollmien-Schlichting instability and cross-flow instability play a role [39]. Near the lower and the upper symmetry lines transition happens via the Tollmien-Schlichting path, for the remainder via the cross-flow path.

This picture, however, depends on both the angle of attack and the Reynolds number. For the lower angle of attack $\alpha = 5°$ the transition depends nearly fully on simultaneously excited Tollmien-Schlichting and cross-flow waves. The region with predominantly cross-flow transition appears only at larger angles of attack. For a low Reynolds number case ($Re_L = 1.5 \cdot 10^6$, $M_\infty = 0.03$) it was found that for $\alpha = 5°$ transition happens completely via the Tollmien-Schlichting path. For $\alpha = 10°$ in that case cross-flow amplification was observed during the iterative prediction process, but finally did not play a role.

In all cases studied in [39] the locations of the determined transition lines agree quite well with the experimentally found lines [40].[8] For the higher Reynolds number cases transition is predicted slightly further upstream, but the forms of the transition lines agree fairly well with the experimentally found ones.

We close this section with an observation how laminar-turbulent transition influences—at least locally—the three-dimensionality of attached viscous flow. Consider first Fig. 7.21 in Sub-Section 7.4.3, page 167. There we see that at the upper symmetry line the streamlines of the external inviscid flow converge towards the symmetry line. They are slightly curved with the convex side toward that line. The skin-friction lines of the laminar flow first converge towards the symmetry line. At $x^1 \approx 0.3$, however, they diverge and turn strongly curved away from it (location a) in Fig. 7.21). They finally meet the skin-friction lines coming from the lower side of the ellipsoid and squeeze each other off the surface (location b) in Fig. 7.21).

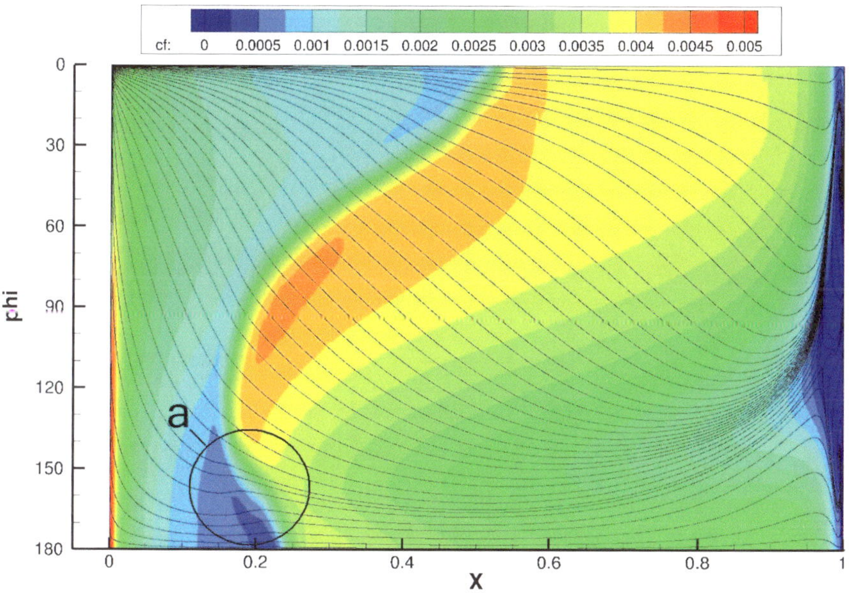

Fig. 10.35. Surface-coordinate parameter map (right half of ellipsoid) [41]: skin-friction lines, and distribution of the skin-friction coefficient c_f. Note: the lower symmetry line is at the top of the figure (phi = 0), the upper symmetry line at the bottom (phi = 180).

[8] Regarding the experimental data see also the discussion and the references given in Section 7.4.3.

Now consider Fig. 10.35. There initially we observe the same picture, location a), although we have only the skin-friction lines at hand and the angle of attack is larger, too. However, close to the transition line the tendency of the skin-friction lines to curve away from the upper symmetry line is broken in favor of a convergence to that line. Only at $x \approx 0.5$ the lines again curve away from the symmetry line.

The reason for this behavior is the following. Once the viscous flow becomes turbulent, the turbulence fluctuations transport momentum (and also energy) vertically towards the body surface, Section 9.2.3. The ensuing time-averaged turbulent boundary-layer profile is fuller than the laminar one, Fig. 9.4 on page 212. In our case it is the stream-wise profile, which is affected. This means that the turbulent flow becomes less curved by the pressure field than the laminar flow. Hence the whole viscous flow attains the tendency to follow more the external inviscid streamlines, which, however, are not indicated in our example. Whereas this effect is well expressed in the surface-coordinate parameter map, it is barely discernible in the true surface plots shown above.

References

1. Hirschel, E.H.: Grenzschichtberechnung für einen Hubschrauberrumpf. MBB-UFE122-AERO-MT-547, Ottobrunn, Germany (1981)
2. Hirschel, E.H.: Computation of Three-Dimensional Boundary Layers on Fuselages. J. Aircraft 21, 23–29 (1984)
3. Polz, G., Quentin, J., Amtsberg, J., Kühn, A.: Windkanaluntersuchungen zur optimalen Gestaltung von Transporthubschrauberzellen hinsichtlich Leistungsbedarf und Stabilitätsverhalten. MBB-UD-299/80 (1981)
4. Hirschel, E.H.: Das Verfahren von Cousteix-Aupoix zur Berechnung von turbulenten, dreidimensionalen Grenzschichten. MBB-UFE122-AERO-MT-484, Ottobrunn, Germany (1983)
5. Schwarz, T.: DLR-Institute of Aerodynamics and Flow Technology: Personal communication (2011)
6. Perry, A.E., Hornung, H.: Some Aspects of Three-Dimensional Separation. Part II: Vortex Skeletons. Z. Flugwiss. und Weltraumforsch (ZFW) 8, 155–160 (1984)
7. Hirschel, E.H., Bretthauer, N., Röhe, H.: Theoretical and Experimental Boundary-Layer Studies on Car Bodies. J. Int. Assoc. Vehicle Design 5(5), 567–584 (1984)
8. Schlichting, H., Truckenbrodt, E.: Aerodynamics of the Aeroplane, 2nd revised edn. McGraw Hill Higher Education, New York (1979)
9. Hirschel, E.H., Fornasier, L.: Flowfield and Vorticity Distribution Near Wing Trailing Edges. AIAA-Paper 1984-0421 (1984)
10. Hirschel, E.H.: Vortex Flows: Some General Properties, and Modelling, Configurational and Manipulation Aspects. AIAA-Paper 96-2514 (1996)
11. Hirschel, E.H.: Considerations of the Vorticity Field on Wings. In: Haase, W. (ed.) Recent Contributions to Fluid Mechanics, pp. 129–137. Springer, Heidelberg (1982)
12. Mangler, K.W., Smith, J.H.B.: Behaviour of the Vortex Sheet at the Trailing Edge of a Lifting Wing. The Aeronautical J. of the Royal Aeronaut. Soc. 74, 906–908 (1970)

13. Kolbe, D.C., Boltz, F.W.: The Forces and Pressure Distributions at Subsonic Speeds on a Plane Wing Having 45° of Sweepback, an Aspect Ratio of 3, and a Taper Ratio of 0.5. NACA RM A51G31 (1951)
14. Hirschel, E.H., Sacher, P.: A Comparative Theoretical Study of the Boundary-Layer Development on Forward Swept Wings. In: Nangia, R.K. (ed.) Proc. Int. Conf. Forward Swept Wings, 1982, Bristol. Univ. of Bristol, U.K. (1983)
15. Fornasier, L.: HISSS—A Higher-Order Subsonic/Supersonic Singularity Method for Calculating Linearized Potential Flow. AIAA-Paper 1984-1646 (1984)
16. Sytsma, H.S., Hewitt, B.L., Rubbert, P.E.: A Comparison of Panel Methods for Subsonic Flow Computation. AGARD-AG-241 (1979)
17. Anderson Jr., J.D.: Fundamentals of Aerodynamics, 5th edn. McGraw Hill, New York (2010)
18. Prandtl, L.: Tragflügeltheorie, I. und II. Mitteilung. Nachrichten der Kgl. Ges. Wiss. Göttingen, Math.-Phys. Klasse, 451–477 (1918) und 107–137 (1919)
19. Vassberg, J.C., DeHaan, M.A., Rivers, S.M., Wahls, R.A.: Development of a Common Research Model for Applied CFD Validation Studies. AIAA-Paper 2008-6919 (2008)
20. `http://aaac.larc.nasa.gov/tsab/cfdlarc/aiaa-dpw/`
21. Vassberg, J.C., Tinoco, E.N., Mani, M., Rider, B., Zickuhr, T., Levy, D.W., Brodersen, O.P., Eisfeld, B., Crippa, S., Wahls, R.A., Morrison, J.H., Mavriplis, D.J., Murayama, M.: Summary of the Fourth AIAA CFD Drag Prediction Workshop. AIAA-Paper 2010-4547 (2010)
22. Brodersen, O.P., Crippa, S., Eisfeld, B., Keye, S., Geisbauer, S.: DLR Results from the Fourth AIAA CFD Drag Prediction Workshop. AIAA-Paper 2010-4223 (2010)
23. Brodersen, O.P.: DLR-Institute of Aerodynamics and Flow Technology: Personal communication (2011)
24. Gerhold, T.: Overview of the Hybrid RANS Code TAU. In: Kroll, N., Fassbender, J. (eds.) MEGAFLOW—Numerical Flow Simulation for Aircraft Design. Notes on Numerical Fluid Mechanics and Multidisciplinary Design, NNFM, vol. 89, pp. 81–92. Springer, Heidelberg (2005)
25. Spalart, P.R., Allmaras, S.R.: A One Equation Turbulence Model for Aerodynamic Flows. AIAA-Paper 1992-0439 (1992)
26. Haase, W., Aupoix, B., Bunge, U., Schwamborn, D. (eds.): FLOMANIA—A European Initiative on Flow Physics Modelling. Results of the European-Union funded project, 2002–2004. NNFM, vol. 94. Springer, Heidelberg (2006)
27. Hirschel, E.H., Weiland, C.: Selected Aerothermodynamic Design Problems of Hypersonic Flight Vehicles. Progress in Astronautics and Aeronautics, AIAA, Reston, Va., vol. 229. Springer, Heidelberg (2009)
28. von Kármán, T.: Aerodynamics—Selected Topics in the Light of their Historical Development. Cornell University Press, Ithaca (1954)
29. Gaster, M.: A Simple Device for Preventing Turbulent Contamination on Swept Leading Edges. J. Royal Aeronautical Soc. 69, 788 (1965)
30. Kilian, T.: DLR-Institute of Aerodynamics and Flow Technology: Personal communication (2012)
31. Seitz, A., Kruse, M., Wunderlich, T., Bold, J., Heinrich, L.: The DLR Project LamAiR: Design of a NLF Forward Swept Wing for Short and Medium Range Transport Application. AIAA-Paper 2011-3526 (2011)

32. Kruse, M., Wunderlich, T., Heinrich, L.: A Conceptual Study of a Transonic NLF Transport Aircraft with Forward Swept Wings. AIAA-Paper 2012-3208 (2012)
33. Hirschel, E.H.: Evaluation of Results of Boundary-Layer Calculations with Regard to Design Aerodynamics. AGARD R-741, 5-1–5-29 (1986)
34. Hirschel, E.H.: Basics of Aerothermodynamics. Progress in Astronautics and Aeronautics, AIAA, Reston, Va, vol. 204. Springer, Heidelberg (2004)
35. Riedelbauch, S.: Aerothermodynamische Eigenschaften von Hyperschallströmungen über strahlungsadiabate Oberflächen (Aerothermodynamic Properties of Hypersonic Flows past Radiation-Cooled Surfaces). Doctoral Thesis, Technische Universität München, Germany. Also DLR-FB 91-42 (1991)
36. Riedelbauch, S., Hirschel, E.H.: Aerothermodynamic Properties of Hypersonic Flow over Radiation-Adiabatic Surfaces. J. of Aircraft 30(6), 840–846 (1993)
37. Désidéri, J.-A., Glowinski, R., Periaux, J. (eds.): Hypersonic Flows for Reentry Problems, vol. 1&2. Springer, Heidelberg (1991)
38. Cebeci, T., Cousteix, J.: Modeling and Computation of Boundary-Layer Flows, 2nd edn. Horizons Publ., Springer, Long Beach, Heidelberg (2005)
39. Krimmelbein, N., Radespiel, R.: Transition Prediction for Three-Dimensional Flows Using Parallel Computation. Computers & Fluids 38, 121–136 (2009)
40. Kreplin, H.-P., Vollmers, H., Meier, H.U.: Wall Shear Stress Measurements on an Inclined Prolate Spheroid in the DFVLR 3 m x 3 m Low Speed Wind Tunnel Göttingen. DFVLR-AVA report IB 22-84 A 33 (1985)
41. Krimmelbein, N.: DLR-Institute of Aerodynamics and Flow Technology: Personal communication (2011)

11

Solutions of the Problems

Problems of Chapter 2

Problem 2.1

With eq. (B.9) for the sought boundary-layer thicknesses and the corresponding constants from Table (B.1) we obtain the results in the following Table.

Table 11.1. Estimated boundary-layer thicknesses δ and displacement thicknesses δ_1 at (bold) $x = 20$ m and $x = 40$ m.

x [m]	δ_{lam} $[cm]$	$\delta_{1,lam}$ $[cm]$	δ_{turb} $[cm]$	$\delta_{1,turb}$ $[cm]$
1	0.19	0.066	1.59	0.20
5	0.43	0.15	5.76	0.72
10	0.60	0.21	10.03	1.25
20	**0.85**	**0.29**	**17.45**	**2.18**
40	**1.21**	**0.42**	**30.39**	**3.80**
50	1.35	0.46	36.33	4.55

Problem 2.2

For concave surface portions such coordinates would lead to a cross-over, if the radius of curvature is small compared to the boundary-layer thickness. Hence one would need to employ hybrid grids.

Problem 2.3

The two components of the external inviscid velocity vector are $u_e = 1$, $v_e = x$.

a) We write the Euler equations for two-dimensional flow taking $\rho = 1$ and insert the velocity components and their derivatives:

E.H. Hirschel, J. Cousteix, and W. Kordulla, *Three-Dimensional Attached Viscous Flow*,
DOI: 10.1007/978-3-642-41378-0_11,

$$u_e \frac{\partial u_e}{\partial x} + v_e \frac{\partial u_e}{\partial y} = -\frac{\partial p}{\partial x} = 0,$$
$$u_e \frac{\partial v_e}{\partial x} + v_e \frac{\partial v_e}{\partial y} = -\frac{\partial p}{\partial y} = 1.$$

b) From the above Euler equations we obtain $\partial p / \partial y = -1$. Integrating this yields $p = -y + c_1$.

c) The streamline function reads

$$\frac{dy}{dx} = \frac{v_e}{u_e} = x.$$

Integrating this yields $y = 0.5\, x^2 + c_2$. The external inviscid streamlines hence are parabolas.

Problem 2.4

We find the needed relations in Appendix B.2 and obtain the following results.

1. The stagnation temperature of the inviscid flow is equal to the atmospheric temperature, $T_t = 300$ K.
   The static temperature at the edge of the boundary layer is given by the isentropic relation:
   $$T_e = \frac{T_t}{1 + \frac{\gamma-1}{2} M_e^2}$$
   i.e., for $M_e = 2$ we get $T_e = 166.67$ K.
2. At thermal equilibrium, the wall temperature is equal to the recovery temperature T_r:
   $$T_r = T_e(1 + r\frac{\gamma - 1}{2} M_e^2) = T_t \frac{1 + r\frac{\gamma-1}{2} M_e^2}{1 + \frac{\gamma-1}{2} M_e^2}$$
   where r is the recovery factor. In laminar flow we have $r = 0.86$ and in turbulent flow, we have $r = 0.9$, Appendix B.2. We deduce the value of the plate temperature for laminar flow to $T_r = 281,34$ K and for turbulent flow to $T_r = 286,67$ K.
3. For a nozzle with a Mach number $M = 5$, the static temperature T_e would be equal to 50 K. At such low temperature condensation occurs during the expansion process. To produce flows at large Mach numbers in wind tunnels, it is therefore required to have higher stagnation temperatures by heating the flow upstream of the nozzle.

Problem 2.5

With the static temperature at $H = 10$ km taken from Table D.2 in Appendix D we obtain the following results:

1. The atmospheric temperature is the static temperature $T_\infty = 223.252$ K. Then, we have
$$T_{t_\infty} = T_\infty \left(1 + \frac{\gamma - 1}{2} M_\infty^2\right) = 401.85 \text{ K}.$$
2. The assumption is permitted, because the vibrational excitation is still very small at that temperature, Sub-Section 3.1.1. Hence to use $\gamma = 1.4$ is tolerable.
3. Through the shock wave in front of the body, the stagnation temperature is conserved. Then, in the inviscid part of the flow, the stagnation temperature is equal to the free stream stagnation temperature. In particular, at the edge of the boundary layer, we have
$$T_{t_e} = T_{t_\infty}$$
We deduce the value of the static temperature at the edge of the boundary layer
$$T_e = T_{t_e} \frac{1}{1 + \frac{\gamma-1}{2} M_e^2}$$
As the wall is adiabatic and neglecting radiation effects, the wall temperature is given by
$$T_w = T_r = T_e \left(1 + r\frac{\gamma - 1}{2} M_e^2\right)$$
Finally, we have
$$T_w = T_\infty \left(1 + \frac{\gamma - 1}{2} M_\infty^2\right) \frac{1 + r\frac{\gamma-1}{2} M_e^2}{1 + \frac{\gamma-1}{2} M_e^2}$$
.
At the stagnation point, $M_e = 0$, hence $T_p = 401.85$ K.
4. At a point on the body where the boundary layer edge Mach number is $M_e = 3$, we have $T_w = 376.02$ K.
5. No.

Problem 2.6

From Table D.2 we obtain for the altitude $H = 10$ km the following data: $T_\infty = 223.252$ K, $\rho_\infty = 4.135 \cdot 10^{-1}$ kg m^{-3}, $\mu_\infty = 1.458 \cdot 10^{-5}$ N s m^{-2}.

The results are found with the formulas given in Appendix B.3.1.

1. The speed of sound is $a_\infty = \sqrt{\gamma R T_\infty}$ with R being the gas constant, Table D.1. In the power-law approximation of the viscosity the exponent is $\omega = 0.65$. We obtain $a_\infty = 299.53$ m s^{-1} and $v_\infty = 239.63$ m s^{-1}. The unit Reynolds number results to $Re^u_\infty = 6.796 \cdot 10^6$ m^{-1}.
   The recovery temperature is $T_r = 248.97$ K. The reference temperature for case a) is $T^* = 241.77$ K and for case b) $T^* = 266.76$ K.
   At the location $x = 5$ m we obtain
   a) for $T_w = T_r$: $\delta_{turb} = 0.059$ m, $\delta_{1,turb} = 0.00735$ m, $\delta_{vs} = 0.177{\cdot}10^{-3}$ m, $\tau_{w,turb} = 20.77$ N m^{-2}.
   b) for $T_w = T_r + 50$ K: $\delta_{turb} = 0.061$ m, $\delta_{1,turb} = 0.00761$ m, $\delta_{vs} = 0.202{\cdot}10^{-3}$ m, $\tau_{w,turb} = 19.44$ N m^{-2}.
2. The displacement thickness in the alternate formulation is a) $\delta_{1,turb} = 0.00804$ m and b) $\delta_{1,turb} = 0.00877$ m.
3. The thicknesses increase with increasing wall temperature, the skin friction decreases. The proportionalities given in eqs. (2.6) and (2.9) can be reproduced. The alternate formulations for $\delta_{1,turb}$ yield somewhat larger values, which is due to the data base underlying the formula (mainly supersonic and hypersonic flow).

Problem 2.7

For the determination of the skin-friction drag D_{sf} we uncoil the assumed circular cylinder and find a flat plate with length $l = 10$ m and width $b = 3.14$ m. A simple formula for the drag (for $\tau(x)$ see Appendix B.3.1) is

$$D_{sf} = b \int_0^l \tau(x) dx = b\, C \mu_\infty v_\infty \left(\frac{T^*}{T_\infty}\right)^{n(1+\omega)-1} (Re^u_\infty)^{1-n} \int_o^l x^{-n} dx =$$
$$= b\, C \mu_\infty v_\infty \left(\frac{T^*}{T_\infty}\right)^{n(1+\omega)-1} (Re^u_\infty)^{1-n} \frac{l^{1-n}}{1-n},$$

with $C = 0.332$ and $n = 0.5$ for laminar, and $C = 0.0296$ and $n = 0.2$ for turbulent flow.

The respective drag coefficient is

$$C_{D_{sf}} = \frac{D_{sf}}{0.5 \rho_\infty v_\infty^2 b\, l}.$$

In our case we find a) $D_{sf} = 709.7$ N, $C_{D_{sf}} = 1.90{\cdot}10^{-3}$, b) $D_{sf} = 664.25$ N, $C_{D_{sf}} = 1.78{\cdot}10^{-3}$.

The result is that the skin-friction drag decreases if the wall temperature is increased.

Problem 2.8

In that case the wall-normal temperature gradient at the wall is negative. This leads, according to the wall compatibility conditions, Section 4.5, to a point of inflection in the tangential velocity profile.

Problem 2.9

The fuller tangential turbulent profile carries a higher stream-wise momentum than the laminar profile.

Problem 2.10

The thermal state of a surface is defined by the wall temperature and the heat flux in the gas at the wall, respectively the temperature gradient in the gas normal to the wall. (It is tacitly assumed that wall-temperature gradients tangential to the surface can be neglected.) The thermal state of the surface governs the thermal surface effects, both viscous and thermo-chemical, Fig. 2.13.

Problem 2.11

The important viscous thermal surface effects are:

- The boundary-layer thicknesses increase with increasing wall temperature.
- The skin friction decreases with increasing wall temperature. The effect is much stronger for turbulent than for laminar flow.
- Both the temperature gradient in the gas at the wall and the wall temperature influence the stability behavior of the laminar boundary layer.
- An increasing wall temperature increases a given flow separation disposition.

Problems of Chapter 3

Problem 3.1

We obtain T = 1,500 K: $\gamma = 1.31$; T = 1,000 K: $\gamma = 1.34$, $Pr = 0.761$; T = 400 K: $\gamma = 1.38$, $Pr = 0.744$.

Problem 3.2

For $\gamma = 1.33$ we get T_t = 1,388 K and for $\gamma = 1.4$ T_t = 1,640 K.

Problem 3.3

We show the approach for the first term in the y-momentum equation, eq. (3.15):

$$\rho u \frac{\partial v}{\partial x} + \underline{v \frac{\partial \rho u}{\partial x}} = \frac{\partial \rho u v}{\partial x}$$

and obtain with the full eq. (3.13) for the y-momentum equation

$$\frac{\partial}{\partial x}\left(\rho u v - \tau_{xy}\right) + \frac{\partial}{\partial y}\left(\rho v^2 + p - \tau_{yy}\right) + \frac{\partial}{\partial z}\left(\rho w v - \tau_{zy}\right) = 0$$

and for the z-momentum equation

$$\frac{\partial}{\partial x}\left(\rho u w - \tau_{xz}\right) + \frac{\partial}{\partial y}\left(\rho v w - \tau_{yz}\right) + \frac{\partial}{\partial z}\left(\rho w^2 + p - \tau_{zz}\right) = 0.$$

Problem 3.4

We write the y-component of eq. (A.6):

$$q_{e_y}|_{y=0} = \left[\rho(e + \frac{1}{2}V^2)v + q_y + pv - u\tau_{yx} - v\tau_{yy}\right]_{y=0}.$$

With the normal velocity component v being zero at the wall we obtain

$$q_{e_y}|_{y=0} = \left[q_y - u\tau_{yx}\right]_{y=0}$$

and with eq. (3.20) eventually

$$q_{e_y}|_{y=0} = -k\frac{\partial T}{\partial y}|_{y=0} - \mu u_w \frac{\partial u}{\partial y}|_{y=0},$$

where u_w is the slip velocity.

Problem 3.5

A solution of the dimensionless equations for given Parameters M and Re as well as Pr and —in the boundary conditions— T_w/T_{ref} is valid for all flows with the same parameters.

Problem 3.6

1. The continuity equation for the instantaneous flow is

$$\frac{\partial \tilde{u}}{\partial x} + \frac{\partial \tilde{v}}{\partial y} + \frac{\partial \tilde{w}}{\partial z} = 0.$$

2. If we assume that the instantaneous flow is two-dimensional, the continuity equation is

$$\frac{\partial \tilde{u}}{\partial x} + \frac{\partial \tilde{v}}{\partial y} = 0.$$

We apply the averaging process to obtain the continuity equation for the average flow:

$$\frac{\partial u}{\partial x} + \frac{\partial v}{\partial y} = 0.$$

The equation for the fluctuating flow is obtained by taking the difference between the equation for the instantaneous flow and the equation for the average flow

$$\frac{\partial u'}{\partial x} + \frac{\partial v'}{\partial y} = 0.$$

3. To be correct, we must apply the average process to the three-dimensional form of the continuity equation

$$\frac{\partial \tilde{u}}{\partial x} + \frac{\partial \tilde{v}}{\partial y} + \frac{\partial \tilde{w}}{\partial z} = 0$$

and the equation for the fluctuating flow is obtained by taking the difference between the equation for the instantaneous flow and the equation for the average flow:

$$\frac{\partial u'}{\partial x} + \frac{\partial v'}{\partial y} + \frac{\partial w'}{\partial z} = 0.$$

If the flow is two-dimensional, the continuity equation for the average flow simplifies to

$$\frac{\partial u}{\partial x} + \frac{\partial v}{\partial y} = 0,$$

but the continuity equation for the fluctuating flow does not simplify because the fluctuations are always three-dimensional even if the average flow is two-dimensional.

Problem 3.7

If the flow is irrotational, the fluctuating flow is also irrotational so that[1]

$$\frac{\partial u_i'}{\partial x_j} - \frac{\partial u_j'}{\partial x_i} = 0.$$

We deduce

$$\langle u_i' \left(\frac{\partial u_i'}{\partial x_j} - \frac{\partial u_j'}{\partial x_i} \right) \rangle = 0.$$

Now from the continuity equation we have:

[1] Einstein's summation convention is applied with $i, j = 1,2,3$.

$$\frac{\partial u_i'}{\partial x_i} = 0$$

and

$$\langle u_i' \frac{\partial u_j'}{\partial x_i} \rangle = \frac{\partial}{\partial x_i} \langle u_i' u_j' \rangle.$$

Finally, we obtain

$$\langle u_i' \left(\frac{\partial u_i'}{\partial x_j} - \frac{\partial u_j'}{\partial x_i} \right) \rangle = \frac{\partial}{\partial x_j} \left(\frac{1}{2} \langle u_i' u_i' \rangle \right) - \frac{\partial}{\partial x_i} \langle u_i' u_j' \rangle = 0$$

or

$$\frac{\partial}{\partial x_i} \langle u_i' u_j' \rangle = \frac{\partial k}{\partial x_j}.$$

The averaged momentum equation is:

$$\begin{aligned} \rho \frac{\partial u_j}{\partial t} + \rho u_i \frac{\partial u_j}{\partial x_i} &= -\frac{\partial p}{\partial x_j} + \frac{\partial \tau_{ij}}{\partial x_i} - \frac{\partial \langle u_i' u_j' \rangle}{\partial x_i} \\ &= -\frac{\partial p}{\partial x_j} + \frac{\partial \tau_{ij}}{\partial x_i} - \frac{\partial k}{\partial x_j}, \end{aligned}$$

where τ_{ij} is the viscous stress. We define a modified pressure

$$P = p + k$$

and the averaged momentum equation reads

$$\rho \frac{\partial u_j}{\partial t} + \rho u_i \frac{\partial u_j}{\partial x_i} = -\frac{\partial P}{\partial x_j} + \frac{\partial \tau_{ij}}{\partial x_i}.$$

This form of the equation is the same as that in laminar flow. Turbulence does not influence the average flow.

Problems of Chapter 4

Problem 4.1

The boundary layer is thin of the order of $O(1/\sqrt{Re_{ref}})$ and hence also the wall-normal coordinate and velocity component within that layer.

Problem 4.2

Formulate the two answers on the basis of the text on page 81 and on eq. (4.35).

Problem 4.3

Very close to the wall, the variation of the tangential velocity component u is nearly linear. We have

$$\tau_w = \left(\mu \frac{\partial u}{\partial y}\right)_{y=0}$$

and, taking into account the no-slip condition at the wall, we get the linearity relation

$$u(y) = \frac{\tau_w}{\mu} y.$$

Using the wall variables and the friction velocity u_τ, we obtain from this equation:[2]

$$u^+ = y^+.$$

In the turbulent part of the wall layer we assume that the time scale of the average flow is equal to the time scale of turbulence which means that there is a complete interaction between the average flow and turbulence. We assume that the turbulent length scale is y, i.e. a typical length scale of turbulent eddies is y.

The hypothesis that the time scales are the same gives

$$\frac{\partial u}{\partial y} = \frac{u_\tau}{\chi y},$$

where χ is a constant. Using the wall variables, this equation becomes

$$\frac{\partial u^+}{\partial y^+} = \frac{1}{\chi y^+}.$$

This equation is valid only in the turbulent part of the wall layer, not very close to the wall. Then, the integrated form of this equation is the law of the wall:

$$u^+ = \frac{1}{\chi} \ln y^+ + c,$$

where the constant c depends on what happens very close to the wall. The logarithmic law of the wall is valid in the turbulent part of the wall layer, see also page 85 and the following page.

Problem 4.4

In the turbulent part of the wall layer, we assume that the time scale of the average flow is equal to the time scale of turbulence which means that there is a full interaction between the average flow and turbulence. We

[2] With $y^+ \approx 5$ the thickness of the viscous sub-layer is defined, Appendix B.3.1, where also an explicit relation for that thickness is given.

assume further that the turbulent length scale is y, i.e. a typical length scale of turbulent eddies is y.

The hypothesis that the time scales are the same gives (see Problem 4.3)

$$\frac{\partial u}{\partial y} = \frac{u_\tau}{\chi y},$$

where χ is a constant. Using the wall variables, this equation becomes

$$\frac{\partial u^+}{\partial y^+} = \frac{1}{\chi y^+}.$$

This equation is valid only in the turbulent part of the wall layer, not very close to the wall, see above.

1. Then, the integrated form of this equation is

$$u^+ = \frac{1}{\chi} \ln y^+ + B_1,$$

where the constant B_1 depends on the wall conditions, i.e. the wall roughness. We hence have $B_1 = B_1(k^+)$. The logarithmic law of the wall is valid in the turbulent part of the wall layer.

2. If we define

$$B_2 = \frac{1}{\chi} \ln k^+ + B_1(k^+),$$

the logarithmic law of the wall becomes

$$u^+ = \frac{1}{\chi} \ln \frac{y^+}{k^+} + B_2 = \frac{1}{\chi} \ln \frac{y}{k} + B_2.$$

3. This logarithmic law can be compared to the logarithmic law of the wall on a smooth wall

$$u^+ = \frac{1}{\chi} \ln y^+ + c.$$

We define

$$B_3 = c - B_2$$

and

$$\Delta u^+ = \frac{\Delta u}{u_\tau} = \frac{1}{\chi} \ln k^+ + B_3.$$

In the case of a rough wall, the logarithmic law of the wall above can then be written in the form

$$u^+ = \frac{1}{\chi} \ln y^+ + c - \Delta u^+.$$

The relation between Δu^+ and k^+ has been determined experimentally for different types of roughnesses.

Problem 4.5

From eq. (4.46), taking into account eq. (4.50), we see that five characteristics in direction normal to the wall exist. To this come two characteristics from the energy equation. The streamlines themselves are an infinite number of sub-characteristics.

Problem 4.6

We write the eqs. (4.57) and (4.58) with the full convective operators:

$$\frac{\partial^2 u}{\partial y^2}|_{y=0} = \left\{ \frac{1}{\mu} \left[\rho u \frac{\partial u}{\partial x} + \rho v \frac{\partial u}{\partial y} + \rho w \frac{\partial u}{\partial z} + \frac{\partial p}{\partial x} - \frac{\partial \mu}{\partial T} \frac{\partial T}{\partial y} \frac{\partial u}{\partial y} \right] \right\}_{y=0},$$

$$\frac{\partial^2 w}{\partial y^2}|_{y=0} = \left\{ \frac{1}{\mu} \left[\rho u \frac{\partial w}{\partial x} + \rho v \frac{\partial w}{\partial y} + \rho w \frac{\partial w}{\partial z} + \frac{\partial p}{\partial z} - \frac{\partial \mu}{\partial T} \frac{\partial T}{\partial y} \frac{\partial w}{\partial y} \right] \right\}_{y=0}.$$

With slip-flow it holds for the tangential velocity components: $u|_{y=0} \neq 0$, and $w|_{y=0} \neq 0$. Depending on the algebraic signs of the gradients $\partial u/\partial x|_{y=0}$, $\partial u/\partial z|_{y=0}$, ..., the second derivatives $\partial^2 u/\partial y^2$ and $\partial^2 w/\partial y^2$ will be influenced by the slip-flow components. If the surface-tangential gradients are zero, there is no influence.

Problems of Chapter 5

Problem 5.1

Eq. (3.55) shows that for $Pr < 1$ the thermal boundary layer is thicker than the flow boundary layer. Hence the integration domain is to be defined by the edge of the thermal boundary layer. This edge can be found with the condition $|T(y) - T_e| \leqq \epsilon$.

Problem 5.2

Rewrite eq. (5.1) by dropping the z-term:

$$\frac{\partial}{\partial x}[\rho_e u_e(\delta_1 - \delta_{1x})] = 0$$

and integrate in x-direction in order to find

$$[\rho_e u_e(\delta_1 - \delta_{1x})]_x - [\rho_e u_e(\delta_1 - \delta_{1x})]_{x=0} = 0.$$

At $x = 0$ we have a zero boundary-layer thickness and find hence $\delta_1 = \delta_{1x}$ for $x > 0$. In the blunt-body case the boundary layer thickness at the stagnation point is positive, however u_e is zero and we get the same result.

Problem 5.3

1. The balance of mass flow in the control volume $\mathcal{D}$ for the real flow and the equivalent inviscid flow gives

$$\rho v_e \Delta x - \rho u_e \frac{d\delta}{dx}\Delta x + \int_0^\delta \rho u dy + \Delta x \frac{d}{dx}\int_0^\delta \rho u dy = \int_0^\delta \rho u dy,$$

$$\rho v_e \Delta x - \rho u_e \frac{d\delta}{dx}\Delta x + \rho u_e \delta + \Delta x \frac{d(\rho u_e \delta)}{dx} = \rho v_w \Delta x + \rho u_e \delta.$$

2. Eliminating v_e between the two equations yields

$$v_w = -\frac{d}{dx}\int_0^\delta u dy + \frac{d(u_e \delta)}{dx}$$

or

$$v_w = \frac{d}{dx}\int_0^\delta (u_e - u) dy.$$

With

$$\delta_1 = \int_0^\delta (1 - \frac{u}{u_e}) dy$$

we obtain the equivalent inviscid source velocity

$$v_w = \frac{d(u_e \delta_1)}{dx}.$$

3. Let $y = f(x)$ be the equation of a streamline. The balance of mass flow in the control volume $\mathcal{V}$ gives

$$\rho u_e f + \Delta x \frac{d(\rho u_e f)}{dx} = \rho v_w \Delta x + \rho u_e f.$$

Taking into account the expression of v_w we deduce

$$\frac{d(u_e f)}{dx} = \frac{d(u_e \delta_1)}{dx}$$

or

$$u_e f = u_e \delta_1 + C,$$

where C is a constant depending on the considered streamline. Then the equation of a streamline is

$$y = \delta_1 + \frac{C}{u_e}.$$

By taking $C = 0$, we see that $y = \delta_1$ is a streamline of the equivalent inviscid flow.

Having in mind the coupling between the inviscid flow and the boundary layer, the boundary-layer effect could be represented by displacing the wall along any streamline. However, at the stagnation point ($u_e = 0$), there is only one regular streamline on which the inviscid flow can rest. This line is defined by $y = \delta_1$. If we took $C \neq 0$, with $u_e \to 0$, we would have $y \to +\infty$ or $y \to -\infty$.

4. A schematic sketch of streamlines in the fictitious flow is given in Fig. 11.1. The line $y = \delta_1$ is a streamline but the line $y = \delta$ is not a streamline. Along the line $y = \delta$, the velocity components of the real flow and of the fictitious flow are identical. Now, the boundary layer is fed by fluid coming from the external flow. Along the line $y = \delta$, the streamlines cut the line $y = \delta$ and enter the boundary layer.

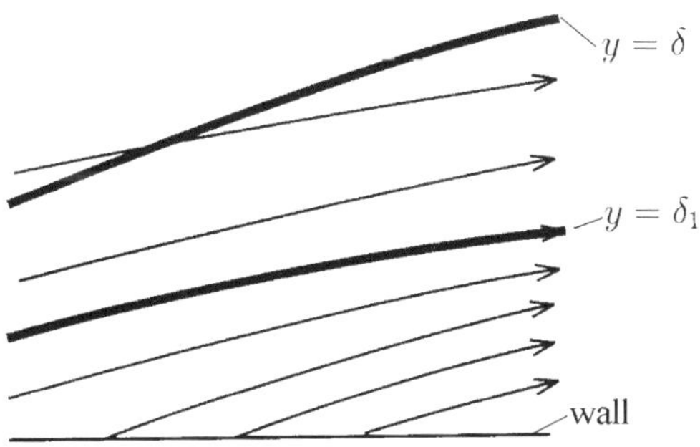

Fig. 11.1. Patterns of streamlines in the equivalent inviscid flow in the boundary-layer domain. The boundary-layer edge is not a streamline.

Problem 5.4

Consider the control volume in Fig. 11.2. The flow comes from the left-hand side, we use a left-handed coordinate system. We allow for a wall-source term v_0, which would be negative for suction or positive for wall-normal blowing.

The mass-flow balance reads, with a as upper bound of the integrals encompassing both the boundary-layer thickness δ and the displacement thickness δ_1 and with <> standing for a mean value:

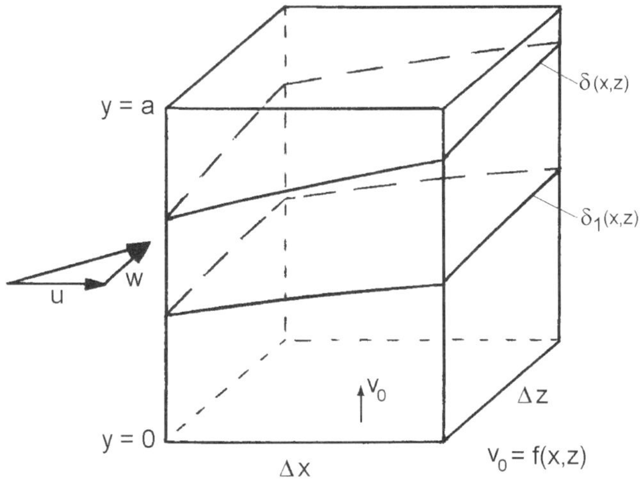

Fig. 11.2. Control volume for the determination of the displacement thickness of three-dimensional boundary layers.

$$\begin{aligned}
&\triangle z \int_0^a u\,dy - \triangle z \int_0^a (u + \frac{\partial u}{\partial x}\triangle x)dy + \\
&+\triangle x \int_0^a w\,dy - \triangle x \int_0^a (w + \frac{\partial w}{\partial z}\triangle z)dy + v_0 = \\
&= \triangle z \int_0^a u_e dy - \triangle z \int_0^a (u_e + \frac{\partial u_e}{\partial x}\triangle x)dy + \\
&+\triangle x \int_0^a w_e dy - \triangle x \int_0^a (w_e + \frac{\partial w_e}{\partial z}\triangle z)dy - \\
&-\triangle z \int_0^{<\delta_1>} u_e dy + \triangle z \int_0^{<\delta_1 + \frac{\partial \delta_1}{\partial x}\triangle x>} (u_e + \frac{\partial u_e}{\partial x}\triangle x)dy - \\
&-\triangle x \int_0^{<\delta_1>} w_e dy + \triangle x \int_0^{<\delta_1 + \frac{\partial \delta_1}{\partial z}\triangle z>} (w_e + \frac{\partial w_e}{\partial z}\triangle z)dy .
\end{aligned}$$

After cancelling out the inflow integrals with the first terms of the outflow integrals this can be reformulated by using

$$\int_0^{<\delta_1 + \frac{\partial \delta_1}{\partial z}\triangle z>} = \int_0^{<\delta_1>} + \int_{<\delta_1>}^{<\delta_1 + \frac{\partial \delta_1}{\partial z}\triangle z>} .$$

We obtain with terms of higher order neglected:

$$\triangle x \triangle z \int_0^a \frac{\partial}{\partial x}(u_e - u)dy + \triangle x \triangle z \int_0^a \frac{\partial}{\partial z}(w_e - w)dy + \triangle x \triangle z\, v_0 =$$

$$= \triangle z\, u_e[< \delta_1 + \frac{\partial \delta_1}{\partial x}\triangle x - \delta_1 >] + \triangle x \triangle z \int_0^{<\delta_1>} \frac{\partial u_e}{\partial x}\, dy +$$

$$+ \triangle x\, w_e[< \delta_1 + \frac{\partial \delta_1}{\partial z}\triangle z - \delta_1 >] + \triangle x \triangle z \int_0^{<\delta_1>} \frac{\partial w_e}{\partial z}\, dy.$$

With the quantities δ_{1x} and δ_{1z} being

$$\delta_{1x} = \int_0^{a=\delta} (1 - \frac{u}{u_e})dy$$

and

$$\delta_{1z} = \int_0^{a=\delta} (1 - \frac{w}{w_e})dy,$$

we arrive eventually at

$$\frac{\partial}{\partial x}\left[u_e(\delta_1 - \delta_{1x})\right] + \frac{\partial}{\partial z}\left[w_e(\delta_1 - \delta_{1z})\right] = v_0.$$

This is a linear partial differential equation of 1st order for δ_1. The variables u_e, w_e, δ_{1x}, and δ_{1z} are locally given functions. The streamlines of the external inviscid flow are sub-characteristics. Initial data must be provided. The integration of the equation has to take into account the Courant-Friedrichs-Lewy (CFL) condition like it is the case with the integration of the three-dimensional boundary-layer equations.

Problems of Chapter 6

Problem 6.1

The larger the aspect ratio, the smaller is the induced drag. The small chord depth means small displacement thicknesses at the trailing edge and hence small pressure drag. This holds also for the slender fuselage. The smaller the wetted surface, the smaller is the skin-friction drag.

Problem 6.2

Both the wing sweep and the supercritical airfoil are means to shift the drag-divergence Mach number to higher subsonic flight Mach numbers.

Problem 6.3

We write the Breguet equation in terms of the flight Mach number M_∞ with a_∞ being the speed of sound:

$$R = M_\infty\, a_\infty \frac{C_L}{C_D}\, I_{sp} \ln\left(1 + \frac{m_F}{m_E + m_P}\right).$$

We observe that the range R increases with the flight Mach number M_∞. However, M_∞ must remain smaller than the drag-divergence Mach number, otherwise the increase of the drag and the decrease of the lift (in terms of the coefficients C_D and C_L)—the other parameters assumed to remain constant—would cancel the effect of the higher flight Mach number. Hence, in principle the product $M_\infty\, C_L/C_D$ has to be maximum.

Problem 6.4

1. The integral equations are

$$\frac{d}{dx}(\delta - \delta_1) + \frac{\delta - \delta_1}{u_e}\frac{du_e}{dx} = c_E,$$

$$\frac{d\delta_2}{dx} + \delta_2 \frac{H+2}{u_e}\frac{du_e}{dx} = \frac{c_f}{2}.$$

It is assumed that $H^* = \dfrac{\delta - \delta_1}{\delta_2}$ is a function of $H = \delta_1/\delta_2$ only:

$$H^* = \frac{(\alpha H^2 + H)}{H-1}, \quad \alpha = 0.631.$$

In order to transform the above two equations, we write

$$\delta - \delta_1 = \delta_2 H^*,$$

whence $\dfrac{d}{dx}(\delta - \delta_1) = H^* \dfrac{d\delta_2}{dx} + \delta_2 H^{*'} \dfrac{d}{dx}\left(\dfrac{\delta_1}{\delta_2}\right), \quad H^{*'} = \dfrac{dH^*}{dH},$

and $\dfrac{d}{dx}(\delta - \delta^*) = (H^* - H H^{*'})\dfrac{d\delta_2}{dx} + H^{*'}\dfrac{d\delta_1}{dx}.$

Thus, we obtain the system "S":

$$(H^* - H H^{*'})\frac{d\delta_2}{dx} + H^{*'}\frac{d\delta_1}{dx} + \frac{\delta - \delta_1}{u_e}\frac{du_e}{dx} = c_E,$$

$$\frac{d\delta_2}{dx} + \delta_2 \frac{H+2}{u_e}\frac{du_e}{dx} = \frac{c_f}{2}.$$

2. The external velocity distribution $u_e(x)$ is given (direct mode). System "S" writes

$$(H^* - H H^{*'})\frac{d\delta_2}{dx} + H^{*'}\frac{d\delta_1}{dx} = c_E - \frac{\delta - \delta_1}{u_e}\frac{du_e}{dx},$$

$$\frac{d\delta_2}{dx} = \frac{c_f}{2} - \delta_2 \frac{H+2}{u_e}\frac{du_e}{dx}.$$

The determinant of this system is

$$\Delta = -H^{*'} = -\frac{\alpha H^2 - 2\alpha H - 1}{(H-1)^2}.$$

This determinant is zero when $H = 1 \pm \sqrt{1 + \frac{1}{\alpha}}$, that is $H = -0.6$ and $H = 2.6$. The significant solution is $H = 2.6$ (Fig. 11.3). This point corresponds to boundary-layer separation.

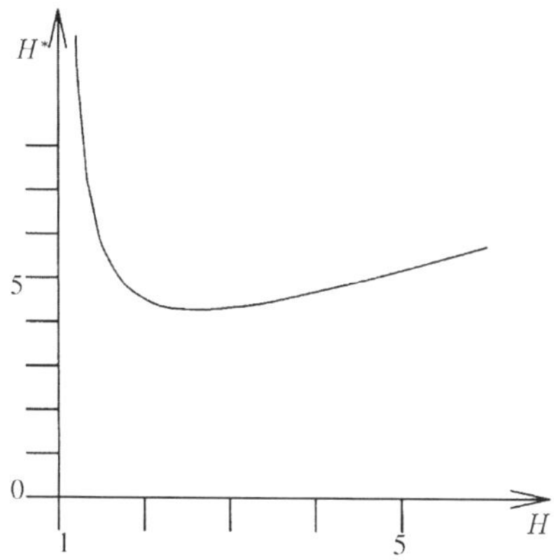

Fig. 11.3. Function $H^*(H)$.

3. The continuity equation can be written as

$$H^* \frac{d\delta_2}{dx} + \delta_2 \frac{dH^*}{dx} + \frac{\delta - \delta_1}{u_e} \frac{du_e}{dx} = c_E.$$

By expressing $\frac{d\delta_2}{dx}$ by means of the momentum equation, we obtain

$$\delta_2 \frac{dH^*}{dx} = c_E - H^* \frac{c_f}{2} + \delta_2 H^* \frac{H+1}{u_e} \frac{du_e}{dx}.$$

Generally, $\frac{dH^*}{dx}$ is not zero at point x_D where $H = 2.6$. Thus, downstream of x_D, H^* becomes less than the minimum of $H^*(H)$ and no solution exists for H. At a separation point, the system of equations becomes singular in the sense that it is not possible to continue the integration of the boundary-layer equations in direct mode downstream of this point.

4. We assume now that $\delta_1(x)$ is a prescribed function (inverse mode). The external velocity $u_e(x)$ becomes unknown. System "S" writes

$$(H^* - HH^{*'}) \frac{d\delta_2}{dx} + \frac{\delta - \delta_1}{u_e} \frac{du_e}{dx} = c_E - H^{*'} \frac{d\delta_1}{dx},$$

$$\frac{d\delta_2}{dx} + \delta_2 \frac{H+2}{u_e} \frac{du_e}{dx} = \frac{c_f}{2}.$$

The determinant of this system is

$$\begin{aligned}\Delta' &= \delta_2 \frac{H+2}{u_e}(H^* - HH^{*'}) - \frac{\delta - \delta_1}{u_e} \\ &= \frac{\delta_2}{u_e}\left[(H+2)(H^* - HH^{*'}) - H^*\right].\end{aligned}$$

We have

$$H^{*'} = \frac{\alpha H^2 - 2\alpha H - 1}{(H-1)^2}, \quad H^* - HH^{*'} = \frac{(\alpha+1)H^2}{(H-1)^2},$$

whence

$$\Delta' = \frac{\delta_2}{u_e} \frac{H^3 + 3\alpha H^2 + H^2 + H}{(H-1)^2}.$$

For $H > 1$, the determinant is never zero: $\Delta' \neq 0$. In the inverse mode, the boundary-layer equations are not singular, in particular not at the separation point.

Problem 6.5

1. As in Problem 6.4 the boundary layer integral equations are written as

$$\begin{aligned}(H^* - HH^{*\prime})\frac{d\delta_2}{dx} + H^{*\prime}\frac{d\delta_1}{dx} + \frac{\delta - \delta_1}{u_e}\frac{du_e}{dx} &= c_E, \\ \frac{d\delta_2}{dx} + \delta_2 \frac{H+2}{u_e}\frac{du_e}{dx} &= \frac{c_f}{2}.\end{aligned}$$

2. With the hypotheses of this problem, the mass flow through a cross-section of the diffuser is

$$\begin{aligned}Q &= 2\int_0^h \rho u \, dy = \\ &= 2\left[\int_0^h (\rho u - \rho u_e)\, dy + \int_0^h \rho u_e \, dy\right] = \\ &= 2\left[-\rho u_e \delta_1 + \rho u_e h\right].\end{aligned}$$

Thus, the conservation of mass in the diffuser writes

$$u_e(h - \delta_1) = \text{const.}$$

By differentiating with respect to x, this equation can be written as

$$-u_e \frac{d\delta_1}{dx} + (h - \delta_1)\frac{du_e}{dx} + u_e \frac{dh}{dx} = 0.$$

3. The calculation of the flow in the diffuser consists of solving the following system of differential equations:

$$
\begin{aligned}
(H^* - HH^{*\prime})\frac{d\delta_2}{dx} + H^{*\prime}\frac{d\delta_1}{dx} + \frac{\delta - \delta_1}{u_e}\frac{du_e}{dx} &= c_E, \\
\frac{d\delta_2}{dx} + \delta_2\frac{H+2}{u_e}\frac{du_e}{dx} &= \frac{c_f}{2}, \\
- u_e\frac{d\delta_1}{dx} + (h - \delta_1)\frac{du_e}{dx} &= -u_e\frac{dh}{dx},
\end{aligned}
$$

where the main unknowns are $\frac{d\delta_2}{dx}$, $\frac{d\delta_1}{dx}$ and $\frac{du_e}{dx}$.
The determinant of the system of equations is

$$
\Delta = \begin{vmatrix} H^* - HH^{*\prime} & H^{*\prime} & \frac{\delta - \delta_1}{u_e} \\ 1 & 0 & \delta_2\frac{H+2}{u_e} \\ 0 & -u_e & h - \delta_1 \end{vmatrix}.
$$

This determinant is zero if $\frac{h}{\delta_2} = f(H)$ with $f = \frac{(H^* - HH^{*\prime})(H+1)}{H^{*\prime}}$ and $H^* = \frac{\alpha H^2 + H}{H - 1}$.

4. For positive values of $\frac{h}{\delta_2}$, the function f has a local minimum for $H = 5.37$ and the corresponding value of f is $f = 28.75$, Fig. 11.4.

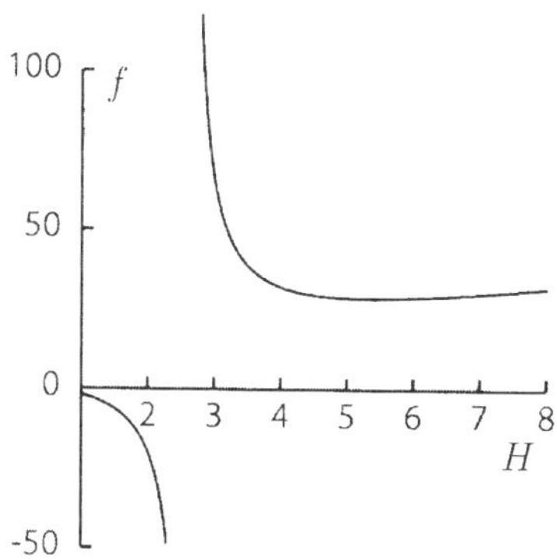

Fig. 11.4. Function $f(H)$.

Thus, if $\frac{h}{\delta_2} < 28.75$, the determinant is always different from zero and the system is non-singular.

Problems of Chapter 7

Problem 7.1

Solve $det(A - \lambda I) = 0$, where A is given by eq. (7.18), I is the $n \times n$ identity matrix and $\lambda_i (n = 1, 2, 3)$ are the eigenvalues:

$$det(A - \lambda I) = 0 = (\frac{1}{2}\frac{\partial p}{\partial y} - \lambda)\left[(\frac{\partial \tau_x}{\partial x} - \lambda)(\frac{\partial \tau_z}{\partial z} - \lambda) - \frac{\partial \tau_x}{\partial z}\frac{\partial \tau_z}{\partial x}\right].$$

We find immediately

$$\lambda_2 = \frac{1}{2}\frac{\partial p}{\partial y}.$$

To find the other two, we rewrite the remainder of the above equation:

$$0 = \left\{\lambda^2 - \frac{1}{2}\left(\frac{\partial \tau_x}{\partial x} + \frac{\partial \tau_z}{\partial z}\right)\right\}^2 - \frac{1}{4}\left\{(\frac{\partial \tau_x}{\partial x} - \frac{\partial \tau_z}{\partial z})^2 + 4\frac{\partial \tau_x}{\partial_z}\frac{\partial \tau_z}{\partial x}\right\}.$$

From this equation λ_1 and λ_3 are found.

Problem 7.2

The sketch shows two half-saddles which fulfills rule 2.

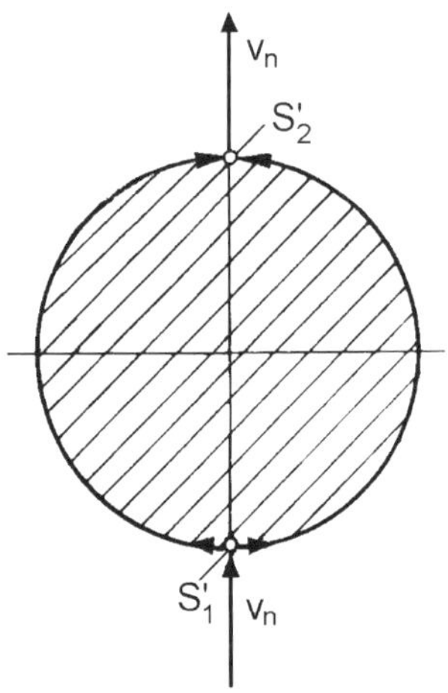

Fig. 11.5. Cross-section of the circular forebody, Fig. 7.8. The velocity components v_n are the flow components due to the angle of attack.

Problem 7.3

We look at the chord cross-section. At the wings leading edge a half-saddle is present, like at the trailing edge. This holds also for the separation bubble at the location where the flow leaves the surface and where it reattaches. In the middle of the bubble sits a center which is counted as a node.

Hence we have in eq. (7.25) one node and four half-saddles which fulfills the rule.

Problem 7.4

Simply because on one side of the attachment line the shape of the stream lines changes from convex to concave, whereas the shape of those on the other side do not change their curvature.

Problem 7.5

The lines coincide in two-dimensional, and in quasi-two-dimensional flows like in the infinite-swept-wing case.

Problem 7.6

The prerequisite for plane-of-symmetry flow is that the flow occurs along a geodesic. The geodesic can be defined as a curve whose tangent vectors always remain parallel if they are conveyed along it, Appendix A.2.3.

Problem 7.7

Write down the summary and compare with Sub-Section 7.4.1.

Problem 7.8

Write down the rules and compare with Section 7.3.

Problems of Chapter 8

Problem 8.1

For small angles ψ we can write with the help of a Taylor expansion:

$$\sin \psi = \frac{x}{R},$$

hence

$$u_e(x) = c_1 u_\infty \frac{x}{R},$$

where R is the radius of the respective shape.

The Taylor expansion of $u_e(x)$ around x_0 yields

$$u_e(x) = u_e|_{x=0} + x\frac{du_e}{dx}|_{x=0} + \cdots,$$

hence

$$\frac{du_e}{dx}|_{x=0} \approx \frac{u_e(x)}{x}.$$

In the stagnation point we have

$$c_{p,s} = \frac{p_s - p_\infty}{0.5\rho_\infty u_\infty^2},$$

where $c_{p,s}$ in the stagnation point for perfect gas is a function of M_∞, eq. (7.29). p_s is the stagnation pressure in x_0.

For incompressible flow this reads

$$c_{p,s} = 1 = \frac{p_s - p_\infty}{0.5\rho_\infty u_\infty^2}.$$

Combining the second and the fourth equation with this one, we get the desired result

$$\frac{du_e}{dx}|_{x=0} = c_1\frac{u_\infty}{R} = \frac{c_1}{R}\sqrt{\frac{2(p_s - p_\infty)}{\rho_\infty}},$$

with $c_1 = 1{,}5$ for the sphere and $c_1 = 2$ for the circular cylinder.

For the infinite swept circular cylinder we obtain:

$$\frac{du_e}{dx}|_{x=0,\varphi_0>0} = cos\varphi_0\left[\frac{c_1}{R}\sqrt{\frac{2(p_s - p_\infty)}{\rho_\infty}}\right]_{\varphi_0=0}.$$

At the attachment point the x-momentum equation holds:

$$\rho_e u_e\frac{\partial u_e}{\partial x} = -\frac{\partial p}{\partial x}.$$

Because $u_e(x = 0)$ is zero, $(\partial p/\partial x)|_{x=0}$ vanishes, too.

Problem 8.2

Three-dimensional boundary-layer methods solve parabolic equations and hence are space-marching methods. For the computations spatial initial conditions for the surface-tangential directions are needed in addition to the external inviscid flow and wall boundary conditions. If a computation does not begin at the forward stagnation point of a configuration, the said assumptions permit to start a solution at other suitable locations.

Navier-Stokes/RANS methods as a rule use time-marching methods even to obtain steady-state solutions. They cover the whole flow field on the configuration under consideration and hence need no spatial initial conditions. However, if the Navier-Stokes/RANS are parabolized, space-marching equations result. These require initial and boundary conditions as in the boundary-layer case.

Problem 8.3

At the stagnation point the tangential velocities v^{*1} and v^{*2} are zero, the normal velocity v^{*3} is not equal zero. The pressure p has a maximum, the gradients in x^1- and x^2-direction are zero. This also holds for the other two scalar entities, the temperature and the density. For the gradients of the tangential velocities in these directions holds $\partial v^{*1}/\partial x^1 \neq 0$, $\partial v^{*2}/\partial x^2 \neq 0$, but $\partial v^{*1}/\partial x^2 = 0$, $\partial v^{*2}/\partial x^1 = 0$.

Problem 8.4

At the symmetry line ($x^2 = 0$) we have $\partial a_{11}/\partial x^2 = 0$, $\partial a_{22}/\partial x^2 = 0$, and further $a_{12} = 0$, $\partial a_{12}/\partial x^1 = 0$, but in the general case $\partial a_{12}/\partial x^2 \neq 0$.

For the remaining metric factors, Appendix A.2.4, we hence obtain $k_{12} = 0$, $k_{17} = 0$, $k_{21} = 0$, $k_{23} = 0$, $k_{26} = 0$, $k_{42} = 0$, but $k_{13} \neq 0$, $k_{43} \neq 0$.

Of these metric factors the following ones are connected to terms which anyway vanish because $v^2 = 0$ or $\partial p/\partial x^2 = 0$: k_{12}, k_{13}, k_{17}, k_{23}, k_{26}, k_{42}, k_{43}.

Problem 8.5

We take the boundary-layer equations for our case from Section 8.3:

$$\frac{\partial v^1}{\partial x^1} + \frac{\partial v^3}{\partial x^3} = 0,$$

$$v^1 \frac{\partial v^1}{\partial x^1} + v^3 \frac{\partial v^1}{\partial x^3} = -\frac{1}{\rho}\frac{\partial p}{\partial x^1} + \nu \frac{\partial^2 v^1}{\partial (x^3)^2},$$

$$v^1 \frac{\partial v^2}{\partial x^1} + v^3 \frac{\partial v^2}{\partial x^3} = \nu \frac{\partial^2 v^2}{\partial (x^3)^2}.$$

1. The external inviscid flow at the boundary-layer edge satisfies the following equations:

$$v_e^1 \frac{\partial v_e^1}{\partial x^1} = -\frac{1}{\rho}\frac{\partial p}{\partial x^1},$$

$$v_e^1 \frac{\partial v_e^2}{\partial x^1} = 0.$$

   The second equation shows that v_e^2 is constant all over the wing and is equal to its value at infinity:

$$v_e^2 = u_\infty \sin \varphi_0,$$

   where u_∞ is the free-stream velocity and φ_0 is the sweep angle of the wing.

2. If $\partial p/\partial x^1 = 0$, v_e^1 is also a constant with $v_e^1 = u_\infty \cos\varphi_0$. If $x^2 = f(x^1)$ is the equation of an external streamline, we have

$$\frac{dx^2}{dx^1} = \frac{v_e^2}{v_e^1} = const. = \tan\varphi_0 .$$

This means that the slope of the external streamlines is constant. The external streamlines are straight lines. Their direction on the wing is the free-stream direction.

3. With $\partial p/\partial x^1 = 0$, the boundary-layer equations are

$$\begin{aligned} \frac{\partial v^1}{\partial x^1} + \frac{\partial v^3}{\partial x^3} &= 0, \\ v^1 \frac{\partial v^1}{\partial x^1} + v^3 \frac{\partial v^1}{\partial x^3} &= \nu \frac{\partial^2 v^1}{\partial (x^3)^2}, \\ v^1 \frac{\partial v^2}{\partial x^1} + v^3 \frac{\partial v^2}{\partial x^3} &= \nu \frac{\partial^2 v^2}{\partial (x^3)^2}. \end{aligned}$$

These equations show that the velocity components $v^1(x^3)$ and $v^3(x^3)$ can be determined independently of the function $v^2(x^3)$.

Now, we can prove that

$$v^2 = v^1 \frac{v_e^2}{v_e^1}$$

is a solution of the x^2-momentum equation.

First, we observe that if we substitute this relation in the x^2-momentum equation we recover exactly the x^1-momentum equation because v_e^2 and v_e^1 are constants. Second, we observe that the boundary conditions are satisfied: $v^2 = 0$ when $x^3 = 0$ because $v^1(x^3 = 0) = 0$ and $v^2 \to v_e^2$ when $x^3 \to \infty$ because $v^1 \to v_e^1$ when $x^3 \to \infty$.

4. When $\partial p/\partial x^1 = 0$, the solution for v^1 and v^3 is the same as the solution for the two-dimensional flat plate, because the equations are the same and the boundary conditions are the same. In addition, the solution for v^2 is

$$v^2 = v^1 \frac{v_e^2}{v_e^1} .$$

Now it is easy to see that the velocity profile is contained in a plane normal to the wall and parallel to the external velocity. The evolution of the boundary-layer along an external streamline is the same as for the two-dimensional flat plate.

5. For turbulent flow, the boundary-layer equations for our infinite swept wing become with eqs. (A.46) and (A.47):

$$\frac{\partial v^1}{\partial x^1} + \frac{\partial v^3}{\partial x^3} = 0,$$
$$v^1 \frac{\partial v^1}{\partial x^1} + v^3 \frac{\partial v^1}{\partial x^3} = -\frac{1}{\rho}\frac{\partial p}{\partial x^1} + \frac{\partial}{\partial x^3}\left[\nu \frac{\partial v^1}{\partial x^3} - < v^{*1'} v^{*3'} >\right],$$
$$v^1 \frac{\partial v^2}{\partial x^1} + v^3 \frac{\partial v^2}{\partial x^3} = \frac{\partial}{\partial x^3}\left[\nu \frac{\partial v^2}{\partial x^3} - < v^{*2'} v^{*3'} >\right].$$

For the case $\partial p/\partial x^1 = 0$, it is not possible to say whether the solution is the same as for the two-dimensional flat plate, because the solution depends on the employed turbulence model. The equations for v^1 and v^2 can be coupled due to that model. Behind this is the fact, that in reality the turbulence is non-isotropic, page 14. If the employed turbulence model does not take that into account, the equations are not coupled.

Problem 8.6

We assume leading-edge oriented orthogonal coordinates with the coordinate x^1 being the surface-tangential coordinate normal to the leading edge and x^2 that in span direction.

The outer boundary conditions are given by the external inviscid flow field with $v_e^2 = u_\infty \sin\varphi_0 =$ const. over the wing's surface in x^1-direction, u_∞ being the free-stream velocity and φ_0 the wing's sweep angle. The external inviscid velocity component in the direction normal to the leading edge is $v_e^1(x^1)$. At the attachment line we have $v_e^1(x^1 = 0) = 0$, but $\partial v_e^1/\partial x^1(x^1 = 0) = A_e^1 > 0$. The temperature boundary condition is $T_e(x^1)$. The wall boundary conditions are the no-slip condition with or without suction or surface-normal blowing, all combined with a suitable thermal boundary condition.

The initial conditions on the attachment line ($x^1 = 0, 0 \leqq x^3 \leqq \delta$) are found from a quasi-one-dimensional solution which yields A^1, v^2, v^3, T as functions of x^3. Prerequisite for the solution is that the attachment line lies on a geodesic, or at least very close to it.

Problem 8.7

The cross-section increase in the axisymmetric case means a thinning of the boundary layer and hence an increase of the skin-friction. The cross-section decrease leads to the reverse effect, i.e. the reverse Mangler effect.

Problems of Chapter 9

Problem 9.1

a) The curve should look like that in Fig. 9.1 with a sharp increase of the wall shear stress at the transition location. b) In the laminar branch the characteristic thickness is the boundary-layer thickness ($\delta_{char} = \delta_{lam}$) and in the turbulent branch it is the thickness of the viscous sub-layer ($\delta_{char} = \delta_{vs}$). c) $\delta_{lam} \propto x^{0.5}$, $\delta_{turb} \propto x^{0.8}$, $\delta_{vs} \propto x^{0.2}$, d) The curve should look like that in Fig. B.1 of Appendix B.

Problem 9.2

The curves should look like those in Figs. 9.2 and 9.3. The equations are the Orr-Sommerfeld equation, eq. (9.16), and the Rayleigh equation, eq. (9.18). The basic assumption regarding the boundary-layer mean flow is the parallel-flow assumption. The consequence is that the stability theory is a local theory, i.e., the statement of stability or instability regards only a given tangential flow profile $u(y)$.

Problem 9.3

1. When the external inviscid flow has the form

$$u_e = kx^m,$$

the Falkner-Skan self similarity solutions for the laminar boundary layer tell us that the Reynolds number Re_{δ_2} is given by

$$Re_{\delta_2} = C\sqrt{Re_x},$$

where C is a function of m.

From the proposed transition criterion, transition occurs when the curve of the evolution of the Reynolds number $Re_{\delta_{2_{tr}}}$ intersects the transition criterion, i.e.

$$1.535\, Re_{x_{tr}}^{0.444} = C\sqrt{Re_{x_{tr}}},$$

whence

$$Re_{x_{tr}} = \left(\frac{1.535}{C}\right)^{1/0.056}.$$

With $m = 1/6$, the Falkner-Skan solution gives $C = 0.509$. We obtain the result $Re_{x_{tr}} = 3.64 \cdot 10^8$ and $Re_{\delta_{2_{tr}}} = 9{,}705$.

For the flat plate, we have $C = 0.664$, and obtain $Re_{x_{tr}} = 3.155 \cdot 10^6$ and $Re_{\delta_{2_{tr}}} = 1{,}179$.

2. We observe from the results that a negative pressure gradient (the flow is accelerated) delays the transition location significantly compared to the zero pressure gradient flow (flat plate). This is in agreement with experiment.

3. a) When using an empirical transition criterion, one must make sure that the considered flow case belongs to the class of flow cases on which the empirical criterion is based.

b) When applying Michel's criterion a glancing intersection results, Fig. 11.6. The criterion in fact is a smoothed mean value of experimentally found data which do not lie on the sharp line seen in the figure. Hence the glancing intersection introduces an uncertainty into the result.

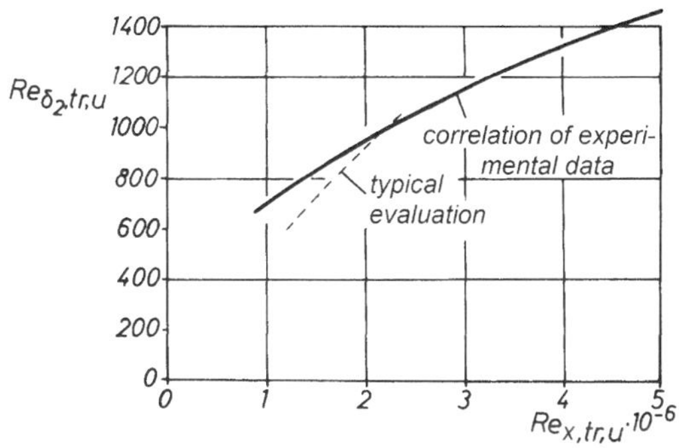

Fig. 11.6. Michel's criterion (full line) and typical evaluation trace (broken line) (from reference [24] of Chapter 9).

Problem 9.4

The curves should look like those in Fig. 4.3. The wall condition is the wall compatibility condition. The wall compatibility condition regards both the boundary-layer equation and the Navier-Stokes/RANS equations.

Problem 9.5

The shape is affected by the tangential pressure gradient, suction or surface-normal blowing, wall heating or cooling. If $u_{yy}(y)$ changes its sign, a point of inflection appears in $u(y)$. The general stability behavior is the following: a negative pressure gradient (accelerated flow) stabilizes, an adverse one destabilizes, suction stabilizes, blowing destabilizes, heat transfer out of the wall into the flow destabilizes, heat transfer out of the flow into the wall stabilizes. All these items affect in particular the function $u_{yy}(y)$, and if $u_{yy}(y)$ changes its sign, a point of inflection appears in $u(y)$.

Problem 9.6

1. We determine first the cross-flow instability Reynolds number $Re_{\delta_{cf}}$. Its definition is

$$Re_{\delta_{cf}} = -\frac{v_{t_e}}{\nu} \int_0^\delta \frac{v_n}{v_{t_e}} \, dy.$$

The velocity component v_n is expressed in the x, z axis-system

$$v_n = -u_e \sin\psi + w_e \cos\psi,$$

whence

$$\frac{v_n}{v_{t_e}} = -\frac{u}{u_e}\frac{u_e}{v_{t_e}} \sin\psi + \frac{w}{w_e}\frac{w_e}{v_{t_e}} \cos\psi = (g' - f')\frac{u_e w_e}{(v_{t_e})^2}$$

and

$$Re_{\delta_{cf}} = -\frac{1}{\nu}\frac{u_e w_e}{v_{t_e}} \int_0^\delta (g' - f') \, \mathrm{d}y = -\frac{1}{\sqrt{\nu k}}\frac{u_e w_e}{v_{t_e}} \int_0^\infty (g' - f') \, \mathrm{d}\eta =$$
$$= \frac{0.3786}{\sqrt{\nu k}}\frac{u_e w_e}{v_{t_e}}.$$

In addition, we have

$$u_e = kx = 2\frac{v_\infty \cos\varphi_0}{R_N} x, \quad w_e = v_\infty \sin\varphi_0, \quad v_{t_e} = \sqrt{u_e^2 + w_e^2}.$$

From this together we obtain

$$Re_{\delta_{cf}} = 0.5354\sqrt{\frac{v_\infty R_N}{\nu}}\sqrt{\cos\varphi_0}\frac{\dfrac{x}{R_N}}{\sqrt{1 + \dfrac{4}{\mathrm{tg}^{\,2}\varphi_0}\left(\dfrac{x}{R_N}\right)^2}}$$

or

$$Re_{\delta_{cf}} = 0.2677\sqrt{\frac{v_\infty R_N}{\nu}}\sqrt{\sin\varphi_0 \, \mathrm{tg}\,\varphi_0}\frac{\dfrac{x}{R_N}}{\sqrt{\dfrac{\mathrm{tg}^{\,2}\varphi_0}{4} + \left(\dfrac{x}{R_N}\right)^2}}.$$

The leading-edge contamination Reynolds number is defined as $Re_{\delta_{2_{LE}}}$. On the attachment line at the leading edge we have

$$w_e = v_\infty \sin\varphi_0$$

whence

$$
\begin{aligned}
Re_{\delta_{2_{LE}}} &= \frac{v_\infty \sin\varphi}{\nu} \int_0^\delta \frac{w}{w_e}\left(1-\frac{w}{w_e}\right) \mathrm{d}y = \\
&= \frac{v_\infty \sin\varphi_0}{\nu} \sqrt{\frac{\nu}{k}} \int_0^\infty g'(1-g')\, \mathrm{d}\eta = 0.4044 \frac{v_\infty \sin\varphi_0}{\sqrt{k\nu}}.
\end{aligned}
$$

With

$$
k = 2\frac{v_\infty \cos\varphi_0}{R_N}
$$

we get

$$
Re_{\delta_{2_{LE}}} = 0.286 \sqrt{\frac{v_\infty R_N}{\nu}} \sqrt{\sin\varphi_0 \operatorname{tg}\varphi_0}.
$$

2. In the expression of the cross-flow Reynolds number $Re_{\delta_{cf}}$ above, the quantity

$$
\frac{\dfrac{x}{R_N}}{\sqrt{\dfrac{\operatorname{tg}^2\varphi_0}{4} + \left(\dfrac{x}{R_N}\right)^2}}
$$

is an increasing function of $\frac{x}{R_N}$ which tends to 1 when $\frac{x}{R_N} \to \infty$. This means that for finite x the radius R_N must become very small.

Comparing the two final expressions for $Re_{\delta_{cf}}$ and $Re_{\delta_{2_{LE}}}$ results in

$$
\left(Re_{\delta_{cf}}\right)_{\max} = 0.94 Re_{\delta_{2_{LE}}},
$$

i.e. if $Re_{\delta_{2_{LE}}}$ remains lower than 100, $\left(Re_{\delta_{cf}}\right)_{\max}$ will never reach the critical value 150.

This result can be interpreted in the sense that for a small leading-edge radius the flow acceleration in chord direction is very large, see Problem 8.1. If we accept the two criteria, this means that attachment-line contamination does not happen due to the strong acceleration ($k = \partial u_e/\partial x = 2u_\infty/R_N$). On the other hand it means that the strong three-dimensionality of the flow is restricted to the vicinity of the leading edge and cross-flow instability does not happen there.

3. The maximum sweep angle of the wing for a laminar boundary layer on the leading edge is such that $Re_{\delta_{2_{LE}}} = 100$. Numerically, we obtain the results which are given Table 11.2. The data are plotted in Fig. 11.7.

4. The maximum sweep angles for which the boundary layers remains laminar according to $Re_{\delta_{2_{LE}}} = 100$ found from the figure are collected in Table 11.3.

Table 11.2. The Reynolds number $Re_{\delta_{2_{LE}}}$ as function of the sweep angle φ_0 for three chord Reynolds numbers Re_c.

$Re_c =$	$15 \cdot 10^6$	$25 \cdot 10^6$	$35 \cdot 10^6$
φ_0	$Re_{\delta_{2_{LE}}}$	$Re_{\delta_{2_{LE}}}$	$Re_{\delta_{2_{LE}}}$
0°	0	0	0
15°	50.52	65.23	77.18
30°	103.08	133.07	157.46
45°	161.33	208.27	246.43
60°	234.97	303.34	358.92

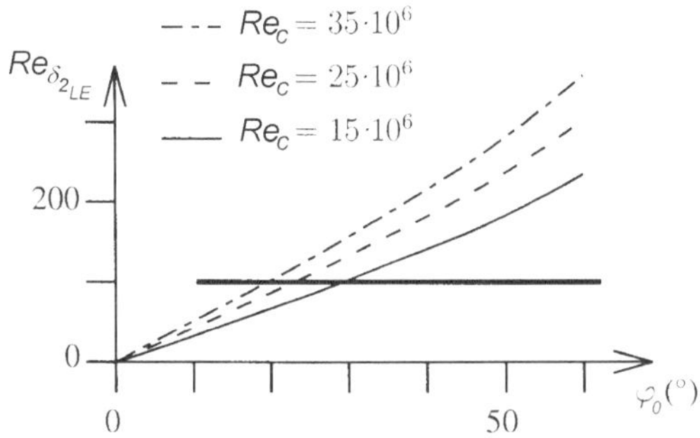

Fig. 11.7. The Reynolds number $Re_{\delta_{2_{LE}}}$ as function of the sweep angle φ_0 for the three chord Reynolds numbers Re_c. The horizontal line denotes $Re_{\delta_{2_{LE}}} = 100$. Below that value leading-edge contamination does not happen.

Table 11.3. The maximum sweep angles for which the boundary layers at the attachment line remain laminar according to $Re_{\delta_{2_{LE}}} = 100$.

Re_c	$\varphi_{0_{max}}$
$15 \cdot 10^6$	29°
$25 \cdot 10^6$	23°
$35 \cdot 10^6$	19°

Problem 9.7

Possible influencing factors are the free-stream turbulence, noise, vibrations, model surface properties, the thermal state of the model surface.

Problem 9.8

The evolution of the displacement thickness of a laminar incompressible flat-plate boundary layer is

$$\delta_1 = 1.721 \frac{x}{\sqrt{Re_x}}.$$

1. At the location $x = 0.075$ m and with $u_e = 100$ m s^{-1} and $\nu = 1.5{\cdot}10^{-5}$ m^2 s^{-1}, we have $Re_x = 5{\cdot}10^5$ and $\delta_1 = 0.182{\cdot}10^{-3}$ m.
2. The displacement thickness can be written:

$$\frac{\delta_1}{\delta} = \int_0^1 \left(1 - \frac{u}{u_e}\right) d\frac{y}{\delta}.$$

With

$$\frac{u}{u_e} = \sin\left(\frac{\pi}{2}\frac{y}{\delta}\right)$$

we obtain

$$\frac{\delta_1}{\delta} = \frac{\pi - 2}{\pi}$$

and the boundary-layer thickness at $x = 0.075$ m: $\delta = 0.502{\cdot}10^{-3}$m.

3. With $d = k$, the transition criterion tells us that transition occurs when (see reference [117] from Chapter 9)

$$\left(\frac{u_k k}{\nu}\right)^{1/2} = 30,$$

that is

$$\frac{u_k k}{\nu} = 900.$$

At the distance $y = k$ from the wall, the velocity in the boundary layer is

$$u_k = u_e \sin\left(\frac{\pi}{2}\frac{k}{\delta}\right).$$

Then, the transition criterion gives

$$u_e \sin\left(\frac{\pi}{2}\frac{k}{\delta}\right)\frac{k}{\nu} = 900,$$

that is

$$\frac{\pi}{2}\frac{k}{\delta}\sin\left(\frac{\pi}{2}\frac{k}{\delta}\right) = 900\,\frac{\nu}{u_e\delta}\frac{\pi}{2} = 0.422.$$

Numerically, we have

$$\frac{\pi}{2}\frac{k}{\delta} = 0.675$$

and

$$\frac{k}{\delta} = 0.43.$$

Finally, the minimum height of the roughness which triggers transition at $x = 75$ mm is $k = 0.216{\cdot}10^{-3}$m, i.e. almost half of the boundary-layer thickness.

Appendix A

Equations of Motion in General Formulations

In this appendix we present—for practical applications—the equations of motion in general form. This concerns the Navier-Stokes equations and the boundary-layer equations. The general application background is that stated in Chapters 3 and 4. The boundary-layer equations and the integral relations, Chapter 5, are given for surface-oriented, non-orthogonal curvilinear coordinates. The same holds for the small-cross flow equations. The boundary-layer equations further are given with contravariant formulation of the velocity components. Higher-order boundary-layer equations are not provided, but some literature is given. A note on computation methods closes the chapter.

A.1 Navier-Stokes/RANS Equations in General Coordinates

We collect the transport/RANS equations for viscous flow in general coordinates. The basic equations have been discussed in Chapter 3. The equations in general coordinates can be formulated in different ways, see, e.g., [1]–[3].

We write the equations in time-dependent conservative flux-vector formulation and for three-dimensional Cartesian coordinates:

$$\frac{\partial \underline{Q}}{\partial t} + \frac{\partial(\underline{E} - \underline{E}_{visc})}{\partial x} + \frac{\partial(\underline{F} - \underline{F}_{visc})}{\partial y} + \frac{\partial(\underline{G} - \underline{G}_{visc})}{\partial z} = 0. \tag{A.1}$$

$\underline{Q}$ is the conservation vector, $\underline{E}$, $\underline{F}$, $\underline{G}$ are the convective (inviscid) and $\underline{E_{visc}}$, $\underline{F_{visc}}$, $\underline{G_{visc}}$ the viscous fluxes in x, y, and z direction. The conservation vector $\underline{Q}$ has the form:

$$\underline{Q} = [\rho, \rho u, \rho v, \rho w, \rho e_t], \tag{A.2}$$

where ρ is the density, u, v, w are the Cartesian components of the velocity vector $\underline{V}$, and $e_t = e + 1/2\ V^2$ is the mass-specific total energy with $V = |\underline{V}| = (u^2 + v^2 + w^2)^{0.5}$.

The convective and the viscous fluxes in the three directions read:

$$\underline{E} = \begin{bmatrix} \rho u \\ \rho u^2 + p \\ \rho uv \\ \rho uw \\ \rho u h_t \end{bmatrix}, \; \underline{E}_{visc} = \begin{bmatrix} 0 \\ \tau_{xx} \\ \tau_{xy} \\ \tau_{xz} \\ -q_x + u\tau_{xx} + v\tau_{xy} + w\tau_{xz} \end{bmatrix}, \tag{A.3}$$

$$\underline{F} = \begin{bmatrix} \rho v \\ \rho uv \\ \rho v^2 + p \\ \rho vw \\ \rho v h_t \end{bmatrix}, \; \underline{F}_{visc} = \begin{bmatrix} 0 \\ \tau_{yx} \\ \tau_{yy} \\ \tau_{yz} \\ -q_y + u\tau_{yx} + v\tau_{yy} + w\tau_{yz} \end{bmatrix}, \tag{A.4}$$

$$\underline{G} = \begin{bmatrix} \rho w \\ \rho uw \\ \rho vw \\ \rho w^2 + p \\ \rho w h_t \end{bmatrix}, \; \underline{G}_{visc} = \begin{bmatrix} 0 \\ \tau_{zx} \\ \tau_{zy} \\ \tau_{zz} \\ -q_z + u\tau_{zx} + v\tau_{zy} + w\tau_{zz} \end{bmatrix}. \tag{A.5}$$

Each of the convective flux vectors $\underline{E}$, $\underline{F}$, $\underline{G}$ represents from top to bottom the transport of mass, Sub-Section 3.2.1, momentum, Sub-Section 3.2.2, and, with changed form, that of (total) energy, Sub-Section 3.2.3. In the above $h_t = e_t + p/\rho$ is the total enthalpy.

In the viscous flux vectors $\underline{E}_{visc}$, $\underline{F}_{visc}$, $\underline{G}_{visc}$, the symbols τ_{xx}, τ_{xy} et cetera represent the components of the viscous stress tensor $\underline{\underline{\tau}}$, eqs. (3.75) to (3.80) in Section 3.5, and q_x, q_y and q_z the components of the heat-flux vector, eqs. (3.83) to (3.85). They contain both the molecular and the apparent turbulent transport of momentum.

In the fifth line of eqs. (A.3) to (A.5) each we have the components of the energy-flux vector $\underline{q}_e$

$$\underline{q}_e = \rho(e + \frac{1}{2}V^2)\underline{V} + \underline{q} + p\underline{V} - \underline{\underline{\tau}} \cdot \underline{V}. \tag{A.6}$$

This vector allows to rewrite the energy equation in (A.1) in conservative energy flux-vector formulation [1]:[1]

$$\frac{\partial}{\partial t}\left[\rho(e + \frac{1}{2}V^2)\right] = -\left(\underline{\nabla} \cdot \underline{q}_e\right). \tag{A.7}$$

The left-hand side represents the rate of increase of energy in the unit volume with time, the right-hand side the gain of energy.

[1] Note the signs of the components of the viscous stress tensor in [1] which are different from our signs.

To compute the flow past configurations with general geometries, the above equations are transformed from the physical space x, y, z into the computation space ξ, η, ζ:

$$\begin{aligned} \xi &= \xi(x, y, z), \\ \eta &= \eta(x, y, z), \\ \zeta &= \zeta(x, y, z). \end{aligned} \tag{A.8}$$

This transformation, which goes back to H. Viviand, [4], and M. Vinokur, [5], regards only the geometry, and not the velocity components. This is in contrast to the approach for the boundary-layer equations in curvilinear coordinates, Sub-Section A.2, where both are transformed. ξ usually defines the main-stream direction, η some lateral direction, and ζ the wall-normal direction, however in general not in the sense of locally monoclinic coordinates, Appendix C.

The transformation results in:

$$\frac{\partial \underline{\widehat{Q}}}{\partial t} + \frac{\partial(\underline{\widehat{E}} - \underline{\widehat{E}}_{visc})}{\partial \xi} + \frac{\partial(\underline{\widehat{F}} - \underline{\widehat{F}}_{visc})}{\partial \eta} + \frac{\partial(\underline{\widehat{G}} - \underline{\widehat{G}}_{visc})}{\partial \zeta} = 0, \tag{A.9}$$

which exhibits the same form as the original formulation, eq. (A.1).

The transformed conservation vector and the convective flux vectors are now:

$$\begin{aligned} \underline{\widehat{Q}} &= J^{-1} \underline{Q}, \\ \underline{\widehat{E}} &= J^{-1}[\xi_x \underline{E} + \xi_y \underline{F} + \xi_z \underline{G}], \\ \underline{\widehat{F}} &= J^{-1}[\eta_x \underline{E} + \eta_y \underline{F} + \eta_z \underline{G}], \\ \underline{\widehat{G}} &= J^{-1}[\zeta_x \underline{E} + \zeta_y \underline{F} + \zeta_z \underline{G}], \end{aligned} \tag{A.10}$$

with J^{-1} being the Jacobi determinant of the transformation (A.8). The transformed viscous flux vectors have the same form as the transformed convective flux vectors.

The fluxes, eqs. (A.3) to (A.5), are transformed analogously, however we do not give the details, and refer instead to, for instance, [6, 2].

A.2 Boundary-Layer Equations in General Coordinates

We give the dimensionless boundary-layer equations for steady flow in surface-oriented non-orthogonal curvilinear coordinates. We depart from the notation used in Chapter 4 and employ the right-handed x, y, z-system with x and y being the surface-parallel non-orthogonal curvilinear coordinates, and z the surface-normal coordinate. Accordingly u and v are the surface tangential velocity components, whereas w is the velocity component normal to the surface. This holds for Appendices A.2.1 and A.2.2. In Appendix A.2.4 another convention is used. For metric properties see Appendix C, where also cross-section coordinates for fuselage-like configurations and percent-line coordinates for swept wings are presented.

A.2.1 First-Order Equations in Non-orthogonal Curvilinear Coordinates

Assumed is that the boundary-layer thickness δ is small compared to the smallest radius of curvature R_{min} of the body surface under consideration:

$$\delta \ll R_{min}. \tag{A.11}$$

The coordinate system then needs to be defined only on the body surface, Appendix C.1.

The first-order boundary-layer equations in non-orthogonal curvilinear coordinates read:

– continuity equation:

$$\frac{\partial}{\partial x}(n_{01}\rho u) + \frac{\partial}{\partial y}(n_{02}\rho v) + n_{03}\frac{\partial}{\partial z}(\rho w) = 0. \tag{A.12}$$

– x-momentum equation:

$$\begin{aligned}\rho(n_{11}u\frac{\partial u}{\partial x} + n_{12}v\frac{\partial u}{\partial y} + w\frac{\partial u}{\partial z} + n_{13}u^2 + n_{14}uv + n_{15}v^2) = \\ = n_{16}\frac{\partial p}{\partial x} + n_{17}\frac{\partial p}{\partial y} + \frac{\partial \tau_x}{\partial z}.\end{aligned} \tag{A.13}$$

– y-momentum equation:

$$\begin{aligned}\rho(n_{21}u\frac{\partial v}{\partial x} + n_{22}v\frac{\partial v}{\partial y} + w\frac{\partial v}{\partial z} + n_{23}u^2 + n_{24}uv + n_{25}v^2) = \\ = n_{26}\frac{\partial p}{\partial x} + n_{27}\frac{\partial p}{\partial y} + \frac{\partial \tau_y}{\partial z}.\end{aligned} \tag{A.14}$$

– energy equation for calorically and thermally perfect gas:

$$\begin{aligned}c_p\rho(n_{31}u\frac{\partial T}{\partial x} + n_{32}v\frac{\partial T}{\partial y} + w\frac{\partial T}{\partial z}) = -\frac{1}{Pr_{ref}}\frac{\partial q_z}{\partial z} + \\ + E_{ref}\left\{[n_{33}u\frac{\partial p}{\partial x} + n_{34}v\frac{\partial p}{\partial y}] + \tau_x\frac{\partial u}{\partial z} + n_{35}\tau_x\frac{\partial v}{\partial z} + \tau_y\frac{\partial v}{\partial z}\right\}.\end{aligned} \tag{A.15}$$

The boundary-layer shear-stress components are

$$\tau_x = \mu \frac{\partial u}{\partial z} - < \tilde{\rho} u' w' >, \tag{A.16}$$

$$\tau_y = \mu \frac{\partial v}{\partial y} - < \tilde{\rho} v' w' >, \tag{A.17}$$

and the boundary-layer heat flux

$$q_z = -k \frac{\partial T}{\partial z} + < \tilde{\rho} h' w' > . \tag{A.18}$$

The metric factors n_{bc} are functions of the metric tensor of the surface, eq. (C.11) in Appendix C.1. We write the factors in terms of the Lamé coefficients h_1, h_2, g, with $a = (h_1)^2(h_2)^2 - g^2 = a_{11}a_{22} - (a_{12})^2$ being the determinant of $a_{\alpha\beta}$:[2]

– continuity equation:

$$n_{01} = \frac{\sqrt{a}}{h_1},\ n_{02} = \frac{\sqrt{a}}{h_2},\ n_{03} = \sqrt{a}. \tag{A.19}$$

– x-momentum equation:

$$\begin{aligned}
n_{11} &= \frac{1}{h_1},\ n_{12} = \frac{1}{h_2}, \\
n_{13} &= \frac{h_1 g}{a}\left[\frac{1}{h_1}\frac{\partial h_1}{\partial y} + \frac{g}{(h_1)^3}\frac{\partial h_1}{\partial x} - \frac{1}{(h_1)^2}\frac{\partial g}{\partial x}\right], \\
n_{14} &= \frac{1}{a}\left[h_1 h_2\left(1 + \frac{g^2}{(h_1)^2(h_2)^2}\right)\frac{\partial h_1}{\partial y} - 2g\frac{\partial h_2}{\partial x}\right], \\
n_{15} &= \frac{h_1}{a}\left[\frac{\partial g}{\partial y} - h_2\frac{\partial h_2}{\partial x} - \frac{g}{h_2}\frac{\partial h_2}{\partial y}\right], \\
n_{16} &= -\frac{(h_2)^2 h_1}{a},\ n_{17} = \frac{g h_1}{a}.
\end{aligned} \tag{A.20}$$

– y-momentum equation:

[2] We use the notation a instead of $p = \sqrt{(h_1)^2(h_2)^2 - g^2} = \sqrt{a}$ in [7].

$$
\begin{aligned}
n_{21} &= \frac{1}{h_1}, \; n_{22} = \frac{1}{h_2},\\
n_{23} &= \frac{h_2}{a}\left[\frac{\partial g}{\partial x} - h_1\frac{\partial h_1}{\partial y} - \frac{g}{h_1}\frac{\partial h_1}{\partial x}\right],\\
n_{24} &= \frac{1}{a}\left[h_1 h_2\left(1 + \frac{g^2}{(h_1)^2(h_2)^2}\right)\frac{\partial h_2}{\partial x} - 2g\frac{\partial h_1}{\partial y}\right],\\
n_{25} &= \frac{h_2 g}{a}\left[\frac{1}{h_2}\frac{\partial h_2}{\partial x} + \frac{g}{(h_2)^3}\frac{\partial h_2}{\partial y} - \frac{1}{(h_2)^2}\frac{\partial g}{\partial y}\right],\\
n_{26} &= \frac{g h_2}{a}, \; n_{27} = -\frac{(h_1)^2 h_2}{a}.
\end{aligned}
\tag{A.21}
$$

– energy equation:

$$
n_{31} = \frac{1}{\sqrt{a_{11}}}, \; n_{32} = \frac{1}{\sqrt{a_{22}}}, \; n_{33} = \frac{1}{\sqrt{a_{11}}}, \; n_{34} = \frac{1}{\sqrt{a_{22}}}, \; n_{35} = \frac{2a_{12}}{\sqrt{a_{11}}\sqrt{a_{22}}}.
\tag{A.22}
$$

The pressure-gradient terms in eqs. (A.13) and (A.14) can be replaced by the velocity components of the external inviscid flow field by

– x-momentum equation:

$$
n_{16}\frac{\partial p}{\partial x} + n_{17}\frac{\partial p}{\partial y} = \rho_e\Big(n_{11}u_e\frac{\partial u_e}{\partial x} + n_{12}v_e\frac{\partial u_e}{\partial y} + n_{13}u_e^2 + n_{14}u_e v_e + n_{15}v_e^2\Big).
\tag{A.23}
$$

– y-momentum equation:

$$
n_{26}\frac{\partial p}{\partial x} + n_{27}\frac{\partial p}{\partial y} = \rho_e\Big(n_{21}u_e\frac{\partial v_e}{\partial x} + n_{22}v_e\frac{\partial v_e}{\partial y} + n_{23}u_e^2 + n_{24}u_e v_e + n_{25}v_e^2\Big).
\tag{A.24}
$$

For two-dimensional boundary layers the displacement thickness δ_1 is found by quadratures. For three-dimensional boundary-layers a partial differential equation has to be solved for δ_1, Chapter 5. In our notation it reads:

$$
\frac{\partial}{\partial x}\left[n_{01}\rho_e u_e(\delta_1 - \delta_{1x})\right] + \frac{\partial}{\partial y}\left[n_{02}\rho_e v_e(\delta_1 - \delta_{1y})\right] = n_{03}\rho_0 w_0,
\tag{A.25}
$$

with

$$
\delta_{1x} = \int_{wall}^{\delta}(1 - \frac{\rho u}{\rho_e u_e})dz,
\tag{A.26}
$$

and

$$\delta_{1y} = \int_{wall}^{\delta} (1 - \frac{\rho v}{\rho_e v_e}) dz. \tag{A.27}$$

The equivalent inviscid source distribution, Chapter 5, reads:

$$(n_{03}\rho_0 w_0)_{inv} = \frac{\partial}{\partial x}\left[n_{01}\rho_e u_e \delta_{1x}\right] + \frac{\partial}{\partial y}\left[n_{02}\rho_e v_e \delta_{1y}\right] + n_{03}\rho_0 w_0. \tag{A.28}$$

A.2.2 Small Cross-Flow Equations

The small cross-flow approximation of the boundary-layer equations in three dimensions leads to a set of quasi-two-dimensional equations. Computation methods making use of this approximation need not much more computational effort than two-dimensional methods. The intricacies of coordinate orientation for truly three-dimensional boundary-layer methods, Appendix C, are avoided. However, the characteristic properties of attached viscous flow, Section 4.4, viz. the three-dimensional domains of dependence and influence, are lost.

The small cross-flow approximation makes use of the principle of prevalence, introduced by E.A. Eichelbrenner and A. Oudart [8]:

1. The velocity vector $[u, v, w]$ is a regular function of the coordinates x, y, z throughout the region close to the surface of the body, the boundary-layer.
2. The directions of the streamlines of the viscous flow differ little—close to the wall—from those of the streamlines of the external inviscid flow, i.e., in the boundary layer the stream-wise flow dominates over the cross-flow. Therefore this approximation yields reasonable results whenever the inviscid streamlines are only slightly curved.

To derive the small cross-flow equations we proceed in two steps. In the first step eqs. (A.12) to (A.15) are written in terms of an orthogonal coordinate system, after that the the small cross-flow assumption is introduced.

The external inviscid stream-line coordinate system, Sub-Section 2.2.2, is a triply orthogonal coordinate system: x is the stream-wise coordinate, y the cross-flow coordinate, and z the coordinate normal to the body surface. The velocity vector consists of u as the stream-wise component, v as the cross-flow component, and w as the component normal to the surface.

Because for the orthogonal coordinates the diagonal term in the metric tensor vanishes, $g = a_{12} = 0$, the metric factors reduce considerably. Hence we can write eqs. (A.12) to (A.15) for orthogonal coordinates directly in terms of h_1 and h_2:

– continuity equation:

$$\frac{\partial}{\partial x}(h_2 \rho u) + \frac{\partial}{\partial y}(h_1 \rho v) + h_1 h_2 \frac{\partial}{\partial z}(\rho w) = 0. \tag{A.29}$$

– x-momentum equation:

$$\begin{aligned}\rho(\frac{u}{h_1}\frac{\partial u}{\partial x} + \frac{v}{h_2}\frac{\partial u}{\partial y} + w\frac{\partial u}{\partial z} + \frac{1}{h_1 h_2}\frac{\partial h_1}{\partial y} uv - \frac{1}{h_1 h_2}\frac{\partial h_2}{\partial x} v^2) = \\ = -\frac{1}{h_1}\frac{\partial p}{\partial x} + \frac{\partial \tau_x}{\partial z}.\end{aligned} \tag{A.30}$$

– y-momentum equation:

$$\begin{aligned}\rho(\frac{u}{h_1}\frac{\partial v}{\partial x} + \frac{v}{h_2}\frac{\partial v}{\partial y} + w\frac{\partial v}{\partial z} - \frac{1}{h_1 h_2}\frac{\partial h_1}{\partial y} u^2 + \frac{1}{h_1 h_2}\frac{\partial h_2}{\partial x} uv) = \\ = -\frac{1}{h_2}\frac{\partial p}{\partial y} + \frac{\partial \tau_y}{\partial z}.\end{aligned} \tag{A.31}$$

– energy equation for calorically and thermally perfect gas:

$$\begin{aligned}c_p \rho(\frac{u}{h_1}\frac{\partial T}{\partial x} + \frac{v}{h_2}\frac{\partial T}{\partial y} + w\frac{\partial T}{\partial z}) = -\frac{1}{Pr_{ref}}\frac{\partial q_z}{\partial z} + \\ + E_{ref}\left\{[\frac{u}{h_1}\frac{\partial p}{\partial x} + \frac{v}{h_2}\frac{\partial p}{\partial y}] + \tau_x \frac{\partial u}{\partial z} + \tau_y \frac{\partial v}{\partial z}\right\}.\end{aligned} \tag{A.32}$$

In eqs. (A.30) and (A.31) the terms

$$-\frac{1}{h_1 h_2}\frac{\partial h_1}{\partial y} = -\frac{1}{2a_{11}\sqrt{a_{22}}}\frac{\partial a_{11}}{\partial y} = K_1 \tag{A.33}$$

and

$$-\frac{1}{h_1 h_2}\frac{\partial h_2}{\partial x} = -\frac{1}{2\sqrt{a_{11}}a_{22}}\frac{\partial a_{22}}{\partial x} = K_2 \tag{A.34}$$

are the geodesic curvatures K_1 and K_2, [9], of the curves y = constant (the streamline of the external inviscid flow) and x = constant (the cross-flow direction).

Again the pressure-gradient terms in eqs. (A.30) and (A.31) can be replaced by the velocity components of the external inviscid flow field. Because in the external stream-line oriented coordinates by definition $v_e = 0$, Fig. 2.3 (note the different notation), we obtain now

– x-momentum equation:

$$-\frac{\partial p}{\partial x} = \rho_e u_e \frac{\partial u_e}{\partial x}. \tag{A.35}$$

– y-momentum equation:

$$-\frac{1}{h_2}\frac{\partial p}{\partial y} = -\rho_e \frac{1}{h_1 h_2}\frac{\partial h_1}{\partial y} u_e^2 = K_1 \rho_e u_e^2. \tag{A.36}$$

If the curvature of the external inviscid streamline K_1 is small, the cross-flow velocity v and the terms in the boundary layer are small, too. Note that anyway holds, Fig. 2.3: $v_{z=0} = 0$, $v_{z=\delta} = 0$.

Now the second step. With the postulate that the stream-wise flow is prevalent, we can neglect the terms containing v and/or the terms containing $\partial/\partial y$. Doing this in eqs. (A.29), (A.30), and (A.32) results in a system of equations for the stream-wise direction from which the y-momentum equation is decoupled. This is similar to the infinite-swept wing concept, Section 8.3, where the omission of terms containing $\partial/\partial y$ serves the same purpose. In both cases the cross-flow velocity component v, though here small, can be determined.

We obtain the small cross-flow equations where all dependent variables are functions of the stream-wise coordinate x and the surface-normal coordinate z only:

– continuity equation:

$$\frac{\partial}{\partial x}(h_2 \rho u) + h_1 h_2 \frac{\partial}{\partial z}(\rho w) = 0. \tag{A.37}$$

– x-momentum equation:

$$\rho\left(\frac{u}{h_1}\frac{\partial u}{\partial x} + w\frac{\partial u}{\partial z}\right) = -\frac{1}{h_1}\frac{\partial p}{\partial x} + \frac{\partial \tau_x}{\partial z}. \tag{A.38}$$

– y-momentum equation:

$$\rho\left(\frac{u}{h_1}\frac{\partial v}{\partial x} + w\frac{\partial v}{\partial z} + K_1 u^2 - K_2 u v\right) = K_1 \rho_e u_e^2 + \frac{\partial \tau_y}{\partial z}. \tag{A.39}$$

– energy equation for calorically and thermally perfect gas:

$$\begin{aligned} c_p \rho\left(\frac{u}{h_1}\frac{\partial T}{\partial x} + w\frac{\partial T}{\partial z}\right) &= -\frac{1}{Pr_{ref}}\frac{\partial q_z}{\partial z} + \\ + E_{ref}\left\{\left[\frac{u}{h_1}\frac{\partial p}{\partial x} - v K_1 \rho_e u_e^2\right] + \tau_x \frac{\partial u}{\partial z} + \tau_y \frac{\partial v}{\partial z}\right]\right\}. \end{aligned} \tag{A.40}$$

Of these equations, the first two form the axially symmetric boundary-layer equations, Chapter 8.6.

The relations for the displacement thickness and the equivalent inviscid source distribution formally are the same as in the preceding sub-section.

A.2.3 The Geodesic as Prerequisite for the Plane-of-Symmetry Flow

The boundary layer in a three-dimensional flow is called quasi-two-dimensional, if the describing equations have only two independent variables. The precondition is a suitable coordinate system with, for instance, the x-coordinate in stream-wise direction and the z-coordinate in surface-normal direction. If this is given, the boundary-layer equations can be solved in the x-z plane.

Four cases must be distinguished:[3]

1. The flow can be constituted as function of x and z for any y (the coordinate normal to the x-z plane). The velocity components u and w are finite, the component v is identical zero. All derivatives of all dependent variables are identical zero in y-direction. The flow field extends to infinity in that direction. The flow is two-dimensional and can be computed with a two-dimensional method.
2. The flow is two-dimensional only for one discrete value of the coordinate y, for instance $y = 0$. The y-momentum equation is zero only for this location. We have plane-of-symmetry flow. This is the classical quasi-two-dimensional case, where for $y = 0$ the x-momentum equation is solved, together with the with respect to y differentiated y-momentum equation, the energy equation and the continuity equation.
3. The y-coordinate lies at a constant, but not necessarily right angle to the x-z plane. The dependent variables u, v and w are all finite, but functions of x and z only. All derivatives in y-direction are zero. We have the situation of the infinite swept wing (ISW). The x- and the y-momentum equation are solved for $y =$ constant together with the energy equation and the continuity equation.
4. If the cross-flow is small, we can formulate the small cross-flow equations, Sub-Section A.2.2. These are quasi-two-dimensional equations, too.

We concentrate on case 2 and ask, what is the prerequisite in terms of the metric properties of the body surface under consideration. Case 2 implies that all streamlines within the boundary layer including the inviscid surface streamline and hence also the pressure gradient lie in the plane $y = 0$. This is given only if the coordinate line $z = 0$ at $y = 0$ fulfills the condition that the geodesic curvature K_1—see the preceding sub-section—is identical zero. If this is the case, the coordinate line lies on a geodesic [9].[4]

The proof for this was given first by L.C. Squire in 1957 [7]. An independent, very detailed proof was given in 1981 by D. Schwamborn [10].

---

[3] For practical examples see Chapter 8.

[4] The geodesic can be defined as a curve whose tangent vectors always remain parallel if they are conveyed along it.

We do not go into the details of the proofs of the two authors but refer the interested reader to the cited publications. We note only that the plane-of-symmetry situation is found at fuselages at angles of attack, at the leading edges of swept wings, swept horizontal tail surfaces and swept vertical stabilizers, all with locally prismatic shape and suitable angles of attack and zero yaw angle. This holds also for the attachment line of the infinite swept wing, case 3. The boundary-layer equations for these cases are presented in Chapter 8. They include also the quasi-one-dimensional case of a forward stagnation point.

A.2.4 Equations in Contravariant Formulation

The contravariant formulation permits a mathematically appealing formulation of the boundary-layer equations and a convenient approach to considerations where geometrical properties come into play as well as to higher-order equations. We give here the first-order equations, as they can be found in [11], although with a slightly different nomenclature.[5]

Tensorial concepts were introduced into boundary-layer theory probably first by J. Kux [12]. He treated the three-dimensional boundary-layer equations for incompressible flow, K. Robert those for compressible flow [13]. The tensorial formulation of the gas dynamic equations for Newtonian fluids in non-orthogonal and accelerated coordinate systems was given by K. Robert and R. Grundmann [14].

We use Einstein's index notation. The two surface-parallel coordinates are x^1 and x^2, the surface-normal coordinate is x^3, Fig. 2.6. The physical velocity components accordingly are v^{*1}, v^{*2}, and v^{*3}.

The contravariant formulation regards the velocity components. The contravariant velocity components v^1 and v^2 are related to the physical ones by means of the diagonal elements a_{11} and a_{22} of the metric tensor of the surface:[6]

$$v^1 = \frac{v^{*1}}{\sqrt{a_{11}}}, \; v^2 = \frac{v^{*2}}{\sqrt{a_{22}}}, \; v^3 = v^{*3}. \tag{A.41}$$

The first-order boundary-layer equations in non-orthogonal curvilinear coordinates with contravariant velocity components read:

– continuity equation:

$$\frac{\partial}{\partial x^1}(k_{01}\rho v^1) + \frac{\partial}{\partial x^2}(k_{01}\rho v^2) + k_{01}\frac{\partial}{\partial x^3}(\rho v^3) = 0. \tag{A.42}$$

[5] The reader is reminded of the conditions usually assumed of this book: steady-state flow and thermally and calorically perfect flow.

[6] Note that $a_{33} = 1$, Appendix C.

– x^1-momentum equation:

$$\rho[v^1\frac{\partial v^1}{\partial x^1} + v^2\frac{\partial v^1}{\partial x^2} + v^3\frac{\partial v^1}{\partial x^3} + k_{11}(v^1)^2 + k_{12}v^1v^2 + k_{13}(v^2)^2] = \\ = k_{16}\frac{\partial p}{\partial x^1} + k_{17}\frac{\partial p}{\partial x^2} + \frac{\partial \tau^1}{\partial x^3}. \tag{A.43}$$

– x^2-momentum equation:

$$\rho[v^1\frac{\partial v^2}{\partial x^1} + v^2\frac{\partial v^2}{\partial x^2} + v^3\frac{\partial v^2}{\partial x^3} + k_{21}(v^1)^2 + k_{22}v^1v^2 + k_{23}(v^2)^2] = \\ = k_{26}\frac{\partial p}{\partial x^1} + k_{27}\frac{\partial p}{\partial x^2} + \frac{\partial \tau^2}{\partial x^3}. \tag{A.44}$$

– energy equation for calorically and thermally perfect gas:

$$c_p\rho(v^1\frac{\partial T}{\partial x^1} + v^2\frac{\partial T}{\partial x^2} + v^3\frac{\partial T}{\partial x^3}) = -\frac{1}{Pr_{ref}}\frac{\partial q^3}{\partial x^3} + \\ + E_{ref}\left\{[v^1\frac{\partial p}{\partial x^1} + v^2\frac{\partial p}{\partial x^2}] + k_{41}\tau^1\frac{\partial v^1}{\partial x^3} + k_{42}\tau^1\frac{\partial v^2}{\partial x^3} + k_{43}\tau^2\frac{\partial v^2}{\partial x^3}\right\}. \tag{A.45}$$

The boundary-layer shear-stress components τ^1, τ^2 and the boundary-layer heat flux component q^3 read:

$$\tau^1 = \mu\frac{\partial v^1}{\partial x^3} - \frac{1}{\sqrt{a_{11}}} < \rho v^{*1'} v^{*3'} >, \tag{A.46}$$

$$\tau^2 = \mu\frac{\partial v^2}{\partial x^3} - \frac{1}{\sqrt{a_{22}}} < \rho v^{*2'} v^{*3'} >, \tag{A.47}$$

$$q^3 = -k\frac{\partial T}{\partial x^3} + < \rho h' v^{*3'} > . \tag{A.48}$$

The metric factors k_{mn} again are functions of the metric tensor of the surface coordinates, eq. (C.11). We write them now fully in terms of this tensor, not with the Lamé coefficients, Appendix C.1:

– continuity equation:

$$k_{01} = \sqrt{a} = \sqrt{a_{11}a_{22} - (a_{12})^2}. \tag{A.49}$$

– x^1-momentum equation:

$$
\begin{aligned}
k_{11} &= \frac{1}{2a}\left[a_{22}\frac{\partial a_{11}}{\partial x^1} - a_{12}\left(2\frac{\partial a_{12}}{\partial x^1} - \frac{\partial a_{11}}{\partial x^2}\right)\right],\\
k_{12} &= \frac{1}{a}\left[a_{22}\frac{\partial a_{11}}{\partial x^2} - a_{12}\frac{\partial a_{22}}{\partial x^1}\right],\\
k_{13} &= \frac{1}{2a}\left[a_{22}\left(2\frac{\partial a_{12}}{\partial x^2} - \frac{\partial a_{22}}{\partial x^1}\right) - a_{12}\frac{\partial a_{22}}{\partial x^2}\right],\\
k_{16} &= -\frac{a_{22}}{a},\ k_{17} = \frac{a_{12}}{a}.
\end{aligned}
\tag{A.50}
$$

– x^2-momentum equation:

$$
\begin{aligned}
k_{21} &= \frac{1}{2a}\left[a_{11}\left(2\frac{\partial a_{12}}{\partial x^1} - \frac{\partial a_{11}}{\partial x^2}\right) - a_{12}\frac{\partial a_{11}}{\partial x^1}\right],\\
k_{22} &= \frac{1}{a}\left[a_{11}\frac{\partial a_{22}}{\partial x^1} - a_{12}\frac{\partial a_{11}}{\partial x^2}\right],\\
k_{23} &= \frac{1}{2a}\left[a_{11}\frac{\partial a_{22}}{\partial x^2} - a_{12}\left(2\frac{\partial a_{12}}{\partial x^2} - \frac{\partial a_{22}}{\partial x^1}\right)\right],\\
k_{26} &= \frac{a_{12}}{a} = k_{17},\ k_{27} = -\frac{a_{11}}{a}.
\end{aligned}
\tag{A.51}
$$

– energy equation:

$$
k_{41} = a_{11},\ k_{42} = 2\,a_{12},\ k_{43} = a_{22}. \tag{A.52}
$$

The pressure-gradient terms in eqs. (8.29) and (8.30) can be replaced by the velocity components of the external inviscid flow field by

– x^1-momentum equation:

$$
k_{16}\frac{\partial p}{\partial x^1} + k_{17}\frac{\partial p}{\partial x^2} = \rho_e\Big(v_e^1\frac{\partial v_e^1}{\partial x^1} + v_e^2\frac{\partial v_e^1}{\partial x^2} + k_{11}(v_e^1)^2 + k_{12}v_e^1 v_e^2 + k_{13}(v_e^2)^2\Big). \tag{A.53}
$$

– x^2-momentum equation:

$$
k_{26}\frac{\partial p}{\partial x^1} + k_{27}\frac{\partial p}{\partial x^2} = \rho_e\Big(v_e^1\frac{\partial v_e^2}{\partial x^1} + v_e^2\frac{\partial v_e^2}{\partial x^2} + k_{21}(v^1)_e^2 + k_{22}v_e^1 v_e^2 + k_{23}(v_e^2)^2\Big). \tag{A.54}
$$

The partial differential equation for δ_1 in contravariant notation reads:

$$
\frac{\partial}{\partial x^1}\left[k_{01}\rho_e v_e^1(\delta_1 - \delta_{1x^1})\right] + \frac{\partial}{\partial x^2}\left[k_{01}\rho_e v_e^2(\delta_1 - \delta_{1x^2})\right] = k_{01}\rho_0 v_0^3, \tag{A.55}
$$

with

$$\delta_{1x^1} = \int_{wall}^{\delta} (1 - \frac{\rho v^1}{\rho_e v_e^1}) dx^3, \tag{A.56}$$

and

$$\delta_{1x^2} = \int_{wall}^{\delta} (1 - \frac{\rho v^2}{\rho_e v_e^2}) dx^3. \tag{A.57}$$

The equivalent inviscid source distribution is:

$$(k_{01}\rho_0 v_0^3)_{inv} = \frac{\partial}{\partial x^1}\left[k_{01}\rho_e v_e^1 \delta_{1x^1}\right] + \frac{\partial}{\partial x^2}\left[k_{01}\rho_e v_e^2 \delta_{1x^2}\right] + k_{01}\rho_0 v_0^3. \tag{A.58}$$

A.2.5 Higher-Order Equations—The SOBOL Method

Aspects of higher-order effects in attached viscous flow were discussed in Chapter 6. Many sets of governing equations describing higher-order effects including the modelling of turbulence can be found in the literature. No attempt is made here to give a review.

We only cite a formulation line of second-order/higher-order three-dimensional boundary-layer equations which had its beginnings in the work environment of the first and the third author of this book. The formulation line originated with K. Robert [13], who introduced a shell-theory concept, the shifters. These are used to project curvature properties of the body surface into the computation domain above the surface.

The formulation line is discussed in [11], too. It led to the development of the finite-difference second-order boundary-layer (SOBOL) method by F. Monnoyer [15]–[17]. SOBOL, [18], during the 1990s found much application in the European aerospace field, namely in the HERMES aerothermodynamics, see also Section 1.3, page 12, of this book. It was also employed by several European aerospace research establishments and companies.

A.3 A Note on Computation Methods

In this section a note is given on computation methods. It is not intended to give overviews over the the different classes of methods. Major publications will be mentioned, the matter of grid generation will be touched.

A.3.1 Navier-Stokes/RANS Methods

Many Navier-Stokes/RANS methods are under development and in use at universities, research establishments and industry. Two publications give overviews of these methods in Europe, [19], and in particular at the European research establishments [20]. Details of the German MEGADESIGN and MegaOpt initiatives can be found in [21].

A retrospective on 40 years of numerical fluid mechanics and aerodynamics with contributions from world-wide all countries with major involvement in the field is given in [22]. The development and application of CFD, for instance, at Boeing is traced in [23]. Workshops on special topics like drag prediction, see Section 10.3, assessments of turbulence models, see Section 9.6, review of high-lift CFD capabilities [24] and so on, give overviews also of the involved Navier-Stokes/RANS methods. Commercial codes are coming more and more in use [25].

Grid generation today is intertwined with the respective numerical code. The road went from structured grids with long generation times to unstructured grids which permit to arrive in one or two days at an entire grid. This has to go from the CAD model of the vehicles surface to the surface grid, see, e.g., [26]. The surface grid has to be lodged with a parametrization of the surface CAD model for grid refinement and coarsening purposes.

Grids for complete airplanes now encompass tens of millions of points. Self organization of grids—with automatic clustering of grid points in regions with large gradients of the dependent variables—in conjunction with hybrid Cartesian grids was proposed in, for instance, [27]. In view of the still increasing computer power due to massive parallel computer architectures meanwhile fully Cartesian grids are put forward [28].

With massive parallel computer architectures code portability issues arise in particular at industry. The prospect of the Virtual Product as a high-fidelity mathermatical/numerical representation of the physical properties and functions of the manoeuvering and real-elastic flight vehicle together with its propulsion system rests on very high computer performance and far-reaching code portability [29].

A.3.2 Boundary-Layer Methods

Three-dimensional boundary-layer methods are available at universities, research establishments and industry. In 1984 at an AGARD event in Brussels representatives of France, Germany, The Netherlands, the UK, and the USA gave overviews of work in their countries on three-dimensional boundary-layer problems and solution methods [30]. Newer overviews of solution methods are not known. The SOBOL method was mentioned above.

Boundary-layer methods basically come as finite-difference methods and integral methods, for the latter see, e.g., the method of the second author of this book [31]. They have in common—because they are space-marching methods—that the spatial CFL condition must be obeyed, Section 4.4. This led for finite-difference methods to the construction of difference molecules with the goal to achieve a domain of dependence as large as possible. The zig-zag or stair-case scheme, for instance, permits a 135° domain [32]. A different approach is that with the characteristic box scheme [33]. In particular for boundary-layer calculations along swept-wing leading edges a double zig-zag scheme was devised [10, 34]. A summarizing discussion of these and other schemes can be found in [16].

A.3.3 Similarity Solutions

Similarity solutions for incompressible and compressible three-dimensional laminar boundary layers have been developed in the 1960s by several authors, see the overview in [32]. Some of them were shown not to fulfill the continuity and/or the Euler equations at the boundary-layer edge [35]. Whereas similarity solutions are not of value for engineering prediction problems, they are well suited for checks of the accuracy of boundary-layer and even of Navier-Stokes methods. The illustrating examples in Sections 2.3 and 4.4 of the present book are based on the external inviscid flow fields underlying the similarity solutions given in [36].

References

1. Bird, R.B., Stewart, W.E., Lightfoot, E.N.: Transport Phenomena, 2nd edn. John Wiley & Sons, New York (2002)
2. Hirsch, C.: Numerical Computation of Internal and External Flows. Fundamentals of Numerical Discretization, vol. 1. John Wiley & Sons, New York (1991)
3. Anderson Jr., J.D.: Computational Fluid Dynamics. McGraw Hill, New York (1995)
4. Viviand, H.: Conservative Forms of Gas Dynamic Equations. La Recherche Aerospatiale 1, 65–68 (1974)
5. Vinokur, M.: Conservative Equations of Gas-Dynamics in Curvilinear Coordinate Systems. J. Comp. Phys. 14, 105–125 (1974)
6. Pulliam, T.H., Steger, J.L.: Implicite Finite-Difference Simulations of Three-Dimensional Compressible Flows. AIAA J. 18(2), 159–167 (1980)
7. Squire, L.C.: The Three-Dimensional Boundary-Layer Equations and some Power Series Solutions. A.R.C. Technical Report, R. & M. No. 3006 (1957)
8. Eichelbrenner, E.A., Oudart, A.: Méthode de calcul de la couche limite tridimensionnelle. Application à un corps fuselé inclin'e sur le vent. O.N.E.R.A. Publication 76 (1955)
9. Aris, R.: Vectors, Tensors and the Basic Equations of Fluid Mechanics. Dover, New York (1989)
10. Schwamborn, D.: Laminare Grenzschichten in der Nähe der Anlegelinie an Flügeln und flügelähnlichen Körpern mit Anstellung (Laminar Boundary Layers in the Vicinity of the Attachment Line at Wings and Wing-Like Bodies at Angle of Attack). Doctoral thesis, RWTH Aachen, Germany, 1981, also DFVLR-FB 81-31 (1981)
11. Hirschel, E.H., Kordulla, W.: Shear Flow in Surface-Oriented Coordinates. NNFM, vol. 4. Vieweg, Braunschweig Wiesbaden (1981)
12. Kux, J.: Über dreidimensionale Grenzschichten an gekrümmten Wänden (About Three-Dimensional Boundary Layers at Curved Surfaces). Doctoral thesis, Universität Hamburg, Germany, Institut fúr Schiffbau der Universität Hamburg, Bericht Nr. 273 (1971)

13. Robert, K.: Higher-Order Boundary Layer Equations for Three-Dimensional, Compressible Flow. DLR-FB 77-36, pp. 205–215 (1976), also ESA-TT-518, 273–288 (1979)
14. Robert, K., Grundmann, R.: Basic Equations for Non-Reacting Newtonian Fluids in Curvilinear. Non-Orthogonal and Accelerated Coordinate Systems. DLR-FB 76-47 (1976)
15. Monnoyer, F.: The Effect of Surface Curvature on Three-Dimensional, Laminar Boundary Layers. Doctoral thesis, Université Libre de Bruxelles, Belgium (1985)
16. Monnoyer, F.: Calculation of Three-Dimensional Viscous Flow on General Configurations Using Second-Order Boundary-Layer Theory. ZFW 14, 95–108 (1990)
17. Monnoyer, F.: Extension et application de la théorie de la couche limite (Extension and Application of the Boundary-Layer Theory). Habilitation thesis, l'Université de Valenciennes et du Hainaut-Cambrésis, France (1996)
18. Monnoyer, F.: SOBOL Mk 2.7 Handbook. UVHC-LMF-NT-001, l'Université de Valenciennes et du Hainaut-Cambrésis, France (1994)
19. Vos, J.B., Rizzi, A., Darracq, D., Hirschel, E.H.: Navier-Stokes Solvers in European Aircraft Design. Progress in Aerospace Sciences 38, 601–697 (2002)
20. Rossow, C.-C., Cambier, L.: European Numerical Aerodynamic Simulation Systems. In: Hirschel, E.H., Krause, E. (eds.) 100 Volumes of 'Notes on Numerical Fluid Mechanics'. NNFM, vol. 100, pp. 189–208. Springer, Heidelberg (2009)
21. Kroll, N., Schwamborn, D., Becker, K., Rieger, H., Thiele, F. (eds.): MEGADESIGN and MegaOpt. NNFM, vol. 107. Springer, Heidelberg (2009)
22. Hirschel, E.H., Krause, E. (eds.): 100 Volumes of 'Notes on Numerical Fluid Mechanics'. NNFM, vol. 100. Springer, Heidelberg (2009)
23. Johnson, F.T., Tinoco, E.N., Yu, N.Y.: Thirty years of Development and Application of CFD at Boeing Commercial Airplanes, Seattle. AIAA-Paper 2003-3439 (2003)
24. Rumsey, C.L., Ying, S.X.: Prediction of High Lift: Review of Present CFD Capability. Progress in Aerospace Sciences 38, 145–180 (2002)
25. Boysan, H.F., Choudhury, D., Engelman, M.S.: Commercial CFD in the Service of Industry: the First 25 Years. In: Hirschel, E.H., Krause, E. (eds.) 100 Volumes of 'Notes on Numerical Fluid Mechanics'. NNFM, vol. 100, pp. 451–461. Springer, Heidelberg (2009)
26. Deister, F., Tremel, U., Hirschel, E.H., Rieger, H.: Automatic Feature-Based Sampling of Native CAD Data for Surface Grid Generation. In: New Results in Numerical and Experimental Fluid Mechanics IV. NNFM, vol. 87, pp. 374–381. Springer, Heidelberg (2004)
27. Deister, F., Hirschel, E.H.: Self-Organizing Hybrid Cartesian Grid/Solution System with Multigrid. AIAA-Paper 2002-0112 (2003)
28. Ishida, T., Takahashi, S., Nakahashi, K.: Efficient and Robust Cartesian Mesh Generation for Building-Cube Method. J. of Computational Science and Technology 2, 435–446 (2008)
29. Hirschel, E.H., Weiland, C.: Issues of Multidisciplinary Design. In: Hirschel, E.H., Krause, E. (eds.) 100 Volumes of 'Notes on Numerical Fluid Mechanics'. NNFM, vol. 100, pp. 255–270. Springer, Heidelberg (2009)
30. N.N.: Three-Dimensional Boundary Layers. Rep. AGARD FDP Round Table Discussion, Brussels, Belgium, May 24, 1984. AGARD-R-719 (1985)

31. Cousteix, J.: Analyse théorique et moyens de prévision de la couche limite turbulente tridimensionelle. Doctoral thesis, University of Paris VI, Paris, France (1974). Also: Theoretical Analysis and Prediction Methods for a Three-Dimensional Turbulent Boundary-Layer. ESA TT-238 (1976)
32. Krause, E., Hirschel, E.H., Bothmann, T.: Die numerische Integration der Bewegungsgleichungen dreidimensionaler laminarer kompressibler Grenzschichten. DGLR-Fachbuchreihe, Band 3, Braunschweig, Germany, 03-1–03-49 (1968)
33. Cebeci, T., Khattab, K., Stewartson, K.: Three-Dimensional Laminar Boundary Layers and the Ok of Accessibility. J. Fluid Mechanics 107, 57–87 (1981)
34. Schwamborn, D.: Boundary Layers on Finite Wings and Related Bodies with Consideration of the Attachment-Line Region. In: Viviand, H. (ed.) Proc. 4th GAMM-Conference on Numerical Methods in Fluid Mechanics, Paris, France, October 7-9, 1981. NNFM, vol. 5, pp. 291–300. Vieweg, Braunschweig Wiesbaden (1982)
35. Kovasznay, L.S.G., Hall, M.G.: Some Impossible Similarity Solutions. AIAA J. 5, 2065–2066 (1967)
36. Yohner, P.L., Hansen, A.G.: Some Numerical Solutions of Similarity Equations for Three-Dimensional Incompressible Boundary-Layer Flows. NACA TN 4370 (1958)

Appendix B

Approximate Relations for Boundary-Layer Properties

Relations for the approximate determination of boundary-layer properties are useful for two reasons. The first one is that they permit to make quickly first assessments of properties of attached viscous flow. The second one is that they allow qualitative checks of flow situations. The prerequisites in our case are that the flow three-dimensionality is only weak, and that the locality principle is not violated. To this may be added that the flow must be transition insensitive.

B.1 Introduction

For approximate determinations and qualitative checks the generalized reference-temperature/enthalpy formulations of G. Simeonides are optimally suited [1]. Although they are intended originally for hypersonic flow problems, they can be used also in the whole flow parameter domain considered in this book. The formulations are valid for attached laminar and turbulent flow and they permit—due to the use of the reference-temperature/enthalpy concept—to take into account wall-temperature and Mach number effects.

We introduce in the next section Simeonides' generalized formulations in terms of the reference-temperature concept. In the following sections the relations for flat surface portions, stagnation points, and attachment lines at swept cylinders are given. In the last section the determination of the virtual origin of boundary layers at junctions is presented. The location of the virtual origin is required, if laminar-turbulent transition is to be regarded, and likewise, if the flow path leads from one to another well defined configuration part. This situation is given, for instance at a deflected aerodynamic trim or control surface, and also at engine inlet ramps.

B.2 The Reference-Temperature Concept

The reference-temperature concept permits the wall temperature and Mach-number effects to be accounted for approximately and in a simple way to enable the determination of boundary layer properties [2]. When used as the reference-enthalpy concept, it enables high enthalpy flows to be treated

[3]. In our case the reference-temperature concept is employed, because the temperature range is assumed not to exceed 1,500 K, Section 3.1.

The reference-temperature/enthalpy concept is discussed in some detail in [4]. It is not an exact but a well-proven approximate concept. Basically it works with boundary layer relations established for incompressible flow. These are applied with the inviscid flow data at the body surface, which are interpreted as being those at the boundary layer edge. Density and viscosity then are interpreted as function of an appropriate reference temperature. In [5], for instance, the viability of the approach is demonstrated.

The characteristic Reynolds number for a boundary-layer like attached viscous flow at the location x is postulated to read

$$Re_x^* = \frac{\rho^* v_e x}{\mu^*}. \tag{B.1}$$

Density ρ^* and viscosity μ^* are reference data, characteristic of the boundary layer. They are determined with the local pressure p and the reference temperature T^*, with v_e being the external inviscid flow velocity.

The reference temperature T^* is empirically composed of the boundary layer edge temperature T_e, the wall temperature T_w, and the recovery temperature T_r [2]

$$T^* = 0.28 T_e + 0.5 T_w + 0.22 T_r. \tag{B.2}$$

The recovery or adiabatic wall temperature T_r is defined by[1]

$$T_r = T_e + r^* \frac{v_e^2}{2c_p} = T_e(1 + r^* \frac{\gamma - 1}{2} M_e^2). \tag{B.3}$$

Here r^* is the recovery factor, which is a function of the Prandtl number Pr at the reference temperature T^*. The Prandtl number depends rather weakly on the temperature, Sub-Section 3.1.2. Usually it is sufficient to assume $r^* = r = \text{const}$. For laminar flow the recovery factor can be taken as $r = \sqrt{Pr}$, and for turbulent flow $r = \sqrt[3]{Pr}$. With the Prandtl number at low temperatures, $Pr \approx 0.74$, we get $r_{lam} \approx 0.86$ and $r_{turb} \approx 0.90$.

Introducing the boundary layer edge data as reference flow data into eq. (B.1) yields

$$Re_x^* = \frac{\rho_e v_e x}{\mu_e} \frac{\rho^*}{\rho_e} \frac{\mu_e}{\mu^*} = Re_{e,x} \frac{\rho^*}{\rho_e} \frac{\mu_e}{\mu^*}, \tag{B.4}$$

with $Re_{e,x} = \rho_e v_e x / \mu_e$. This relation can be simplified. If we apply it to boundary layer like flows, we can write, because $p \approx p_e \approx p_w$

$$\frac{\rho^*}{\rho_e} = \frac{T_e}{T^*}. \tag{B.5}$$

[1] The total temperature T_t is found with $r^* = 1$.

If, for simplicity, we further introduce the power-law expression for the viscosity given in Sub-Section 3.1.2, we obtain

$$\frac{\mu^*}{\mu_e} = \frac{(T^*)^{\omega^*}}{(T_e)^{\omega_e}}. \tag{B.6}$$

Only if T^* and T_e are both in the same temperature interval, ω^* and ω_e are equal and we get:

$$\frac{\mu^*}{\mu_e} = \left(\frac{T^*}{T_e}\right)^{\omega}. \tag{B.7}$$

Introducing eqs. (B.5) and (B.7) into eq. (B.4) reduces the latter to

$$Re_x^* = Re_{e,x}\left(\frac{T_e}{T^*}\right)^{1+\omega}. \tag{B.8}$$

B.3 Generalized Reference-Temperature Relations

In the following sub-sections we give approximate boundary-layer relations in generalized form [1]. The relations are valid for both laminar and turbulent flow in a Reynolds number range up to 10^7. We assume that the temperature does not exceed 1,500 K. Although derived originally for hypersonic flow problems, the relations can be used also for low speed problems.[2]

In the following relations we have for laminar flow the exponent $n = 0.5$ and for turbulent flow $n = 0.2$. If the power-law expressions for the viscosity are used, and T^* and T_e are both in the same temperature interval (see above), the exponents of the viscosity laws ω^* and ω_e are equal, and a given relation can be further reduced. For a number of the relations we give also the reduced form.

B.3.1 Flat Plate

We consider a sufficient flat surface portion as a flat plate. If the surface portion is inclined against the free stream or if it is the surface of a deflected control surface or an inlet ramp, the more general relations from [1] must be applied. The Blasius boundary-layer relation [6] and the 1/7-power law turbulent boundary layer-relation [7], respectively, are the basis for the following formulations. The 1/7-power law is detested by some authors. It permits, however, for engineering purposes simple and fast estimates of properties of turbulent boundary layers.

[2] Of course it makes sense, as with all approximate relations, to check the range of applicability and to establish the error range. This holds in particular if the relations are of empirical or semi-empirical character.

Boundary-Layer Thicknesses. The boundary-layer thicknesses of the Blasius and the 1/7-power law—incompressible—boundary layers can be written in generalized form:

$$\delta_i = C_i \frac{x^{1-n}}{(Re^u_\infty)^n}, \tag{B.9}$$

with the unit Reynolds number being $Re^u_\infty = \rho_\infty v_\infty/\mu_\infty$, and $i = 0$ for the boundary-layer thickness δ (where we leave away the lower index 0), $i = 1$ for the displacement thickness δ_1, $i = 2$ for the momentum thickness δ_2.

The constants C_i for the different thicknesses and the exponent n, both for laminar and turbulent flow are given in Table B.1.

Table B.1. Constants C_i in eq. (B.9). The exponent in that equation is $n = 0.5$ for laminar flow and $n = 0.2$ for turbulent flow. Also given is the shape factor $H_{12} = \delta_1/\delta_2$.

	δ	δ_1	δ_2	H_{12}
laminar BL: C_i =	5	1.721	0.664	2.591
turbulent BL: C_i =	0.37	0.046	0.036	1.278

Redefining the original Re^u_∞ with the relation eq. (B.4) as Re^{u*}_∞ and introducing the reference temperature, eq. (B.9) becomes

$$\delta_i = C_i \frac{x^{1-n}}{(Re^u_\infty)^n} \left(\frac{\rho^* \mu_\infty}{\rho_\infty \mu^*}\right)^{-n}, \tag{B.10}$$

respectively

$$\delta_i = C_i \frac{x^{1-n}}{(Re^u_\infty)^n} \left(\frac{T^*}{T_\infty}\right)^{n(1+\omega)}. \tag{B.11}$$

Alternative formulations are for:

– the **thickness of the laminar boundary layer** [8]

$$\delta_{lam} = \delta_{lam,ic} \sqrt{\frac{T_\infty}{T^*} \frac{\mu^\star}{\mu_\infty}} \left(1 + 0.442 \frac{\gamma - 1}{2} M^2_\infty + 0.386 \frac{T_w - T_r}{T_\infty}\right), \tag{B.12}$$

with that one for the incompressible flow being

$$\delta_{lam,ic} = 5 \frac{x^{0.5}}{\sqrt{Re^u_\infty}}, \tag{B.13}$$

– the **displacement thickness** δ_1 of **laminar flow** [8]

$$\delta_{1,lam} = \delta_{1,lam,ic} \sqrt{\frac{T_\infty}{T^*} \frac{\mu^\star}{\mu_\infty}} \left(1 + 1.284 \frac{\gamma - 1}{2} M_\infty^2 + 1.121 \frac{T_w - T_r}{T_\infty}\right), \tag{B.14}$$

with that for the incompressible flow being

$$\delta_{1,lam,ic} = 1.721 \frac{x^{0.5}}{\sqrt{Re_\infty^u}}, \tag{B.15}$$

– the **displacement thickness** δ_1 of **turbulent flow** [8]

$$\delta_{1,turb} = \delta_{1,turb,ic} \left(\frac{T_\infty}{T^*}\right)^{0.8} \left(\frac{\mu^*}{\mu_\infty}\right)^{0.2} \left(1 + 0.286 M_\infty^2 + 0.871 \frac{T_w - T_r}{T_\infty}\right), \tag{B.16}$$

with that for turbulent incompressible flow with a constant different from that given in Table B.1

$$\delta_{1,turb,ic} = 0.0504 \frac{x}{(Re_{e,x})^{0.2}}. \tag{B.17}$$

– and for the **momentum thickness** δ_2 [1]

$$\delta_2 = C_2 \frac{x^{1-n}}{1-n} \left(\frac{\rho^* \mu^*}{\rho_\infty \mu_\infty}\right)^n \left(\frac{\rho^*}{\rho_\infty}\right)^{1-2n} \left(\frac{1}{Re_\infty^u}\right)^n, \tag{B.18}$$

with $C_2 = 0.332$ for laminar flow and $C_2 = 0.0296$ for turbulent flow.

The **characteristic thicknesses of laminar and turbulent boundary layers** can be used to explain quite a number of phenomena in laminar and turbulent attached viscous flow,[3] see [4, 9], and Sub-Section 2.3.3, page 43 of the present book. They also must be taken into account in grid generation for numerical methods.

The characteristic thicknesses govern the wall shear stress and the heat flux in the gas at the wall of attached viscous flow of boundary layer type. In the laminar flow domain, the characteristic thickness Δ_{lam} is approximately the boundary layer thickness δ_{lam}, the 99 per cent thickness. In the turbulent domain Δ_{turb} is the **thickness of the viscous sub-layer** δ_{vs}, not the thickness δ_{turb}. δ_{vs} is much smaller than the boundary layer thickness δ_{turb} as indicated schematically in Fig. B.1. However, at and in the vicinity of singular points and lines the characteristic thickness can not be approximated by δ_{lam} or δ_{vs}.

[3] The characteristic thicknesses can also be used for the scaling of, for instance, the thermal state of the surface from a smaller to a larger high-speed flight vehicle [4].

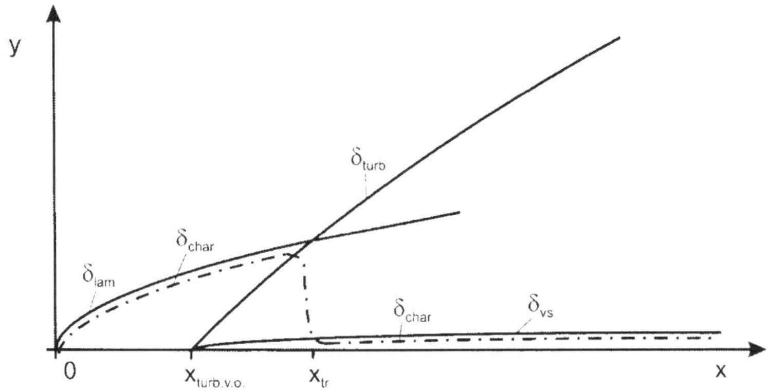

Fig. B.1. Schematic of the characteristic thickness δ_{char} in a boundary layer [9]. The location of the virtual origin of the turbulent boundary layer, Sub-Section B.4, is denoted with $x_{turb,\,v.o.}$, the transition location with x_{tr}.

The thickness δ_{lam} is that given above. The thickness δ_{vs} usually is expressed in terms of the non-dimensional wall distance y^+ [6, 10]:

$$y^+ = \frac{y \rho u_\tau}{\mu}, \tag{B.19}$$

with the friction velocity u_τ being

$$u_\tau = \sqrt{\frac{\tau_w}{\rho_w}}. \tag{B.20}$$

With $y^+ \approx 5$ defining the thickness of the viscous sub-layer [6], accurate flow computations usually require that the first grid line away from the surface is located at $y^+ \lessapprox 1$. Because the velocity profile is linear in that regime, it is sufficient to put two to three grid lines below $y^+ \approx 5$.

The above holds, if the transport equations of turbulence are integrated down to the body surface, which is the case with the so-called low-Reynolds number formulations. If, however, a law-of-the-wall formulation is used, the first grid line can be located at $y^+ \approx 50$–100 and the distance down to the wall is bridged with the law-of-the-wall.

The explicit relation for the thickness of the viscous sub-layer usually is not given in the boundary-layer literature. Exceptions are for instance the book by E.R.G. Eckert and R.M. Drake [11] and the report by G. Simeonides [12]. The relation of the latter reads for the flat plate:[4]

$$\delta_{vs} = 33.78 \frac{x^{0.2}}{(Re^u_\infty)^{0.8}} \left(\frac{\rho_\infty}{\rho^*} \frac{\mu^*}{\mu_\infty} \right)^{0.8}, \tag{B.21}$$

[4] Note the different dependencies of δ_{vs} and δ_{turb} on x and Re^u_∞: $\delta_{vs} \propto x^n (Re^u_\infty)^{1-n}$ compared to $\delta_{turb} \propto x^{1-n} (Re^u_\infty)^{-n}$ (eq.(B.9)), where $n = 0.2$. The authors of [11] give $n = 0.1$ for the δ_{vs}-relation.

respectively:

$$\delta_{vs} = 33.78 \frac{x^{0.2}}{(Re^u_\infty)^{0.8}} \left(\frac{T^*}{T_\infty} \right)^{0.8(1+\omega)} . \tag{B.22}$$

Wall Shear Stress and Heat Flux. For the wall shear stress over a flat plate we get in generalized form, with $C = 0.332$ for laminar flow and $C = 0.0296$ for turbulent flow

$$\tau_w = C \mu_\infty v_\infty x^{-n} \left(\frac{T_\infty}{T^*} \right)^{1-n} \left(\frac{\mu^*}{\mu_\infty} \right)^n (Re^u_\infty)^{1-n}, \tag{B.23}$$

respectively

$$\tau_w = C \mu_\infty v_\infty x^{-n} \left(\frac{T^*}{T_\infty} \right)^{n(1+\omega)-1} (Re^u_\infty)^{1-n}. \tag{B.24}$$

Eq. (B.23) can also be written as non-dimensional coefficient c_f:

$$\frac{\tau_w}{0.5 \rho_\infty v_\infty^2} = c_f = 2\, C\, x^{-n} \left(\frac{T_\infty}{T^*} \right)^{1-n} \left(\frac{\mu^*}{\mu_\infty} \right)^n (Re^u_\infty)^{-n}. \tag{B.25}$$

The heat flux in the gas at the wall reads, again with $C = 0.332$ for laminar flow and $C = 0.0296$ for turbulent flow

$$q_{gw} = C x^{-n} k_\infty Pr^{1/3} (T_r - T_w) \left(\frac{T_\infty}{T^*} \right)^{1-n} \left(\frac{\mu^*}{\mu_\infty} \right)^n (Re^u_\infty)^{1-n}, \tag{B.26}$$

respectively

$$q_{gw} = C x^{-n} k_\infty Pr^{1/3} (T_r - T_w) \left(\frac{T^*}{T_\infty} \right)^{n(1+\omega)-1} (Re^u_\infty)^{1-n}. \tag{B.27}$$

B.3.2 Stagnation Point

In [1] no explicit relation for the **thickness of the boundary layer** δ at a stagnation point is given. We take the formula of F. Homann, eq. (8.12) for axisymmetric stagnation-point flow[5]

$$\delta|_{x=0} = 2.8 \sqrt{\frac{\mu_e}{\rho_e} \frac{1}{\frac{du_e}{dx}|_{x=0}}}. \tag{B.28}$$

[5] The reader may compare this with other relations given in the literature, e.g., [13].

The velocity gradient can be expressed as

$$\frac{du_e}{dx}|_{x=0} = \frac{k}{R_N}\sqrt{\frac{2(p_s - p_\infty)}{\rho_s}}, \tag{B.29}$$

see Chapter 8, Problem 8.1, where R_N is the nose radius, with $k = 1.5$ for the sphere. The subscript 's' denotes the stagnation point. This relation holds for incompressible flow, but can be used to a degree also for compressible flow, see also [4].

With this expression for the velocity gradient the boundary-layer thickness at the stagnation point becomes:

$$\delta|_{x=0} = 2.8\frac{\sqrt{R_N}}{\sqrt{\frac{k\rho_e}{\mu_e}\left[\frac{2(p_s-p_\infty)}{\rho_s}\right]^{0.5}}}. \tag{B.30}$$

The data ρ_e, μ_e are interpreted as reference data ρ_{ref}, μ_{ref} which can be replaced by reference-temperature data ρ^*, μ^* in order to obtain a dependence on the wall temperature. We find finally, after replacing the subscript 's' by 'e' for the 'edge value' in the stagnation point:

$$\delta|_{x=0} = 2.8\frac{\sqrt{R_N}}{\sqrt{\frac{k\rho_e}{\mu_e}\left[\frac{2(p_e-p_\infty)}{\rho_e}\right]^{0.5}}}\left(\frac{T^*}{T_e}\right)^{0.5(1+\omega)}. \tag{B.31}$$

This relation shows not only that δ increases with increasing T^*, but also with increasing radius R.

For the **heat flux in the gas at the wall** q_{gw} we find at the sphere, respectively the circular cylinder (2-D case), for perfect gas with the generalized reference-temperature formulation:

$$q_{gw} = k_\infty Pr^{1/3} g_{sp}\frac{1}{R_N}T_r(1 - \frac{T_w}{T_r}), \tag{B.32}$$

where:

$$g_{sp} = C\left(\frac{p_e}{p_\infty}\right)^{0.5}\left(\frac{T^*}{T_\infty}\right)^{0.5(\omega-1)}\left(\frac{R_N}{u_\infty}\frac{du_e}{dx}|_{x=0}\right)^{0.5}(Re_{\infty,R})^{0.5}, \tag{B.33}$$

with $C = 0.763$ for the sphere and $C = 0.57$ for the circular cylinder.

B.3.3 Attachment-Line at a Swept Cylinder

In [1] no explicit relation for the **thickness of the boundary layer** δ for this is given. We consider the **wall shear stress** τ_w along the attachment line of an infinite swept circular cylinder, '$_{scy}$', assuming perfect-gas flow. This wall shear stress is constant in the direction of the attachment line:

$$\tau_{w,scy} = \rho_\infty u_\infty^2 f_{scy}, \tag{B.34}$$

with

$$f_{scy} = \\ = C\left(\frac{w_e}{u_\infty}\right)^{2(1-n)}\left(\frac{\rho^*}{\rho_\infty}\right)^{1-n}\left(\frac{\mu^*}{\mu_\infty}\right)^n\left(\frac{R}{u_\infty}\frac{du_e}{dx}|_{x=0}\right)^n\frac{1}{(Re_{\infty,R})^n}. \tag{B.35}$$

R is the radius of the cylinder. The gradient of the inviscid external velocity normal to the attachment line $du_e/dx|_{x=0}$ reads [4]:

$$\frac{du_e}{dx}|_{x=0} = \cos(\varphi)\frac{1.33}{R}\sqrt{\frac{2(p_s - p_\infty)}{\rho_s}}, \tag{B.36}$$

where φ is the sweep angle of the cylinder and $w_e = u_\infty \sin\varphi$ the inviscid external velocity along the attachment line.

For laminar flow $C = 0.57$, $n = 0.5$. As reference-temperature values are recommended, [1], $\rho^* = \rho_e^{0.8}\rho_w^{0.2}$ and $\mu^* = \mu_e^{0.8}\mu_w^{0.2}$.

For turbulent flow $C = 0.0345$, $n = 0.21$. The reference-temperature values in this case, following a proposal from [14] to put more weight on the recovery temperature and less on the wall temperature, are taken at:

$$T^* = 0.30T_e + 0.10T_w + 0.60T_r. \tag{B.37}$$

The density at reference temperature conditions, ρ^*, is found with T^* and the external pressure p_e.

An alternative formulation is:

$$\tau_{w,scy} = \frac{\mu_\infty u_\infty}{R} f_{scy}^*, \tag{B.38}$$

with

$$f_{scy}^* = \\ = C\left(\frac{w_e}{u_\infty}\right)^{2(1-n)}\left(\frac{p_e}{p_\infty}\right)^{1-n}\left(\frac{T^*}{T_\infty}\right)^{n(1+\omega)-1}\left(\frac{R}{u_\infty}\frac{du_e}{dx}|_{x=0}\right)^n\frac{1}{(Re_{\infty,R})^n}. \tag{B.39}$$

The **heat flux in the gas at the wall** q_{gw} reads:

$$q_{gw,scy} = Pr^{\frac{1}{3}} k_\infty g_{scy}^* \frac{1}{R} T_r \left(1 - \frac{T_w}{T_r}\right), \tag{B.40}$$

with

$$g^*_{scy} = \\ = C \left(\frac{w_e}{u_\infty}\right)^{1-2n} \left(\frac{p_e}{p_\infty}\right)^{1-n} \left(\frac{T^*}{T_\infty}\right)^{n(1+\omega)-1} \left(\frac{R}{u_\infty}\frac{du_e}{dx}|_{x=0}\right)^n (Re_{\infty,R})^{1-n}. \tag{B.41}$$

For laminar flow $C = 0.57$, $n = 0.5$ and $\rho^* = \rho_e^{0.8}\rho_w^{0.2}$, $\mu^* = \mu_e^{0.8}\mu_w^{0.2}$. For turbulent flow $C = 0.0345$, $n = 0.21$. The reference-temperature values are found with eq. (B.37), and ρ^* again with T^* and the external pressure p_e.

B.4 Virtual Origin of Boundary Layers at Junctions

Despite the general character of the above discussed reference-temperature relations, they cannot be applied directly, if different boundary-layer growth rates are present over different surface portions. Simeonides lists five cases, where the boundary layer passes from one generic aerodynamic surface to another one at a junction [1]. These are:

1. Laminar state of flow ⇒ turbulent state of flow (laminar-turbulent transition).
2. Flat plate ⇒ ramp or ramp ⇒ ramp.
3. Blunt nose or leading edge ⇒ conical, cylindrical or plane afterbody.
4. Conical fore body ⇒ cylindrical afterbody.
5. Cylindrical or conical forebody ⇒ flare.

We first treat case 1. Fig. B.2 shows the x_1, y_1-coordinate system in which the laminar boundary layer is defined, as well as the x_2, y_2-coordinate system in which the turbulent boundary layer is defined. The latter is the x_1, y_1-system shifted downstream to the location of the virtual origin $x_{2,v.o.}$ of the turbulent boundary layer.

The junction in this case is the location x_{tr}, where the laminar-turbulent transition is assumed to happen instantly.[6] This location at the same time is denoted $x_{1,j}$ and $x_{2,j}$. The Reynolds number does not change at the transition location, however, the flow properties change due to the transition process.

It is evident that at x_{tr} simply a continuation of the flow is not possible. In [15] therefore a matching procedure is proposed, which essentially leads to a turbulent boundary layer with a virtual origin different from that of the laminar boundary layer.

[6] In reality instant transition never happens, Section 9.2.1. In the frame of the present approximative relations, however, the assumption of instant transition is permitted, as it is usually done in applied aerodynamics.

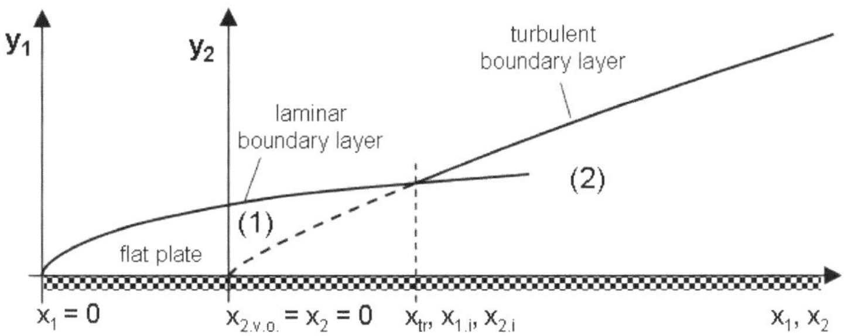

Fig. B.2. Illustration of the virtual origin of a boundary layer at a junction demonstrated by means of flat-plate laminar-turbulent transition.

Proposed is the matching of the momentum deficit of the two boundary layers on both sides of the junction $x_{1,j} = x_{2,j}$

$$(\rho_e u_e^2 \delta_2)|_2 = (\rho_e u_e^2 \delta_2)|_1, \tag{B.42}$$

with δ_2 being the momentum-loss thickness, see above.

The procedure is the following:

- Determine with the flow parameters of the laminar flow (1) the momentum thickness $\delta_2|_1$, eq. (B.18), at the junction $x_1 = x_{1,j}$.
- Determine $\delta_2|_2$ from eq. (B.42) with the turbulent flow parameters (2). For the flat-surface transition case $\rho_e u_e^2|_2 = \rho_e u_e^2|_1$ and hence $\delta_2|_2 = \delta_2|_1$.
- Find the virtual junction coordinate $x_{2,j}$ of the turbulent boundary layer with the inverted eq. (B.18). The effective turbulent-flow coordinate x_2 is then in terms of x_1 and in view of the turbulent flow properties: $x_2 = x_1 + (x_{2,j} - x_{1,j})$. The virtual origin $x_{2,v.o.}$ of the turbulent boundary layer in terms of x_1 lies at $x_1 = x_{1,j} - x_{2,j}$.

In case 2 the situation is different. A flat plate/ramp flow is the idealized flow past a deflected aerodynamic trim or control surface, or past a ramp inlet with one or more ramps.

At the junction of the flat plate and the ramp the local unit Reynolds number Re_e^u changes. Actually it increases, and hence at the ramp the boundary layer becomes thinner with an increase of both the wall shear stress and the wall heat transfer.

The unit Reynolds-number change, Fig. B.3, however, is well defined only if the (inviscid) flat-plate flow is supersonic and if the ramp angle δ, Fig. B.4, is such that it remains supersonic on the ramp.[7]

[7] It is neglected that in this case in an usually small domain around the junction flat plate/ramp a strong interaction between the boundary layer and the shock wave occurs [9].

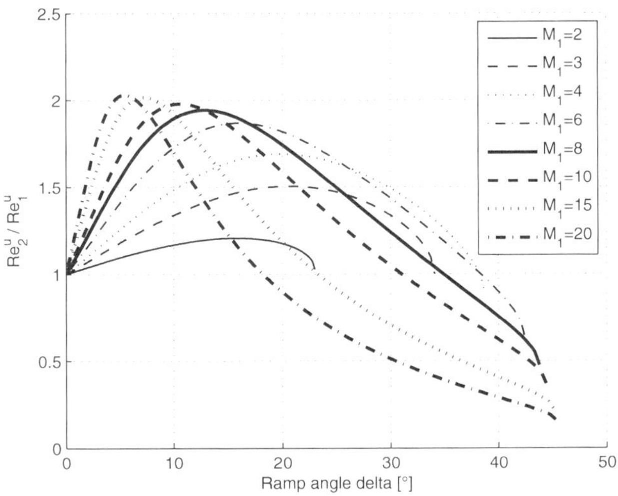

Fig. B.3. Ratio of unit Reynolds numbers across the shock wave of a flat-plate('1')/ramp('2') configuration as function of the ramp angle δ for different flat-plate Mach numbers M_1 and the ratio of the specific heats $\gamma = 1.4$ [9]. The viscosity μ in all cases was computed with the power-law relation, eq. (3.7), hence the figure holds for temperatures $T_1, T_2 \gtrapprox 200$ K.

The unit Reynolds number Re^u_e then always rises across the shock wave of the flat-plate/ramp configuration, except for high flat-plate Mach numbers M_1 at large ramp angles δ.

If the (inviscid) flat-plate flow is subsonic, the establishment of the unit Reynolds-number change unfortunately is problematic. It can only be done, if plateaus of the flow properties upstream and downstream of the junction can, at least approximately, be defined.

The general case is depicted in Fig. B.4. At the junction of the two planar surfaces a jump or rise of the flow parameters, and especially of the unit Reynolds number, happens.

Again eq. (B.42) is employed, but now we have $\rho_e u_e^2|_2 \neq \rho_e u_e^2|_1$. The following steps are the same as in case 1. The effective ramp coordinate x_2 is then in terms of x_1 and the ramp angle η: $x_2 = x_{2,j} + (x_1 - x_{1,j})/\cos\eta$. The virtual origin $x_{2,v.o.}$ of the ramp boundary layer in terms of x_1 lies at $x_1 = x_{1,j} - x_{2,j}/\cos\eta$.

The remaining cases are treated in an analogous way [1].

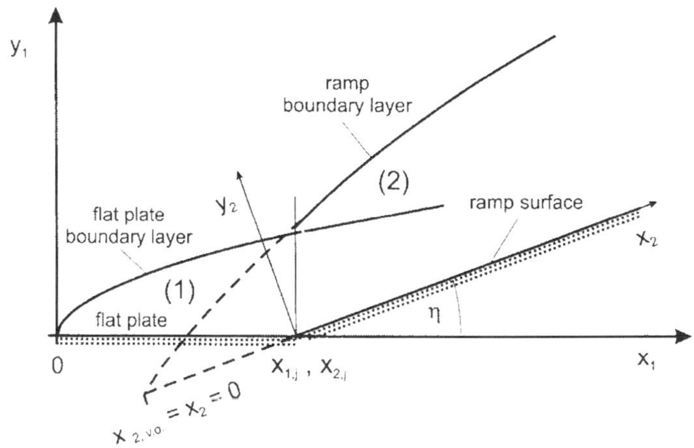

Fig. B.4. Illustration of the virtual origin of a boundary layer at a junction demonstrated by means of a flat plate/ramp configuration [9].

References

1. Simeonides, G.: Generalized Reference-Enthalpy Formulation and Simulation of Viscous Effects in Hypersonic Flow. Shock Waves 8(3), 161–172 (1998)
2. Rubesin, M.W., Johnson, H.A.: A Critical Review of Skin Friction and Heat Transfer Solutions of the Laminar Boundary Layer of a Flat Plate. Trans. ASME 71, 385–388 (1949)
3. Eckert, E.R.G.: Engineering Relations of Friction and Heat Transfer to Surfaces in High-Velocity Flow. J. Aeronautical Sciences 22(8), 585–587 (1955)
4. Hirschel, E.H.: Basics of Aerothermodynamics. Progress in Astronautics and Aeronautics, AIAA, Reston, Va, vol. 204. Springer, Heidelberg (2004)
5. Simeonides, G., Walpot, L.M.G., Netterfield, M., Tumino, G.: Evaluation of Engineering Heat Transfer Prediction Methods in High Enthalpy Flow Conditions. AIAA-Paper 96-1860 (1996)
6. Schlichting, H., Gersten, K.: Boundary Layer Theory, 8th edn. Springer, Heidelberg (2000)
7. Cebeci, T., Cousteix, J.: Modeling and Computation of Boundary-Layer Flows, 2nd edn. Horizons Publ., Springer, Long Beach, Heidelberg (2005)
8. Simeonides, G.: Hypersonic Shock Wave Boundary Layer Interactions over Compression Corners. Doctoral Thesis, University of Bristol, U.K. (1992)
9. Hirschel, E.H., Weiland, C.: Selected Aerothermodynamic Design Problems of Hypersonic Flight Vehicles. Progress in Astronautics and Aeronautics, AIAA, Reston, Va, vol. 229. Springer, Heidelberg (2009)
10. Smits, A.J., Dussauge, J.-P.: Turbulent Shear Layers in Supersonic Flow, 2nd edn. AIP/Springer, New York (2004)
11. Eckert, E.R.G., Drake, R.M.: Heat and Mass Transfer. MacGraw-Hill, New York (1950)
12. Simeonides, G.: On the Scaling of Wall Temperature Viscous Effects. ESA/ESTEC EWP - 1880 (1996)

13. Reshotko, E.: Heat Transfer to a General Three-Dimensional Stagnation Point. Jet Propulsion 28, 58–60 (1958)
14. Poll, D.I.A.: Transition Description and Prediction in Three-Dimensional Flow. AGARD-R-709, 5-1–5-23 (1984)
15. Hayes, J.R., Neumann, R.D.: Introduction to the Aerodynamic Heating Analysis of Supersonic Missiles. In: Mendenhall, M.R. (ed.) Tactical Missile Aerodynamics: Prediction Methodology. Progress in Astronautics and Aeronautics, AIAA, Reston, Va, pp. 63–110 (1992)

Appendix C

Boundary-Layer Coordinates: Metric Properties, Transformations, Examples

First-order boundary-layer equations in surface-oriented locally monoclinic non-orthogonal curvilinear coordinates, also in contravariant formulation, are given in Appendix A.2. The metric factors appearing in the equations are combinations of the elements of the metric tensor of the surface coordinates and partly also of their derivatives, Chapter 8.

We give in this appendix the application-relevant relations which are needed to compute metric properties and the metric tensor, transformations and the like. General basics, derivations and proofs can be found in, e.g., [1]–[4], and for our topic in particular in [5]. The provided relations permit an easy and consistent treatment of the geometrical problems which are connected to surface-oriented non-orthogonal curvilinear coordinates.

We write the relations in index notation, e.g., the boundary-layer coordinates coordinates as (x^1, x^2, x^3). It should be no problem for the reader to change then to the (x, y, z)-coordinates or to the (x, z, y)-coordinates which we employed for the derivation of the boundary-layer equations.

Two examples are discussed in some detail: (1) a fuselage cross-section coordinate system as canonical example of an airliner fuselage, (2) a wing percent-line coordinate system as canonical example of a finite-span swept wing. The metric properties of surface coordinates of axisymmetric bodies, flat wings, unswept wings, infinite swept wings can be derived from them, see also the examples in [5]. Other examples can be found in [6]–[8].

C.1 Metric Properties of Surface Coordinates

First-order boundary-layer theory employs coordinates on the body surface only, Section 2.2. We consider a surface element which is embedded in the Cartesian reference coordinate system, the $x^{1'}, x^{2'}, x^{3'}$-system, Fig. C.1. The unit base vectors of the latter system are $\underline{e}_{1'}$, $\underline{e}_{2'}$, and $\underline{e}_{3'}$.

The boundary-layer coordinate system x^1, x^2, x^3 is defined on the surface. The lines x^2 = const. (the x^1-coordinates) and x^1 = const. (the x^2-coordinates) lie on the surface at $x^3 = 0$. The x^3-coordinate is rectilinear and normal to both, and therefore locally normal to the surface. That is why we call surface-oriented non-orthogonal curvilinear boundary-layer coordinates also "surface-oriented locally monoclinic coordinates".

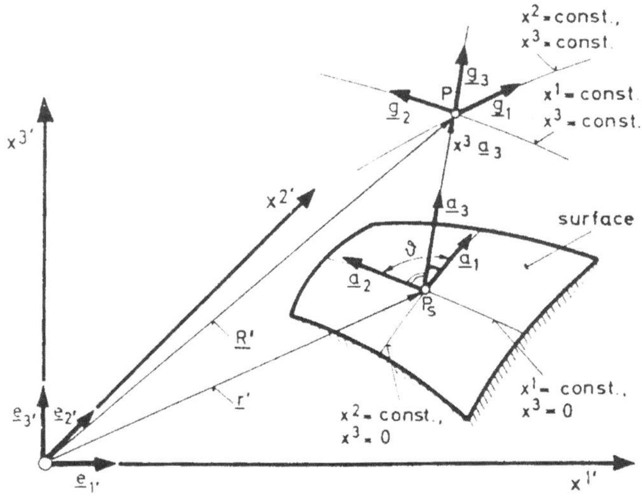

Fig. C.1. Surface element in general coordinates [6].

Both x^1 and x^2—called Gaussian surface parameters—have no length properties in general. Both parameters are not necessarily counted along the coordinate lines.

The coordinate base indicated in Fig. C.1 is called a covariant base. The covariant base vectors $\underline{a}_1$ and $\underline{a}_2$ belonging to the x^1, x^2, x^3-coordinates are defined by

$$\underline{a}_1 = \beta_1^{1'} \underline{e}_{1'} + \beta_1^{2'} \underline{e}_{2'} + \beta_1^{3'} \underline{e}_{3'} \tag{C.1}$$

and

$$\underline{a}_2 = \beta_2^{1'} \underline{e}_{1'} + \beta_2^{2'} \underline{e}_{2'} + \beta_2^{3'} \underline{e}_{3'}, \tag{C.2}$$

where the Cartesian components of the base vectors $\underline{a}_1$ and $\underline{a}_2$

$$\beta_1^{1'} = \frac{\partial x^{1'}}{\partial x^1}, \; \beta_1^{2'} = \frac{\partial x^{2'}}{\partial x^1}, \; ... \; \beta_2^{2'} = \frac{\partial x^{2'}}{\partial x^2}, \; \beta_2^{3'} = \frac{\partial x^{3'}}{\partial x^2} \tag{C.3}$$

are the derivatives of the contour functions

$$x^{1'} = x^{1'}(x^1, x^2), \;\; x^{2'} = x^{2'}(x^1, x^2), \;\; x^{3'} = x^3(x^1, x^2), \tag{C.4}$$

which define the x^1, x^2-coordinates on the surface of the configuration under consideration.

The third base vector $\underline{a}_3$ is a unit vector which points in x^3-direction:

$$\underline{a}_3 = \frac{\underline{a}_1 \times \underline{a}_2}{|\underline{a}_1 \times \underline{a}_2|} = \beta_3^{1'} \underline{e}_{1'} + \beta_3^{2'} \underline{e}_{2'} + \beta_3^{3'} \underline{e}_{3'}. \tag{C.5}$$

The components of the base vector $\underline{a}_3$ are

$$\beta_3^{1'} = \frac{\triangle_1}{\triangle},\ \beta_3^{2'} = \frac{\triangle_2}{\triangle},\ \beta_3^{3'} = \frac{\triangle_3}{\triangle} \tag{C.6}$$

with

$$\triangle_1 = \beta_1^{2'}\beta_2^{3'} - \beta_1^{3'}\beta_2^{2'}, \tag{C.7}$$

$$\triangle_2 = \beta_2^{1'}\beta_1^{3'} - \beta_2^{3'}\beta_1^{1'}, \tag{C.8}$$

$$\triangle_3 = \beta_1^{1'}\beta_2^{2'} - \beta_1^{2'}\beta_2^{1'}, \tag{C.9}$$

$$\triangle = [(\triangle_1)^2 + (\triangle_2)^2 + (\triangle_3)^2]^{1/2} = \sqrt{a}, \tag{C.10}$$

where a is the determinant of the metric tenor, see below.

Once the components of the covariant base vectors are known, every geometrical aspect of the problem at hand can be described. The difficulty lies in the definition of the contour functions. Regarding these functions two examples are given in Sections C.3 and C.4.

The components—the metric coefficients—of the (symmetric) metric tensor of the surface coordinates $a_{\alpha\beta} = a_\alpha \cdot a_\beta$:

$$\underline{\underline{a}} = \begin{pmatrix} a_{11} & a_{12} \\ a_{21} & a_{22} \end{pmatrix} \tag{C.11}$$

are defined as follows

$$\begin{aligned} a_{11} &= (\beta_1^{1'})^2 + (\beta_1^{2'})^2 + (\beta_1^{3'})^2, \\ a_{12} &= \beta_1^{1'}\beta_2^{1'} + \beta_1^{2'}\beta_2^{2'} + \beta_1^{3'}\beta_2^{3'} = a_{21}, \\ a_{22} &= (\beta_2^{1'})^2 + (\beta_2^{2'})^2 + (\beta_2^{3'})^2. \end{aligned} \tag{C.12}$$

The often used Lamé coefficients h_1, h_2, and g are related to the metric coefficients in the following way:

$$h_1 = \sqrt{a_{11}},\ \ h_2 = \sqrt{a_{22}},\ \ g = a_{12}. \tag{C.13}$$

The determinant of the metric tensor of the surface reads

$$a = a_{11}a_{22} - (a_{12})^2. \tag{C.14}$$

The angle ϑ between the coordinate lines x^1 = const. and x^2 = const., Fig. C.1, is found from the scalar product of the base vectors $\underline{a}_1 \cdot \underline{a}_2 = |\underline{a}_1|\,|\underline{a}_2| \cos\vartheta$:

$$\cos\vartheta = \frac{a_{12}}{\sqrt{a_{11}}\sqrt{a_{22}}}. \tag{C.15}$$

For orthogonal coordinates $a_{12} = 0$ and we get $\cos\vartheta = \pi/2$ as expected.

The length elements are, with the asterisk denoting dimensional quantities

$$dx^1 = \frac{dx^{*1}}{\sqrt{a_{11}}},\ dx^2 = \frac{dx^{*2}}{\sqrt{a_{22}}},\ dx^3 = dx^{*3}. \tag{C.16}$$

The coordinates x^1 and x^2 have no length properties, they are Gaussian surface parameters. The general metric—the length element ds—reads [1]:

$$(ds)^2 = a_{11}(dx^1)^2 + 2a_{12}dx^1dx^2 + a_{22}(dx^2)^2 + (dx^3)^2, \tag{C.17}$$

which on the body surface reduces to:

$$ds^2 = a_{11}(dx^1)^2 + a_{22}(dx^2)^2 + 2\,a_{12}\,dx^1\,dx^2. \tag{C.18}$$

The surface element dA and the volume element dV are defined by:

$$dA = \sqrt{a}\,dx^1dx^2, \tag{C.19}$$

$$dV = \sqrt{a}\,dx^1dx^2dx^3. \tag{C.20}$$

For second-order boundary-layer theory the curvature of the surface is taken into account. For the determination of the covariant curvature tensor, the principle surface curvatures and their directions, the metric tensor of the coordinate system off the surface, etc. see, e.g., [5, 9].

C.2 Transformations

The velocity components of the external inviscid flow usually are given in the reference coordinate system and need to be transformed into the boundary-layer system. On the other hand, the wall shear-stress components are to be transformed back into the reference coordinate system. The—rather simple—transformation laws, if contravariant velocity components are used, are given in the following, see also [5, 8].

We begin with the transformation of a vector from the boundary-layer coordinates into the Cartesian reference coordinates. A vector $\underline{F}$ with its components F^1, F^2, F^3 is defined in the covariant base a_i (i = 1,2,3) by

$$\underline{F} = F^1\underline{a}_1 + F^2\underline{a}_2 + F^3\underline{a}_3. \tag{C.21}$$

In the frame of first-order boundary-layer theory only the tangential components of the external inviscid flow vector and, anyway, the wall-shear stress vector are concerned, Fig. C.2.

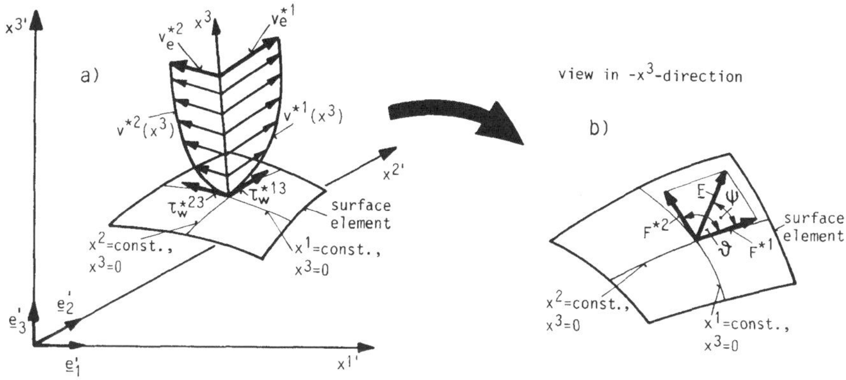

Fig. C.2. Transformation of vectors [8]. a) vector components to be transformed, b) view in negative x^3-direction toward the surface.

The components F^i are contravariant components belonging to the covariant base. They are related to the physical components F^{*i} by

$$F^1 = \frac{F^{*1}}{\sqrt{a_{11}}},\ F^2 = \frac{F^{*2}}{\sqrt{a_{22}}},\ F^3 = F^{*3}. \tag{C.22}$$

The fundamental transformation is from the x^i-system into the Cartesian reference $x^{j'}$-system with:

$$\begin{aligned} F^{1'} &= \beta_1^{1'} F^1 + \beta_2^{1'} F^2 + \beta_3^{1'} F^3, \\ F^{2'} &= \beta_1^{2'} F^1 + \beta_2^{2'} F^2 + \beta_3^{2'} F^3, \\ F^{3'} &= \beta_1^{3'} F^1 + \beta_2^{3'} F^2 + \beta_3^{3'} F^3. \end{aligned} \tag{C.23}$$

The inverse transformation from the $x^{j'}$-system into the x^i-system reads:

$$\begin{aligned} F^1 &= \beta_{1'}^1 F^{1'} + \beta_{2'}^1 F^{2'} + \beta_{3'}^1 F^{3'}, \\ F^2 &= \beta_{1'}^2 F^{1'} + \beta_{2'}^2 F^{2'} + \beta_{3'}^2 F^{3'}, \\ F^3 &= \beta_{1'}^3 F^{1'} + \beta_{2'}^3 F^{2'} + \beta_{3'}^3 F^{3'}. \end{aligned} \tag{C.24}$$

The inverse transformation matrix $\beta_{l'}^k$, the Jacobian, is given in terms of the components of the base vectors $\beta_j^{i'}$ of the surface coordinate system, eq. (C.5), with a being the determinant of the metric tensor, eq. (C.14):

$$\beta^k_{l'} = \begin{pmatrix} \beta^1_{1'} & \beta^1_{2'} & \beta^1_{3'} \\ \beta^2_{1'} & \beta^2_{2'} & \beta^2_{3'} \\ \beta^3_{1'} & \beta^3_{2'} & \beta^3_{3'} \end{pmatrix} =$$
$$= \frac{1}{\sqrt{a}} \begin{pmatrix} (\beta^{2'}_2 \beta^{3'}_3 - \beta^{3'}_2 \beta^{2'}_3) & (\beta^{3'}_2 \beta^{1'}_3 - \beta^{1'}_2 \beta^{3'}_3) & (\beta^{1'}_2 \beta^{2'}_3 - \beta^{2'}_2 \beta^{1'}_3) \\ (\beta^{3'}_1 \beta^{2'}_3 - \beta^{2'}_1 \beta^{3'}_3) & (\beta^{1'}_1 \beta^{3'}_3 - \beta^{3'}_1 \beta^{1'}_3) & (\beta^{2'}_1 \beta^{1'}_3 - \beta^{1'}_1 \beta^{2'}_3) \\ (\beta^{2'}_1 \beta^{3'}_2 - \beta^{3'}_1 \beta^{2'}_2) & (\beta^{3'}_1 \beta^{1'}_2 - \beta^{1'}_1 \beta^{3'}_2) & (\beta^{1'}_1 \beta^{2'}_2 - \beta^{2'}_1 \beta^{1'}_2) \end{pmatrix} . \tag{C.25}$$

We consider as an example the transformation of the vector $\underline{v}'_e$ of the external inviscid flow into the flow vector $\underline{v}_e$ in the boundary-layer coordinates:

$$\begin{aligned} v^1_e &= \beta^1_{1'} v^{1'}_e + \beta^1_{2'} v^{2'}_e + \beta^1_{3'} v^{3'}_e , \\ v^2_e &= \beta^2_{1'} v^{1'}_e + \beta^2_{2'} v^{2'}_e + \beta^2_{3'} v^{3'}_e , \\ v^3_e &= \beta^3_{1'} v^{1'}_e + \beta^3_{2'} v^{2'}_e + \beta^3_{3'} v^{3'}_e . \end{aligned} \tag{C.26}$$

The component v^3_e of course should be zero, but it should be computed in order to check the accuracy of both the inviscid flow data, and the whole geometrical representation of the configuration under consideration.

The physical velocity components, if needed, are:

$$v^{*1}_e = v^1_e \sqrt{a_{11}}, \; v^{*2}_e = v^2_e \sqrt{a_{22}}, \; v^{*3}_e = v^3_e . \tag{C.27}$$

The magnitude of the velocity vector is:

$$\begin{aligned} |\underline{v}_e| &= \left[(\sqrt{a_{11}} v^1_e)^2 + 2 a_{12} v^1_e v^2_e + (\sqrt{a_{22}} v^2_e)^2 \right]^{1/2} = \\ &= \left[(v^{*1}_e)^2 + 2 \cos\vartheta \; v^{*1}_e v^{*2}_e + (v^{*2}_e)^2 \right]^{1/2} . \end{aligned} \tag{C.28}$$

The angle ψ, Fig. C.2, reads

$$tan\psi_e = \frac{\sqrt{a} \; v^2_e}{a_{11} \, v^1_e + a_{12} \, v^2_e} = \frac{\sin\vartheta \; v^{*2}_e}{v^{*1}_e + \cos\vartheta \; v^{*2}_e} . \tag{C.29}$$

C.3 Example 1: Fuselage Cross-Section Coordinate System

Consider the surface coordinates of the body with a quite general shape in Fig. C.3. The body is defined in the $x^{1'}, x^{2'}, x^{3'}$-Cartesian reference coordinate system. The two poles are linked by the $x^{1'}$-coordinate axis. This is not mandatory.

Because most fuselages are defined by cross sections, a cross-section coordinate system is the natural choice, Fig. C.4. Such a system can be employed if the forward stagnation point lies close to the nose point (sufficiently slender fuselage, small angle of attack), see also Section 8.2.

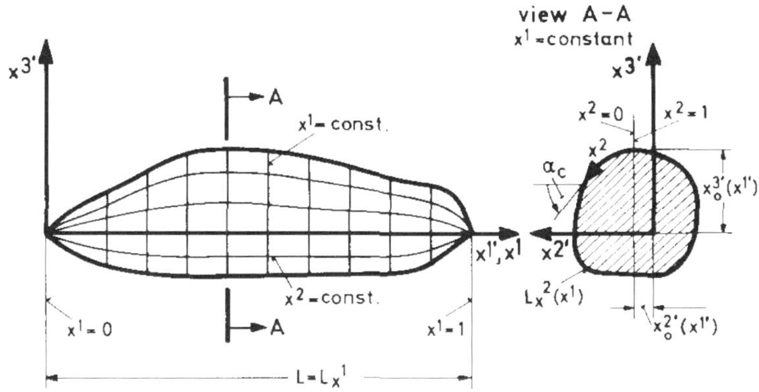

Fig. C.3. Surface coordinates of a general fuselage coordinate system: general conventions [6].

Hence the x^2-coordinate lines on the body—lines where x^1 = const.—are defined by transverse frame cuts (cross-sections parallel to the $x^{2'}, x^{3'}$-plane). They link points of equal fraction of the circumferential length of the cross sections. The circumferential arc length of each transverse frame L_{x^2} is the normalizing length for x^2. The x^1-coordinate is measured along the fuselage axis. Therefore the body length is chosen to be the normalizing length for x^1: $L_{x^1} = L$.

The coordinates are normalized such that:

$$\begin{aligned} x^1 &= 0 \text{ at the front pole,} \\ x^1 &= 1 \text{ at the aft pole,} \\ x^2 &= 0 \text{ at the cross-section apex,} \\ x^2 &= 1 \text{ at the cross-section apex.} \end{aligned} \tag{C.30}$$

We obtain, with $\alpha_c(x^1, x^2)$ being the contour angle, the general contour functions eq. (C.4):

$$\begin{aligned} x^{1'} &= L_{x^1}\, x^1 = L\, x^1, \\ x^{2'} &= x_0^{2'}(x^1) + L_{x^2}(x^1) \int_0^{x^2} \cos\alpha_c\,(x^1, \xi^2)\, d\xi^2, \\ x^{3'} &= x_0^{3'}(x^1) - L_{x^2}(x^1) \int_0^{x^2} \sin\alpha_c\,(x^1, \xi^2)\, d\xi^2. \end{aligned} \tag{C.31}$$

For the canonical fuselage coordinates shown in Fig. C.4, of course we have everywhere $x_0^{2'}(x^1) = 0$.

We obtain in this case directly from the foregoing equations the components of the base vectors $\underline{a}_1$ and $\underline{a}_2$, eq. (C.3):

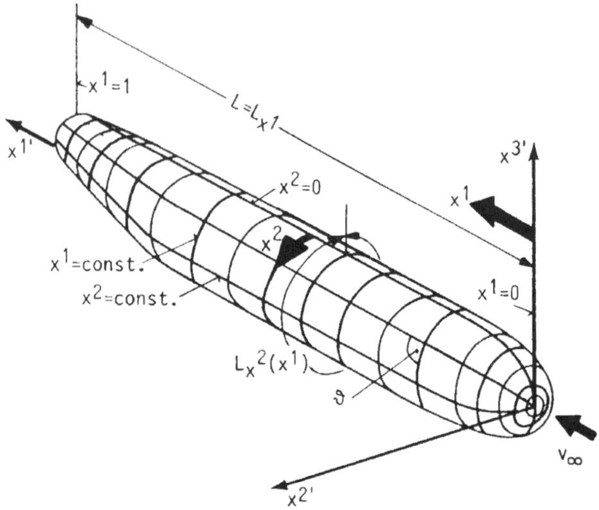

Fig. C.4. Surface coordinates of the canonical fuselage coordinate system: airplane fuselage [8].

$$
\begin{aligned}
\beta_1^{1'} &= L_{x^1} = L,\\
\beta_1^{2'} &= \frac{d}{d\,x^1}\left[L_{x^2}(x^1)\right]\int_0^{x^2}\cos\alpha_c(x^1,\xi^2)\,d\xi^2+\\
&\quad + L_{x^2}(x^1)\int_0^{x^2}\frac{\partial}{\partial\,x^1}\left[\cos\alpha_c\,(x^1,\xi^2)\right]\,d\xi^2,\\
\beta_1^{3'} &= \frac{d}{d\,x^1}\left[x_0^{3'}(x^1)\right]-\frac{d}{d\,x^1}\left[L_{x^2}(x^1)\right]\int_0^{x^2}\sin\alpha_c\,(x^1,\xi^2)\,d\xi^2- \qquad (C.32)\\
&\quad - L_{x^2}(x^1)\int_0^{x^2}\frac{\partial}{\partial\,x^1}\left[\sin\alpha_c\,(x^1,\xi^2)\right]\,d\xi^2,\\
\beta_2^{1'} &= 0,\\
\beta_2^{2'} &= L_{x^2}(x^1)\cos\alpha_c\,(x^1,x^2),\\
\beta_2^{3'} &= -\,L_{x^2}(x^1)\sin\alpha_c\,(x^1,x^2).
\end{aligned}
$$

The components of the base vector $\underline{a}_3$ can be found from these components as shown in Section C.1. If the cross sections are given point-wise and $x^{2'}(x^2)$ and $x^{3'}(x^2)$ are given as spline approximations, some of the base-vector components can be found directly from the spline approximations.

In Fig. C.5 we show the surface-coordinate parameter map of the coordinate system. The fuselage nose and the base are represented by singular lines because all $x^2 =$ const. lines fall together there.

Parameter maps of this kind permit detailed visualizations for the study of flow-field properties. In this book examples in surface-parameter maps of

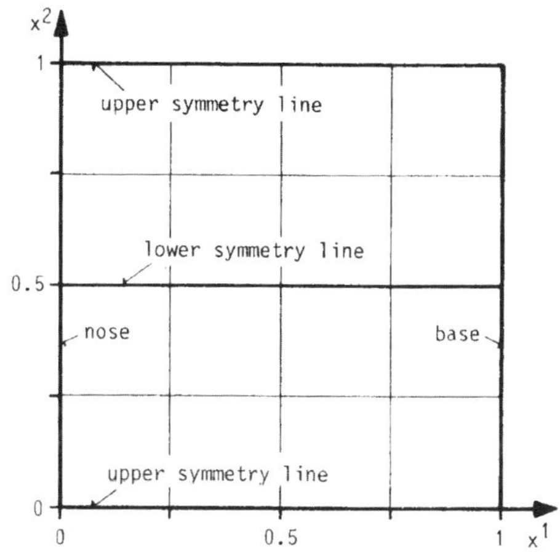

Fig. C.5. Surface-coordinate parameter map of the canonical fuselage coordinate system [8].

skin-friction lines are given in Section 10.5 on page 283, of external inviscid streamlines and skin-friction lines in Section 7.4.3 on page 167 and in the same section of separation lines etc. on page 170.

We derive as an example the surface metric tensor of an ellipsoid. The x^1-coordinate is normalized with $L_{x^1} = L$. The normalizing length for x^2 in each cross section is $L_{x^2} = \pi\, r(x^1)$, because due to the symmetry of the body only one half of it is considered. Thus $0 \leqq x^1 \leqq 1$ and $0 \leqq x^2 \leqq 1$.

The local radius r is, with the diameter $D = 2\, r_{max}$:

$$r(x^1) = D\left[x^1(1 - x^1)\right]^{1/2} \tag{C.33}$$

and its x^1-derivative

$$\frac{dr(x^1)}{dx^1} = \frac{D}{2}\,\frac{1 - 2x^1}{\sqrt{x^1(1 - x^1)}}. \tag{C.34}$$

With $\alpha_c = \pi\, x^2$ being the contour angle we obtain as contour functions instead of eq. (C.31) directly:

$$\begin{aligned} x^{1'} &= L_{x^1}\, x^1 = L\, x^1, \\ x^{2'} &= r(x^1)\, sin(\pi\, x^2), \\ x^{3'} &= r(x^1)\, cos(\pi\, x^2). \end{aligned} \tag{C.35}$$

From this follows the surface metric tensor of the ellipsoid:

$$\underline{\underline{a}} = \begin{pmatrix} L^2 + (\frac{dr(x^1)}{dx^1})^2 & 0 \\ 0 & (L_{x^2}(x^1))^2 \end{pmatrix}. \tag{C.36}$$

C.4 Example 2: Wing Percent-Line Coordinate System

Consider the wing shown in Fig. C.6. The leading edge and the trailing edge are smooth. If a break would be present, for instance at the trailing edge, as is common today for large airplanes, see the configuration considered in Section 10.3, either it would have to be smoothed out, or two separate domains would have to be defined. In the span-wise direction the chord sections of the wing can be quite arbitrary. The wing may be, for instance, twisted.

The coordinate system in the figure is a per-cent line system. It is found by cuts of constant span (x^2 = const. coordinate lines) and and lines of constant chord measured on the wing surface (x^1 = const. coordinate lines). The half-span is denoted s. The x^2-axis is measured along the $x^{2'}$-axis of the Cartesian reference system and not along the x^1 = const. lines. The latter of course is also possible.

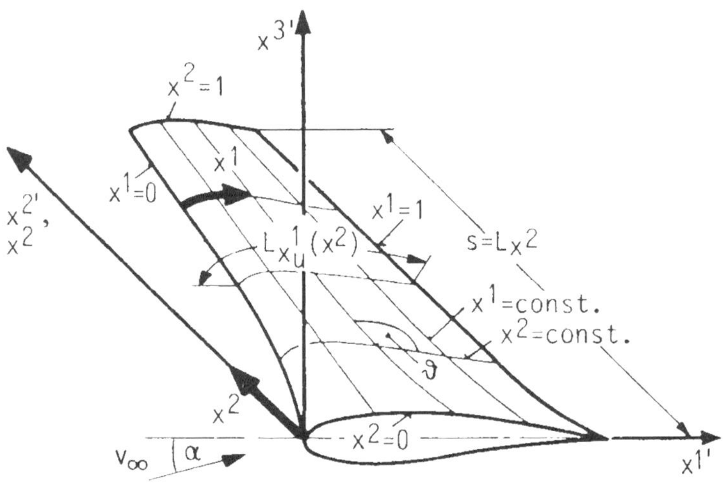

Fig. C.6. Surface coordinates of the canonical wing coordinate system [8].

The coordinates are normalized such that:

$$
\begin{aligned}
x^1 &= 0 \text{ at the leading edge,}\\
x^1 &= 1 \text{ at the trailing edge,}\\
x^2 &= 0 \text{ at the wing root,}\\
x^2 &= 1 \text{ at the wing tip.}
\end{aligned}
\qquad (C.37)
$$

It is assumed now that the chord (airfoil) sections of the wing are given in planes parallel to the $x^{1'}$-$x^{3'}$-plane. In each chord section $x^{2'} =$ const. the coordinates $x^{1'}_{LE}$, $x^{3'}_{LE}$ of the leading edge are known and furthermore the arc lengths $L_{x^{1'}u}$ on the upper side and $L_{x^{3'}l}$ on the lower side. They are measured from the leading edge ($x^1_u = 0$, $x^1_l = 0$) to the trailing edge ($x^1_u = 1$, $x^1_l = 1$).

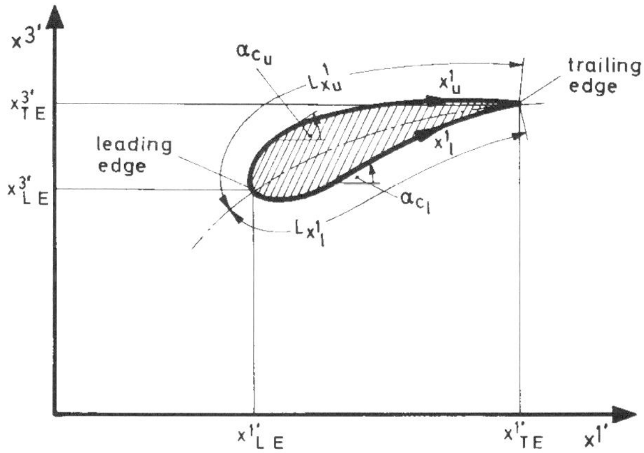

Fig. C.7. Chord section of the wing at $x^2 = x^{2'} =$ const. [8]. The contour angle α_c is measured in planes $x^{2'} =$ constant.

The arc lengths are the local normalizing lengths in x^1-direction for the upper and the lower side of the wing. By introducing the contour angle α_c, the relations eq. (C.4) are found (we show them for the upper side only):

$$
\begin{aligned}
x^{1'}_u &= x^{1'}_{LE}(0, x^2) + L_{x^1_u}(x^2) \int_0^{x^1} \cos\alpha_{c_u}(\xi^1, x^2)\, d\xi^1, \\
x^{2'}_u &= L_{x^2}\, x^2 = s\, x^2, \\
x^{3'}_u &= x^{3'}_{LE}(0, x^2) + L_{x^1_u}(x^2) \int_0^{x^1} \sin\alpha_{c_u}(\xi^1, x^2)\, d\xi^1.
\end{aligned}
\tag{C.38}
$$

We obtain from these equations the components of the base vectors $\underline{a}_1$ and $\underline{a}_2$, eq. (C.3):

$$
\begin{aligned}
\beta_1^{1'} &= L_{x_u^1}(x^2) \cos \alpha_{c_u}(x^1, x^2), \\
\beta_1^{2'} &= 0, \\
\beta_1^{3'} &= L_{x_u^1}(x^2) \sin \alpha_{c_u}(x^1, x^2), \\
\beta_2^{1'} &= \frac{d}{dx^2}\left[x_{LE}^{1'}(0, x^2)\right] + \frac{d}{dx^2}[L_{x_u^1}(x^2)] \int_0^{x^1} \cos \alpha_{c_u}(\xi^1, x^2)\, d\xi^1 + \\
&\quad + L_{x_u^1}(x^2) \int_0^{x^1} \frac{\partial}{\partial x^2}\left[\cos \alpha_{c_u}(\xi^1, x^2)\right] d\xi^1, \qquad \text{(C.39)} \\
\beta_2^{2'} &= L_{x^2} = s, \\
\beta_2^{3'} &= \frac{d}{dx^2}\left[x_{LE}^{3'}(0, x^2)\right] + \frac{d}{dx^2}\left[L_{x_u^1}(x^2)\right] \int_0^{x^1} \sin \alpha_{c_u}(\xi^1, x^2)\, d\xi^1 + \\
&\quad + L_{x_u^1}(x^2) \int_0^{x^1} \frac{\partial}{\partial x^2}\left[\sin \alpha_{c_u}(\xi^1, x^2)\right] d\xi^1.
\end{aligned}
$$

The components of the base vector $\underline{a}_3$ can be found from these components as shown in Section C.1. If the chord sections are given point-wise and $x^{1'}(x^1)$ and $x^{3'}(x^1)$ are given as spline approximations, some of the base-vector components can be found directly from the spline approximations. Then the contour angle $\alpha_{c_u}(x^1, x^2)$ needs not to be known.

In Fig. C.8 we show the surface-coordinate parameter map of the coordinate system of the wing.

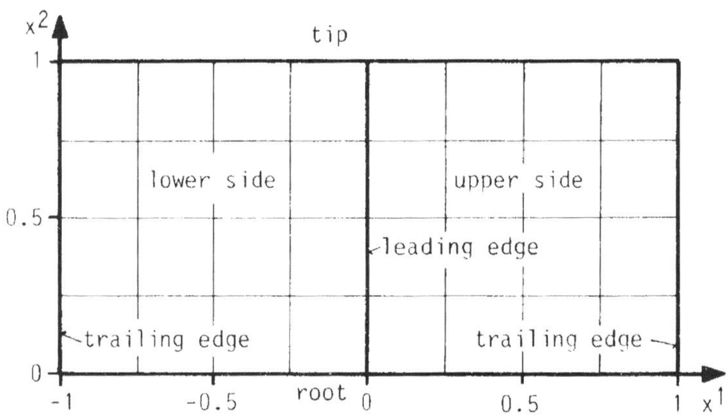

Fig. C.8. Surface-coordinate parameter map of the canonical wing coordinate system [8].

As an example the surface metric tensor of the infinite swept wing with the leading-edge oriented non-orthogonal curvilinear coordinates shown in Fig. 8.4 of Section 8.3 is derived. In this case x^2 is measured along the wing's leading edge. All derivatives in that direction are identically zero, all variables are functions of x^1 only. We normalize the lengths such that $L_{x^1} = L_{x^2} = 1$. With the sweep angle φ_0 we have

$$x^{1'}_{LE} = \sin\varphi_0\, x^2,\ x^{2'}_{LE} = \cos\varphi_0\, x^2 \tag{C.40}$$

and obtain the contour functions for the upper side of the wing, choosing $x^{3'}_{LE} = 0$

$$\begin{aligned} x^{1'}_u &= \sin\varphi_0\, x^2 + \int_0^{x^1} \cos\alpha_{c_u}(\xi^1)\, d\xi^1, \\ x^{2'}_u &= \cos\varphi_0\, x^2, \\ x^{3'}_u &= \int_0^{x^1} \sin\alpha_{c_u}(\xi^1)\, d\xi^1. \end{aligned} \tag{C.41}$$

From this follows the surface metric tensor of the infinite swept wing with leading-edge oriented non-orthogonal curvilinear coordinates:

$$\underline{\underline{a}} = \begin{pmatrix} 1 & \sin\varphi_0\, \cos\alpha_{c_u}(x^1) \\ \sin\varphi_0\, \cos\alpha_{c_u}(x^1) & 1 \end{pmatrix}. \tag{C.42}$$

For the infinite swept wing with orthogonal curvilinear coordinates the metric tensor reduces to:

$$\underline{\underline{a}} = \begin{pmatrix} 1 & 0 \\ 0 & 1 \end{pmatrix}. \tag{C.43}$$

Note that in this case the surface properties, i.e. α_{c_u}, no longer appear in the metric tensor. This leads to the question, whether in computations of the boundary-layer of wings the latter in some cases can be considered as flat from the beginning.[1]

In [5] a hierarchy of governing equations is discussed and the equations with an a-metric with $\alpha_{c_u}(x^1, x^2) = 0$ are called zero-order equations. For bodies of course they are not suited, but for wings of all kinds. There, however, they have a meaning only for non-orthogonal coordinates.

In [10, 11] wing cases were studied in conjunction with boundary-layer stability and transition investigations. We show in particular a result obtained for the ONERA wing M6, lower part of Fig. C.9, where experimental data were available [12]. The wing has the symmetrical airfoil ONERA "D". The measurements were made with a Reynolds number $Re_{\bar{t}} = 3.5 \cdot 10^6$, with $\bar{t}$ being the mean chord length $\bar{t} = 0.63$ m.

Because the wing has a taper ratio $\lambda = 0.56$, the locally-infinite-swept-wing approximation, Section 8.4, was employed. The considered span-wise location was $x^2 = 0.45$, lower part of Fig. C.9.

1 Cases where this certainly should not be done are boundary-layer stability studies.

The experimental results show the existence of a separation bubble for angles of attack $\alpha \gtrapprox 5°$. In the upper part of Fig. C.9 we see a remarkably good agreement between the measured and computed separation location at $\alpha = 5°$ and $10°$ if the contour angle α_c is taken into account (first-order theory). Zero-order theory with $\alpha_c \equiv 0$ yields locations somewhat downstream of these locations. For $\alpha = 15°$ separation of larger extent exists already near the leading edge, resulting in a wrong prediction of the inviscid flow field (panel method) which is seen from the difference of the predicted and measured location of the attachment line.

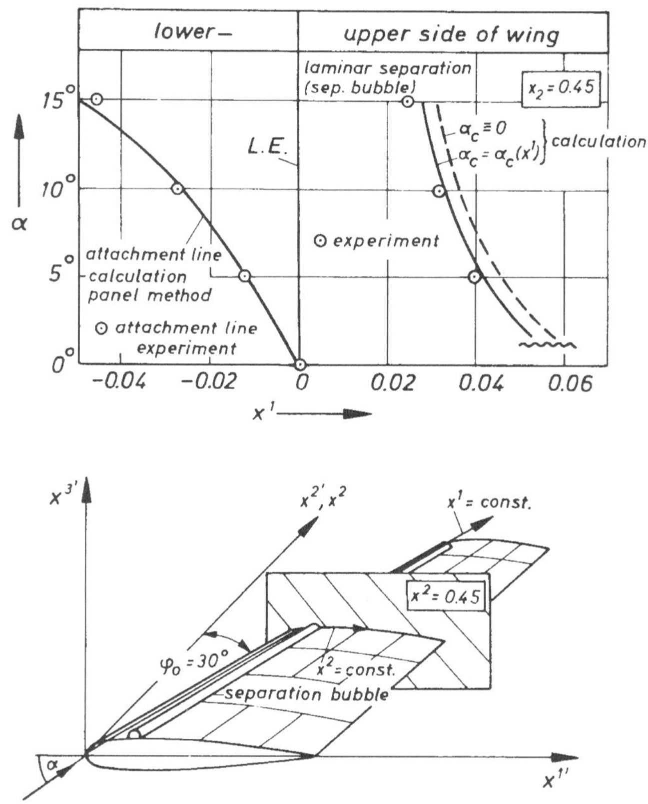

Fig. C.9. Comparison of computed, [11], and and measured, [12], locations of incompressible laminar separation (separation bubble downstream of the leading-edge) on a swept tapered wing [5].

The results—although to be seen with some reservations, because in the experiment the flow was slightly compressible ($v_\infty = 90$ m/s), unstable at $x^1 = 0.02$ and the separation bubble was three-dimensional—show that zero-order boundary-layer solutions are a viable tool to determine viscous effects in attached three-dimensional wing flow without the effort needed to take into account the contour angle. This holds also if the flow is turbulent.

The results also explain why handbook methods or the generalized reference-temperature equations—presented in Appendix B.3—applied in main-stream direction yield reasonably accurate estimations of, for instance, skin friction and boundary-layer thicknesses on simplified wing and fuselage configurations.

References

1. Aris, R.: Vectors, Tensors, and the Basic Equations of Fluid Mechancis. Prentice Hall, Englewood Cliffs (1962); unabridged Dover republication (1989)
2. Flügge, W.: Tensor Analysis and Continuum Mechanics. Springer, New York (1972)
3. Sokolnikoff, I.S.: Tensor Analysis—Theory and Applications to Geometry and Continuum Mechanics, 2nd edn. John Wiley and Sons, New York (1964)
4. Klingbeil, E.: Tensorrechnung für Ingenieure. 2. Auflage, Hochschultaschenbuch, Band 197, Bibliographisches Institut, Mannheim Wien Zürich (1989)
5. Hirschel, E.H., Kordulla, W.: Shear Flow in Surface-Oriented Coordinates. NNFM, vol. 4. Vieweg, Braunschweig Wiesbaden (1981)
6. Hirschel, E.H.: Boundary-Layer Coordinates on General Wings and Bodies. Zeitschrift für Flugwissenschaften und Weltraumforschung (ZFW) 6, 194–202 (1982)
7. Hirschel, E.H., Bretthauer, N., Röhe, H.: Theoretical and Experimental Boundary-Layer Studies of Car Bodies. Int. J. of Vehicle Design 5, 567–584 (1984)
8. Hirschel, E.H.: Evaluation of Results of Boundary-Layer Calculations with Regard to Design Aerodynamics. AGARD R-741, 5-1–5-29 (1986)
9. Monnoyer, F.: Calculation of Three-Dimensional Viscous Flow on General Configurations Using Second-Order Boundary-Layer Theory. ZFW 14, 95–108 (1990)
10. Hirschel, E.H., Jawtusch, V., Grundmann, R.: Berechnung dreidimensionaler Grenzschichten an Pfeilflügeln. In: Jahrestagung der Deutschen Gesellschaft für Luft- und Raumfahrt, München, Germany, DGLR 76-187, September 14-16 (1976)
11. Hirschel, E.H., Jawtusch, V.: Nachrechnung des experimentell ermittelten Übergangs laminar-turbulent an einem gepfeilten Flügel. In: F. Maurer (ed.): Beiträge zur Gasdynamik und Aerodynamik. DLR-FB 77-36, pp. 179–190 (1977)
12. Schmitt, V., Cousteix, J.: Etude de la couche limite tridimensionelle sur une aile en flèche. ONERA Rapport Technique No. 14/1713 AN (1975)

Appendix D

Constants, Atmosphere Data, Units, and Conversions

In this book, units are in general the SI units (Système International d'Unités), see [1, 2], where also the constants can be found. In the following sections we give first constants and air properties, Section D.1, and then a selection of atmosphere data, Section D.2. The basic units, the derived units, and conversions to US units are given in Section D.3.[1]

D.1 Constants and Air Properties

Molar universal gas constant: $R_0 = 8.314472{\cdot}10^3$ kg m^2 s^{-2} kmol^{-1} K^{-1} =
= $4.97201{\cdot}10^4$ lb$_m$ ft^2 s^{-2} (lb$_m$-mol)$^{-1}$ °R^{-1}

Standard gravitational acceleration of earth at sea level: $g_0 = 9.80665$ m s^{-2} = 32.174 ft s^{-2}

Table D.1. Molecular weights, gas constants, and characteristic vibrational temperatures of air constituents for the low temperature domain [3, 4]. * is the U.S. standard atmosphere value [5], + the value from [4].

Gas	Molecular weight M [kg kmol^{-1}]	Specific gas constant R [m^2 s^{-2} K^{-1}]	Characteristic vibrational temperature Θ_{vibr} [K]
air	28.9644* (28.97+)	287.06	
N_2	28.02	296.73	3,390.0
O_2	32.00	259.83	2,270.0

[1] Details can be found, for instance, at `http://physics.nist.gov/cuu/Reference/contents/html`

D.2 Atmosphere Data

Table D.2. Properties of the 15°C U.S. standard atmosphere as function of the altitude [5].

Altitude H [km]	Temperature T [K]	Pressure p [Pa]	Density ρ [kg m^{-3}]	Dynamic viscosity μ [N s m^{-2}]	Thermal conductivity k [W m^{-1} K^{-1}]
0.0	288.150	$1.013 \cdot 10^5$	$1.225 \cdot 10^0$	$1.789 \cdot 10^{-5}$	$2.536 \cdot 10^{-2}$
1.0	281.651	$8.988 \cdot 10^4$	$1.112 \cdot 10^0$	$1.758 \cdot 10^{-5}$	$2.485 \cdot 10^{-2}$
2.0	275.154	$7.950 \cdot 10^4$	$1.007 \cdot 10^0$	$1.726 \cdot 10^{-5}$	$2.433 \cdot 10^{-2}$
3.0	268.659	$7.012 \cdot 10^4$	$9.092 \cdot 10^{-1}$	$1.694 \cdot 10^{-5}$	$2.381 \cdot 10^{-2}$
4.0	262.166	$6.166 \cdot 10^4$	$8.193 \cdot 10^{-1}$	$1.661 \cdot 10^{-5}$	$2.329 \cdot 10^{-2}$
5.0	255.676	$5.405 \cdot 10^4$	$7.364 \cdot 10^{-1}$	$1.628 \cdot 10^{-5}$	$2.276 \cdot 10^{-2}$
6.0	249.187	$4.722 \cdot 10^4$	$6.601 \cdot 10^{-1}$	$1.595 \cdot 10^{-5}$	$2.224 \cdot 10^{-2}$
7.0	242.700	$4.110 \cdot 10^4$	$5.900 \cdot 10^{-1}$	$1.561 \cdot 10^{-5}$	$2.170 \cdot 10^{-2}$
8.0	236.215	$3.565 \cdot 10^4$	$5.258 \cdot 10^{-1}$	$1.527 \cdot 10^{-5}$	$2.117 \cdot 10^{-2}$
9.0	229.733	$3.080 \cdot 10^4$	$4.671 \cdot 10^{-1}$	$1.493 \cdot 10^{-5}$	$2.063 \cdot 10^{-2}$
10.0	223.252	$2.650 \cdot 10^4$	$4.135 \cdot 10^{-1}$	$1.458 \cdot 10^{-5}$	$2.009 \cdot 10^{-2}$
12.0	216.650	$1.940 \cdot 10^4$	$3.119 \cdot 10^{-1}$	$1.421 \cdot 10^{-5}$	$1.953 \cdot 10^{-2}$
14.0	216.650	$1.417 \cdot 10^4$	$2.279 \cdot 10^{-1}$	$1.421 \cdot 10^{-5}$	$1.953 \cdot 10^{-2}$

D.3 Units and Conversions

Basic and derived SI units are listed of the major flow, transport, and thermal entities. In the left column name and symbol are given and in the right column the unit (dimension), with → the symbol used in Appendix E, and in the line below its conversion.

SI Basic Units

length, L — [m], → [L]
1.0 m $= 100.0$ cm $= 3.28084$ ft
$1{,}000.0$ m $= 1.0$ km

mass, m — [kg], → [M]
1.0 kg $= 2.20462$ lb$_m$

time, t	[s] (= [sec]), $\rightarrow$ [t]
temperature, T	[K], $\rightarrow$ [T] 1.0 K = 1.8 °R $\Rightarrow T_{\mathrm{Kelvin}} = (5/9)\,(T_{\mathrm{Fahrenheit}} + 459.67)$ $\Rightarrow T_{\mathrm{Kelvin}} = T_{\mathrm{Celsius}} + 273.15$
amount of substance, mole	[kmol], $\rightarrow$ [mole] 1.0 kmol = 2.20462 $\mathrm{lb_m}$-mol

SI Derived Units

area, A	$[\mathrm{m^2}]$, $\rightarrow$ $[\mathrm{L^2}]$ $1.0\ \mathrm{m^2} = 10.76391\ \mathrm{ft^2}$
volume, V	$[\mathrm{m^3}]$, $\rightarrow$ $[\mathrm{L^3}]$ $1.0\ \mathrm{m^3} = 35.31467\ \mathrm{ft^3}$
speed, velocity, v	$[\mathrm{m\ s^{-1}}]$, $\rightarrow$ $[\mathrm{L\,t^{-1}}]$ $1.0\ \mathrm{m\ s^{-1}} = 3.28084\ \mathrm{ft\ s^{-1}}$
force, F	$[\mathrm{N}] = [\mathrm{kg\,m\,s^{-2}}]$, $\rightarrow$ $[\mathrm{M\,L\,t^{-2}}]$ $1.0\ \mathrm{N} = 0.224809\ \mathrm{lb_f}$
pressure, p	$[\mathrm{Pa}] = [\mathrm{N\,m^{-2}}]$, $\rightarrow$ $[\mathrm{M\,L^{-1}\,t^{-2}}]$ $1.0\ \mathrm{Pa} = 10^{-5}\ \mathrm{bar} = 9.86923{\cdot}10^{-6}\ \mathrm{atm} =$ $= 0.020885\ \mathrm{lb_f\ ft^{-2}}$
density, ρ	$[\mathrm{kg\,m^{-3}}]$, $\rightarrow$ $[\mathrm{M\,L^{-3}}]$ $1.0\ \mathrm{kg\,m^{-3}} = 0.062428\ \mathrm{lb_m\,ft^{-3}} =$ $= 1.94032{\cdot}10^{-3}\ \mathrm{lb_f\,s^2\,ft^{-4}}$
(dynamic) viscosity, μ	$[\mathrm{Pa\,s}] = [\mathrm{N\,s\,m^{-2}}]$, $\rightarrow$ $[\mathrm{M\,L^{-1}\,t^{-1}}]$ $1.0\ \mathrm{Pa\,s} = 0.020885\ \mathrm{lb_f\,s\,ft^{-2}}$
kinematic viscosity, ν	$[\mathrm{m^2\,s^{-1}}]$, $\rightarrow$ $[\mathrm{L^2\,t^{-1}}]$ $1.0\ \mathrm{m^2\,s^{-1}} = 10.76391\ \mathrm{ft^2\,s^{-1}}$
shear stress, τ	$[\mathrm{Pa}] = [\mathrm{N\,m^{-2}}]$, $\rightarrow$ $[\mathrm{M\,L^{-1}\,t^{-2}}]$ $1.0\ \mathrm{Pa} = 0.020885\ \mathrm{lb_f\,ft^{-2}}$
energy, enthalpy, work, quantity of heat	$[\mathrm{J}] = [\mathrm{N\,m}]$, $\rightarrow$ $[\mathrm{M\,L^2\,t^{-2}}]$ $1.0\ \mathrm{J} = 9.47813{\cdot}10^{-4}\ \mathrm{BTU} =$ $= 23.73036\ \mathrm{lb_m ft^2\,s^{-2}} = 0.737562\ \mathrm{lb_f\,s^{-2}}$

(mass specific) internal energy e, enthalpy h	$[\mathrm{J\,kg^{-1}}] = [\mathrm{m^2\,s^{-2}}]$, $\rightarrow [\mathrm{L^2\,t^{-2}}]$ $1.0\ \mathrm{m^2\,s^{-2}} = 10.76391\ \mathrm{ft^2\,s^{-2}}$
(mass) specific heat, c_v, c_p specific gas constant, R	$[\mathrm{J\,kg^{-1}\,K^{-1}}] = [\mathrm{m^2\,s^{-2}\,K^{-1}}]$, $\rightarrow [\mathrm{L^2\,t^{-2}\,T^{-1}}]$ $1.0\ \mathrm{m^2\,s^{-2}\,K^{-1}} = 5.97995\ \mathrm{ft^2\,s^{-2}\,{}^\circ R^{-1}}$
power, work per unit time	$[\mathrm{W}] = [\mathrm{J\,s^{-1}}] = [\mathrm{N\,m\,s^{-1}}]$, $\rightarrow [\mathrm{M\,L^2\,t^{-3}}]$ $1.0\ \mathrm{W} = 9.47813{\cdot}10^{-4}\ \mathrm{BTU\ s^{-1}} =$ $= 23.73036\ \mathrm{lb_m\,ft^2\,s^{-3}}$
thermal conductivity, k	$[\mathrm{W\,m^{-1}\,K^{-1}}] = [\mathrm{N\,s^{-1}\,K^{-1}}]$, $\rightarrow [\mathrm{M\,L\,t^{-3}\,T^{-1}}]$ $1.0\ \mathrm{W\,m^{-1}\,K^{-1}} =$ $= 1.60496{\cdot}10^{-4}\ \mathrm{BTU\,s^{-1}\,ft^{-1}\,{}^\circ R^{-1}} =$ $= 4.018342\ \mathrm{lb_m\,ft\,s^{-3}\,{}^\circ R^{-1}}$
heat flux, q	$[\mathrm{W\,m^{-2}}] = [\mathrm{J\,m^{-2}\,s^{-1}}]$, $\rightarrow [\mathrm{M\,t^{-3}}]$ $1.0\ \mathrm{W\,m^{-2}} = 0.88055{\cdot}10^{-4}\ \mathrm{BTU\,s^{-1}\,ft^{-2}} =$ $= 2.204623\ \mathrm{lb_m\,s^{-3}}$

References

1. Taylor, B.N. (ed.): The International System of Units (SI). US Dept. of Commerce, National Institute of Standards and Technology, NIST Special Publication 330, US Government Printing Office, Washington, D.C. (2001)
2. Taylor, B.N.: Guide for the Use of the International System of Units (SI). US Dept. of Commerce, National Institute of Standards and Technology, NIST Special Publication 811, US Government Printing Office, Washington, D.C. (1995)
3. Hirschfelder, J.O., Curtiss, C.F., Bird, R.B.: Molecular Theory of Gases and Liquids. John Wiley, New York (1964) (corrected printing)
4. Bird, R.B., Stewart, W.E., Lightfoot, E.N.: Transport Phenomena, 2nd edn. John Wiley & Sons, New York (2002)
5. N.N.: U.S. Standard Atmosphere. Government Printing Office, Washington, D.C. (1976)

Appendix E

Symbols, Abbreviations, and Acronyms

Only the important symbols are listed. If a symbol appears only locally or infrequent, it is not included. In general the page number is indicated, where a symbol is defined or appears first. Dimensions are given in terms of the SI basic units: length [L], time [t], mass [M], temperature [T], and amount of substance [mole], Appendix D. For actual dimensions and their conversions see Appendix D.3.

E.1 Latin Letters

A	area, p. 369, $[\mathrm{L}^2]$
A^1, A^2	dependent variables, p. 182
a	speed of sound, p. 62, $[\mathrm{Lt}^{-1}]$
a	determinant of metric tensor, p. 353, $[\mathrm{L}^4]$
$\underline{a}_i$	covariant base vector, p. 352, [–]
$\underline{\underline{a}}$	metric tensor, p. 353, [–]
$a_{\alpha\beta}(\alpha,\beta=1,2)$	components of metric tensor, p. 353, $[\mathrm{L}^2]$
b	wing span, p. 259, [L]
C_D	drag coefficient, p. 124, [–]
C_L	lift coefficient, p. 124, [–]
c_f	skin-friction coefficient, p. 125, [–]
c_p	(mass) specific heat at constant pressure, p. 52, $[\mathrm{L}^2\mathrm{t}^{-2}\mathrm{T}^{-1}]$
c_p	pressure coefficient, p. 152, [–]
$c_{p_{stag}}$	stagnation pressure coefficient, p. 153, [–]
$c_{p_{vac}}$	vacuum pressure coefficient, p. 261, [–]
c_v	(mass) specific heat at constant volume, p. 55, $[\mathrm{L}^2\mathrm{t}^{-2}\mathrm{T}^{-1}]$
$c_{v_{vibr_{O_2}}}$, $c_{v_{vibr_{N_2}}}$	(mass) specific heats of vibration energy at constant volume, p. 52, $[\mathrm{L}^2\mathrm{t}^{-2}\mathrm{T}^{-1}]$
D	diameter, p. 359, [L]
D	drag, p. 250, $[\mathrm{MLt}^{-2}]$
D_i	induced drag, p. 250, $[\mathrm{MLt}^{-2}]$
dA	differential surface element, p. 354, $[\mathrm{L}^2]$
dV	differential volume element, p. 354, $[\mathrm{L}^3]$
ds	differential length element, p. 354, [L]
$dx^i(i=1,2,3)$	differential length elements, p. 354, [–]

$dx^{*i}(i = 1, 2, 3)$	differential physical length elements, p. 354, [L]
E	Eckert number, p. 65, [–]
$\underline{e}_i(i = 1, 2, 3)$	unit base vectors, p. 351, [–]
F	force, p. 369, $[MLt^{-2}]$
g	gravitational acceleration, p. 367, $[\mathrm{Lt}^{-2}]$
H	altitude, p. 35, [L]
H	shape factor, p. 340, [–]
h	(mass-specific) enthalpy, p. 59, $[\mathrm{L}^2\mathrm{t}^{-2}]$
h_t	total enthalpy, p. 102, $[\mathrm{L}^2\mathrm{t}^{-2}]$
h_1, h_2, g	Lamé coefficients, p. 353, [L]
k	thermal conductivity, p. 55, $[\mathrm{MLt}^{-3}\mathrm{T}^{-1}]$
k_{mn}	metric factor, p. 330
L	length, p. 368, [L]
L	lift, p. 250, $[\mathrm{MLt}^{-2}]$
L/D	lift-to-drag ratio, p. 125, [–]
M	Mach number, p. 62, [–]
M_e	boundary-layer edge Mach number, p. 213, [–]
M_i	molecular weight of species i, p. 367, $[\mathrm{Mmole}^{-1}]$
M_∞	flight Mach number, p. 43, [–]
m	mass, p. 368, [M]
n	exponent in boundary-layer relations, p. 340, [–]
n_{bc}	metric factor, p. 323
Pe	Péclet number, p. 64, [–]
Pr	Prandtl number, p. 64, [–]
p	pressure, p. 51, $[\mathrm{ML}^{-1}\mathrm{t}^{-2}]$
p_∞	free-stream pressure, p. 152, $[\mathrm{ML}^{-1}\mathrm{t}^{-2}]$
q	heat flux, p. 370, $[\mathrm{Mt}^{-3}]$
$\underline{q}$	vector of heat conduction, p. 59, [–]
q_x, q_y, q_z	heat fluxes, p. 59, $[\mathrm{Mt}^{-3}]$
q_{gw}	heat flux in the gas at the wall, p. 60, $[\mathrm{Mt}^{-3}]$
q_w	heat flux in the wall, p. 60, $[\mathrm{Mt}^{-3}]$
q_{rad}	thermal radiation heat flux, p. 60, $[\mathrm{Mt}^{-3}]$
q_y	boundary-layer heat flux, p. 79, $[\mathrm{Mt}^{-3}]$
q_∞	free-stream dynamic pressure, p. 152, $[\mathrm{ML}^{-1}\mathrm{t}^{-2}]$
R	gas constant, p. 52, $[\mathrm{L}^2\mathrm{t}^{-2}\mathrm{T}^{-1}]$
R_0	universal gas constant, p. 367, $[\mathrm{ML}^2\mathrm{t}^{-2}\mathrm{mole}^{-1}\mathrm{T}^{-1}]$
R	radius, p. 154, [L]
R_N	nose radius, p. 344, [L]
Re	Reynolds number, p. 63, [–]
Re^u	unit Reynolds number, p. 340, $[\mathrm{L}^{-1}]$
Re^u_e	boundary-layer edge unit Reynolds number, p. 44, $[\mathrm{L}^{-1}]$
r	radius, p. 359, [L]
r	recovery factor, p. 338, [–]
s	half span of wing, p. 360, [L]
T	temperature, p. 51, [T]

T_e	boundary-layer edge temperature, [T]
T_{gw}	temperature of the gas at the wall, p. 40, [T]
T_r	recovery temperature, p. 338, [T]
T_{ra}	radiation-adiabatic temperature, p. 60, [T]
T_t	total temperature, p. 338, [T]
T_w	wall temperature, p. 40, [T]
T_∞	free-stream temperature, p. 35, [T]
T^*	reference temperature, p. 338, [T]
t	time, p. 369, [t]
t, n, z	external inviscid streamline-oriented coordinates, p. 30
V	magnitude of velocity vector, p. 320, [Lt^{-1}]
$\underline{V}$	velocity vector, p. 320, [–]
u, v, w	Cartesian velocity components, [Lt^{-1}]
u_e, v_e	boundary-layer edge velocity, [Lt^{-1}]
u_∞, v_∞	free-stream velocity, flight speed, [Lt^{-1}]
v_n	cross-flow velocity component, p. 31, [Lt^{-1}]
v_t	stream-wise velocity component, p. 31, [Lt^{-1}]
$v^i (i = 1, 2, 3)$	contravariant velocity components, p. 329, [t^{-1}]
$v^{*i} (i = 1, 2, 3)$	physical velocity components, p. 329, [Lt^{-1}]
$v_e^i (i = 1, 2, 3)$	contravariant external inviscid velocity components, p. 356, [t^{-1}]
$v_e^{*i} (i = 1, 2, 3)$	physical external inviscid velocity components, p. 356, [Lt^{-1}]
x, y, z	Cartesian coordinates, [L]
x, y, z	body axis coordinates, [L]
$x^i (i = 1, 2, 3)$	surface-oriented locally monoclinic non-orthogonal curvilinear coordinates, p. 351
$x^{i'} (i' = 1, 2, 3)$	Cartesian reference coordinates, p. 351

E.2 Greek Letters

α	angle of attack, [°]
α_c	contour angle, p. 357, [°]
β	sideslip (yaw) angle, [°]
Γ	circulation, p. 253, [$\mathrm{L}^2\mathrm{t}^{-1}$]
γ	ratio of specific heats, p. 55, [–]
Δ_c	characteristic boundary layer thickness, p. 341, [L]
δ	flow (ordinary) boundary layer thickness, p. 66, [L]
δ	ramp angle, p. 348, [°]
δ_{lam}	laminar boundary-layer thickness, p. 340, [L]
δ_{turb}	turbulent boundary-layer thickness, p. 340, [L]
δ_{vs}	viscous sub-layer thickness, p. 342, [L]
δ_1	displacement thickness, p. 100, [L]
δ_2	momentum-flow displacement thickness, p. 101, [L]

δ_3	energy-flow displacement thickness, p. 102, [L]
δ^*	displacement thickness ($\equiv \delta_1$), p. 100, [L]
ε	vortex-line angle, p. 252, [°]
θ	momentum thickness ($\equiv \delta_2$), p. 100, [L]
ϑ	coordinate angle, p. 353, [°]
μ	viscosity, p. 53, [$\mathrm{ML^{-1}t^{-1}}$]
μ_e	boundary-layer edge viscosity, [$\mathrm{ML^{-1}t^{-1}}$]
ν	kinematic viscosity, p. 64, [$\mathrm{L^2t^{-1}}$]
ρ	density, p. 51, [$\mathrm{ML^{-3}}$]
ρ_e	boundary-layer edge density, [$\mathrm{ML^{-3}}$]
ρ_∞	free-stream density, [$\mathrm{ML^{-3}}$]
$\underline{\underline{\tau}}$	viscous stress tensor, p. 58, [–]
$\tau_{ij}(i,j=1,2,3)$	components of viscous shear stress tensor, p. 58, [–]
τ_w	wall shear stress, skin friction, p. 45, [$\mathrm{ML^{-1}t^{-2}}$]
τ_x, τ_z	boundary-layer shear stress components, p. 79, [$\mathrm{ML^{-1}t^{-2}}$]
φ	sweep angle of leading edge or cylinder, p. 189, [°]
φ	angle of extremum line, p. 152, [°]
ψ	trailing-edge flow shear angle, p. 253, [°]
ω	exponent in the power-law equations of viscosity and heat conductivity, p. 53, [–]
$\underline{\omega}$	vorticity vector, p. 253, [–]
$\omega_{O_2}, \omega_{N_2}$	mass fractions, p. 52, [–]

E.3 Indices

E.3.1 Upper Indices

T	transposed
u	unit
$*$	reference-temperature/enthalpy value
$+$	dimensionless sub-layer entity

E.3.2 Lower Indices

bl, BL	boundary layer
c	compressible
cr	critical
D	drag
e	boundary-layer edge, external (inviscid flow)
gw	gas at the wall
ic	incompressible
inv	inviscid
ip	inflection point
k	thermal conductivity

L lift
L length
LE leading edge
lam laminar
ra radiation adiabatic
rad radiation
ref reference
$stag$ stagnation
T thermal
TE trailing edge
t total
tr transition
$turb$ turbulent
vac vacuum
$vibr$ vibrational
vs viscous sub-layer
w wall
x,y,z Cartesian coordinates
μ viscosity
0 leading edge
∞ infinity

E.4 Other Symbols

$O(\,)$ order of magnitude
$\overline{q}$ time-integrated value of q
$\underline{v}$ vector
$\underline{\underline{t}}$ tensor
$\propto$ proportional to
$\widehat{=}$ corresponds to
$\succ$ succeeds
$< >$ average

E.5 Abbreviations, Acronyms

AGARD Advisory Group for Aerospace Research & Development
AIAA American Institute of Aeronautics and Astronautics
ARA Aircraft Research Association
BDW blunt delta wing
BL boundary layer
CAD computer aided design
CERT Centre d'Etudes et de Recherches de Toulouse
CF cross flow

CFD	computational fluid dynamics
CFL	Courant-Friedrichs-Lewy
CIRA	National Aerospace Research Center Italy
CRM	Common Research Model
DLR	German Aerospace Center
DNS	direct numerical simulation
ERCOFTAC	European Research Community on Flow, Turbulence and Combustion
ESA	European Space Agency
ETW	European Transonic Windtunnel
FOI	Swedish Defence Research Agency
HISSS	Higher-Order Subsonic-Supersonic Singularity (method)
HLF	hybrid laminar flow
HTP	horizontal tail plane
ISW	infinite swept wing
IUTAM	International Union of Theoretical and Applied Mechanics
LE	leading edge
LEBU	large eddy breakup
LES	large eddy simulation
LFC	laminar flow control
LISW	locally infinite swept wing
LTA	large transport airplane
MBB	Messerschmitt-Bölkow-Blohm GmbH
NACA	National Advisory Committee for Aeronautics
NASA	National Aeronautics and Space Administration
NASP	National Aerospace Plane
NLF	natural laminar flow
NLR	National Aerospace Laboratory of the Netherlands
NS	Navier-Stokes
NTF	National Transonic Facility
ONERA	National Aerospace Research Center France
PSE	parabolized stability equations
RAE	Royal Aircraft Establishment
RANS	Reynolds-averaged Navier-Stokes
SOBOL	second-order boundary-layer (method)
TE, TEF	trailing edge, trailing-edge flow
TS	Tollmien-Schlichting
VKI	von Kármán Institute

Permissions

Figures reproduced with permission by

- DLR-Institute of Aerodynamics and Flow Technology, Braunschweig, Germany: lower part of Fig. 10.4,
- J. Häberle and C. Weiland: Figs. 2.14, 2.16,
- S. Hein: Fig. 9.8,
- Ch. Mundt et al.: Fig. 6.7,
- S. Riedelbauch: Figs. 10.25 to 10.30,
- D. Schwamborn: Figs. 7.10, 7.13, 7.17, 7.18,
- Zeitschrift für Flugwissenschaften und Weltraumforschung (ZfW): Fig. 6.5 (K.M. Wanie et al.), and Fig. 7.24 (H.U. Meier et al.).

Figures have directly been provided by many colleagues as acknowledged in the preface of this book.

Name Index

Subject Index

Printed by Printforce, the Netherlands